HEAT TRANSFER

HEAT TRANSFER
Emerging Concepts and Analysis

Erica Carlson

Kruger Brentt
Publishers
2025

Kruger Brentt Publishers UK. LTD.
Company Number 9728962

Regd. Office: 68 St Margarets Road, Edgware, Middlesex HA8 9UU

ISBN: 978-1-78715-380-6

For information on all our publications visit our website at http://krugerbrentt.com/

PREFACE

Heat transfer is a fundamental process that underpins numerous phenomena in nature and engineering, from the cooling of electronic devices to the regulation of temperature in industrial processes and the Earth's climate system. As our understanding of heat transfer continues to evolve, new concepts, theories, and analytical techniques are emerging to address complex challenges and harness the potential of heat transfer for technological innovation and sustainability. Understanding the latest advancements in heat transfer is essential for engineers, researchers, and policymakers seeking to develop efficient thermal management solutions, optimize energy systems, and mitigate the impacts of climate change.

This handbook, titled "Heat Transfer: Emerging Concepts and Analysis," is a comprehensive exploration of the latest developments in heat transfer theory, methods, and applications. Drawing upon the expertise of heat transfer specialists, researchers, and industry practitioners, this volume offers theoretical insights, computational techniques, and practical examples for understanding and analyzing heat transfer processes in diverse systems.

From the fundamentals of conduction, convection, and radiation to the latest advancements in heat exchanger design, thermal management of electronic devices, and renewable energy technologies, the "Heat Transfer: Emerging Concepts and Analysis" covers a wide range of topics essential for mastering the field of heat transfer. Through clear explanations, mathematical derivations, and computational simulations, readers will gain a deep understanding of the principles and methods used to analyze and optimize heat transfer systems.

Moreover, this handbook explores the emerging concepts and trends in heat transfer research, including nanoscale heat transfer, thermal metamaterials, and heat transfer in complex fluids and materials. By examining case studies, research findings, and future directions, we highlight the ways in which emerging concepts are shaping the future of heat transfer science and engineering.

As we confront the challenges of climate change, energy transition, and sustainable development, the importance of advancing our understanding of heat transfer has never been greater. It is our hope that the "Heat Transfer: Emerging Concepts and Analysis" will serve as a valuable resource for students, researchers, and practitioners in the field of heat transfer and related disciplines. Whether you are studying heat transfer phenomena, designing thermal systems, or developing innovative solutions for sustainable energy, this handbook invites you to explore the frontiers of heat transfer and contribute to building a more sustainable and resilient future.

Erica Carlson

CONTENTS

PART 2: BOILING, FREEZING AND CONDENSATION HEAT TRANSFER

PART 1
HEAT TRANSFER IN MICRO SYSTEMS

1 Chapter

ANSFER IN FLUIDIZED BED REACTORS

1. INTRODUCTION: AN OVERVIEW

The fluidization of gas-solid systems is a longstanding topic of fundamental research. The close contact between fluid and particles in an isothermal system enhances heat and mass transmission, making it advantageous for chemical reactions, mixing, drying, and several other applications. Despite significant advantages, some issues remain related with gas-solid fluidization systems. The solid properties, particularly particle size and size distribution, strongly influence particle interaction, mobility, and fluid-particle mixing. Observable macroscopic phenomena of plug formation, channeling, and particle agglomeration were noted with increasing surface gas velocity in a conventional fluidized bed of fine/nano-particles (group C in the Geldart classification) due to significant interparticle forces. The powder–particle fluidized bed (PPFB) concept was initially proposed by Kato et al. in the early 1990s. [1]. It is known to be a useful technique to fluidize group C particles without external aid like acoustic, centrifugal, magnetic, stirring and/or vibrating fields, etc. In the PPFB process, fine powders (group C) are fluidized with coarse particles (group B). The bi- modal fluidized bed system at steady state gives a certain stable hold-up of fine powders in the bed [2]. Many investigators, Sun and Grace [3], and Xue et al. [4] studied bi-modal fluidization in bubbling or turbulent regimes, while, Wei et al. [5] and Du [6, 7] concentrate on CFBs. The investigation of Xue et al. showed that by adding coarse particles fluidization quality of fine particles could improve [4]. Extensive studies of Wei et al. indicated that the addition of coarse particles to a fluid catalytic cracking (FCC) riser decreased the lateral solids mixing and had insignificant influence on axial solids mixing [5]. Du et al. studied the axial and lateral mixing by using tracer particles of different sizes in a FCC riser and found that the axial solids back mixing increased, while radial solids mixing decreased with the increase of particle size and density. Recently, Zeeshan et al. explained bi-modal/bi-particle fluidized bed system hydrodynamics, attrition, mixing behaviour, and its applications [8].

Achieving stable and homogeneous heat transmission in Fluidized Bed Reactors (FBR) without external or internal heat sources poses a significant challenge for designers. Continuous heat supply and removal are essential for the operation of FBR to regulate highly endothermic and exothermic reactions, respectively. The advanced concept of bi-modal particle (Gas-Solid-Solid) fluidization is presented by FLOTU to address the aforementioned reaction barriers in fluidized bed technology. This chapter provides a thorough introduction of GSS-FRB research, accompanied by a practical example of the direct propane dehydrogenation reaction. Direct propane dehydrogenation is hampered by engineering constraints due to equilibrium limitations and the endothermic nature of the process; thus, a continuous supply of heat is necessary. The superiority of GSS-FBR operation was analyzed and evaluated in terms of hydrodynamics and reaction outcomes. The outcomes of the fixed bed micro-reactor were compared with those of the integrated bi-modal particle fluidized bed reactors and defined parametrically. The notable characteristics of this technology were emphasized, advocating for an enhanced design as a viable opportunity for both highly endothermic and exothermic processes using FBR, offering economic and operational advantages.

2. DESIGN AND REACTORS SEQUENCE: INTEGRATED APPROACH

Zeeshan et al. introduced an innovative design of a cofluidized bed reactor to enhance fluidized bed technology for high endothermic or exothermic processes, in which coarse catalyst particles were cofluidized with fine FCC catalyst [9, 10].Coarse particles serve as the principal catalyst for the intended reaction (endothermic/exothermic), whilst fine particles operate as heat transporters or absorbers. In an endothermic process, the minute catalyst particles absorb heat from the regenerator, convey it to the primary catalyst (coarse particles) within a fluidized bed reactor, and may subsequently act as a secondary catalyst in other reactors, including olefins inter-conversion, MTO, and MTP, as depicted in Figure 1.

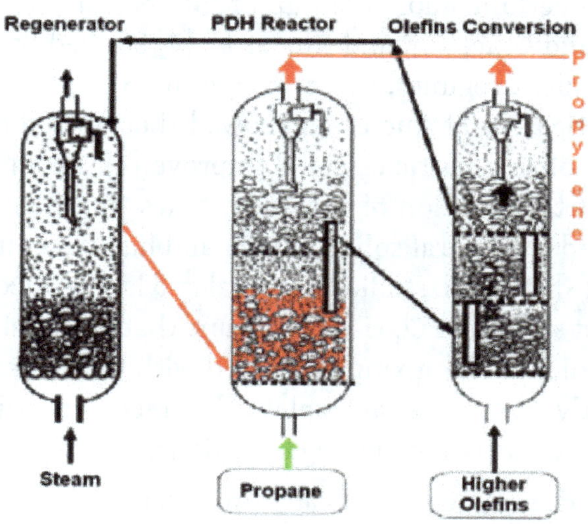

Fig. 1. Proposed process design and reactors sequence

3. CHARACTERISTICS FEATURES OF TECHNOLOGY

The gas-solid-solid (GSS) fluidization system is a distinctive apparatus characterized by its superior mixing quality. The mixing and hydrodynamic behavior of FCC particles with coarse particles were examined in 2D and 3D co-fluidized beds, respectively. The enhancement of mixing and fluidization qualities by the addition of coarse particles is mostly attributed to their mobility, which disrupts the strong interparticle interactions among FCC particles and eliminates bubble wakes, employing a bubble dissociation method intermittently. Details regarding the design of 2D and 3D co-fluidized bed reactors, as well as experimental specifications, can be found elsewhere. [8].

To process the solid particulates in fluidized bed and slurry phase reactors, attrition is an inevitable consequence and is therefore one of the preliminary parameters for the catalyst design for a fluidized bed reactor. The mechanical degradation propensity of the zeolite catalysts (particles) were investigated in a bi-modal distribution environment using a Gas Jet Attrition - ASTM standard fluidized bed test (D-5757) [11]. The experiment was done to investigate the parameters influencing attrition phenomena in bimodal fluidization. In a bimodal fluidization system, two distinct particle types were co-fluidized isothermally. Comprehensive information regarding experimental design is available elsewhere.[12].

3.1 Mixing behaviour

The production, rupture, and expansion of bubbles are crucial for fluidization, thermal exchange, and mixing efficacy. Fluidization transpires mostly due to a dynamic equilibrium between gravitational forces and fluid forces within the bed. In systems where the fluidizing particles exhibit markedly varied densities and sizes, unstable local pockets with a very high void fraction, referred to as bubbles, are formed. This system is referred to as a self-excited nonlinear system. It is established that periodic disturbance of a system parameter can lead to chaos suppression. This study investigates uniform mixing in co-fluidization (GSS) and identifies a distinctive bubble braking phenomenon that primarily facilitates uniform mixing. Following bubble formation, the bubble flow pattern and pressure distribution are influenced by the placement of the blocks.

Fig. 2. Unique bubble braking phenomena of GSS fluidization

In the bi-modal particle system (GSS FBR), the giant bubbles were fragmented into smaller bubbles. This occurs when coarse particles disrupt huge bubble wakes, resulting in the generation of swirling bubbles. The bubble transferred momentum to solids in both axial and lateral directions, so enhancing mixing. The coarse particles provide a high frequency of bubbles and a low pressure amplitude surrounding the orifice compared to an extra distributor, thus facilitating smoother fluidization of the FCC catalyst with enhanced uniform mixing. Nevertheless, coarse particles within the bed exhibit excellent symmetrical flow, and no obstructions were identified. This effect was seen in a 2D co-fluidized bed, as illustrated in Figure 2 [8].

3.2 Hydrodynamics of co-fluidization

Following a fundamental description of the function of bubbles in a bi-modal fluidized bed, the hydrodynamics of FCC particles with coarse particles was examined. The fluidization characteristics and mixing were enhanced overall. Consequently, we concentrate our focus on the comprehensive behavior of the bed.. The fluidization of FCC catalyst shows crack formation at low superficial gas veloCities. With the increase of the gas velocity to around 0.1 m/s, the FCC particles become fluidizing and their fine counter parts air borne at the upper part of the bed. The experimental results of the pressure drop and bed height ratio, of FCC and coarse particles bed (independently) are shown in Figure 3 (a) and (b), respectively. This is the reason that the bed height ratio of FCC particle is become violent with small increase in velocity. The upper part of bed demonstrates very dilute bed, but in turbulent fluidization behavior. The bed surface was severely disturbed by large gas bubble eruptions and fine particle ejections, which lead to enormous elutriation. While, coarse particles also feel difficulty in fluidization at lower gas veloCities, till 0.3 m/s. With the increase of gas velocity, the pressure drop curves do not show a plateau. Figure 3 (b) shows that the expansion of FCC particles bed is much higher than that of the coarse particles bed. This gap can be modified by decreasing the particle size difference. The physical mixture of FCC particles with coarse particles was made in following ratio: The FCC particles were fixed to 100 g, with coarse particles were added as 25, 50 and 75 g. Details about the gross bed behavior, pressure drop and expansion characteristics are shown in Figure 4 (a) and (b). The incorporation of coarse particles in varying quantities into FCC particles has markedly enhanced the overall fluidization behavior. Moreover, the minimum fluidization velocity of the combination diminishes. In a co-fluidization system, measuring ΔP across a single particle type, whether small or coarse, is challenging due to their thorough mixing.

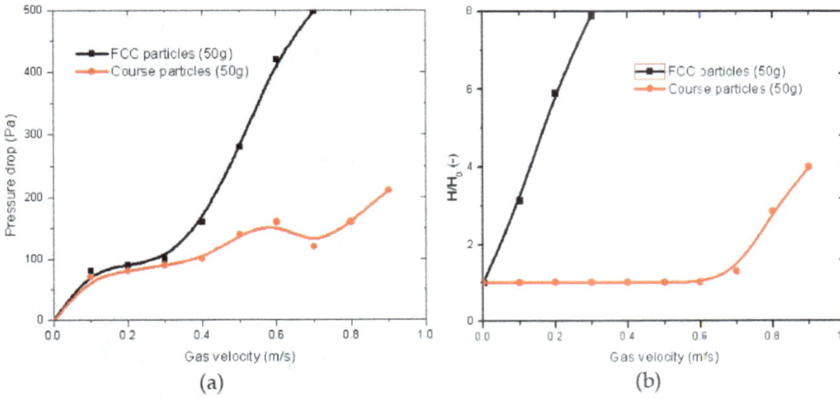

Fig. 3. (a) Bed pressure drop curve of FCC and coarse particles, (b) Bed expansion curve of FCC and coarse particles

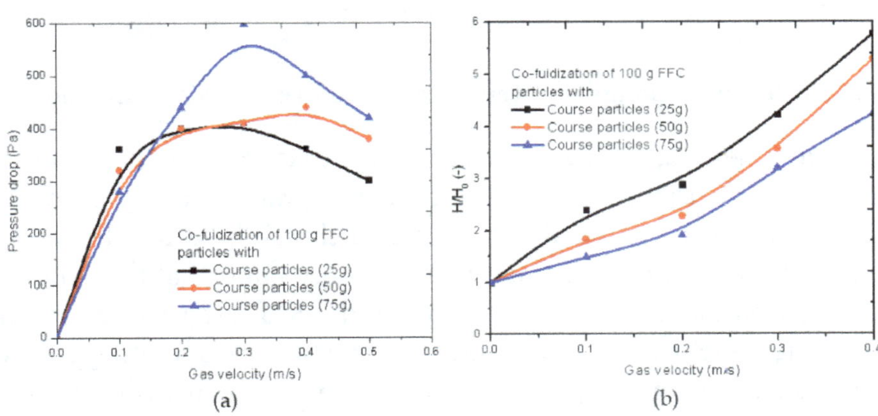

Fig. 4. (a) Bed pressure drop during co-fluidization of FCC and coarse particles, (b) Bed expansion during co-fluidization of FCC and coarse particles

3.3 Attrition study

Numerous experimental research and empirical models were created to delineate the degree of attrition in a fluidized environment. Gwyn et al. [12] established the subsequent empirical relationship (refer to equation 1) by analyzing the shearing duration of attrition, utilizing a single particle system with high-velocity air jets in a fluidized bed. The Gwyn constants K and m are functions of material qualities and size, determined experimentally; further details can be found elsewhere. [12].

$$W = Kt^m \tag{1}$$

However, to date, no study has focused on particles attrition phenomena (systematically explained in Figure 5) in bi-modal particles fluidized bed (co-fluidization environments) and multi-particle sized system, i.e. the most feasible design for handling extremely endothermic or exothermic reactions, while, disadvantage of attrition become serious. Therefore the study focused attention not only on attrition calculations, but explore its incremental phenomena and develop a more generalizes relationship for calculating attrition debris in bi-model particle system.

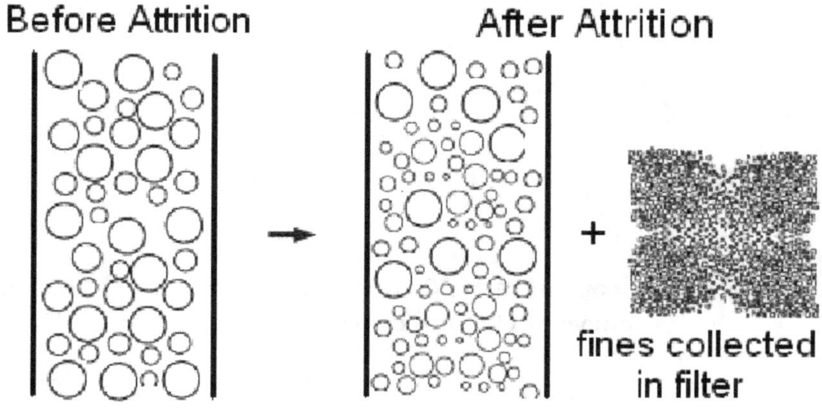

Fig. 5. Systematic image of attrition tube before and after attrition run

AJI in certain cases was as high as 0.28, means that 28 % of the small particles were lost due to attrition. The attrition evaluation was generally pronounced by the attrition rate which depends upon certain thresh hold for the size limit of fines collected. In practice, attrition mass was more important than the collected fine's size distribution, therefore ASTM standard D-5757 attrition test method was selected for the present study and operated under standard operating procedure, in the density range of 1-3 g/cm^3 [12]. The flow regime changes with density variations and fluidization veloCities, but settling chamber provides slipping area to keep the particles airborne in these circumstances, and only let the fines go out. Therefore, ASTM standard fluidized bed was selected and operated at fixed superficial gas velocity i.e. 0.059 m/s according to the standard operating procedure. The precise values of Gwyn constants for each particle (small zeolite catalysts) at standard conditions were assessed in accordance with a definitive protocol.

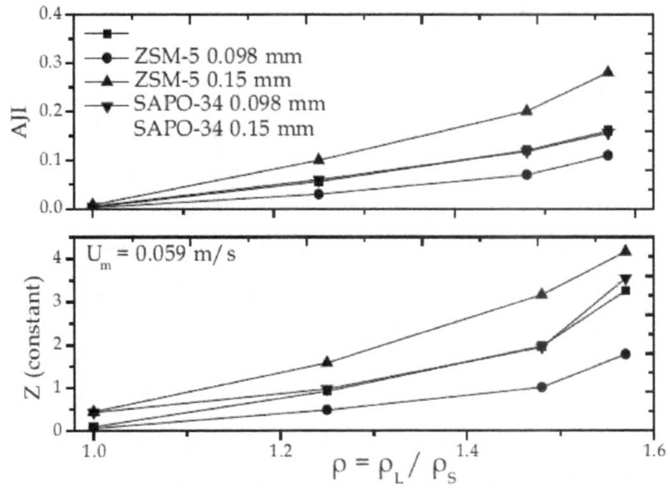

Fig. 6. Effect of relative density on newly defined constant Z and AJI after 5 hr

Effect of relative density was considered to define this new constant Z and its influence can be observed in Figure 6. There it was assumed that S was constant and its value is 1. This newly developed function (equation 2) has the ability to successfully explain small particles attrition of zeolites and other synthetic catalysts used in bimodal fluidized bed reactors. This empirical relationship will become more generalized and accurate because Gwyn constants of each particle (catalyst) were determined individually. Therefore the applicability of this function was not only limited to the measurement of zeolites attrition in a bimodal fluidized bed environment but can be used for other catalysts. The narrow span of deviation i.e. ± 0.50 g in attrition debris after 5.0 hours operation was observed. The model was generalized as changes related to material properties, and by considering these properties the relationship was made generalized. For each type of material first determine the values of their constants for single particle system as explained by Gwyn, and then using proposed model the attrition loss of that sample in a bi-model fluidization at designated conditions can be calculated.

$$x = (\rho)^5 \cdot (S)^{0.44} \cdot (K) \cdot (t)^{1.53m} \tag{2}$$

Furthermore, the magnitude of impact was found to be a function of substantial particle density and size ratios.

4. CASE STUDY

The suggested bi-modal fluidized bed has extensive applications in the chemical industry, particularly for exothermic and endothermic reactions. A comprehensive case study on direct propane dehydrogenation utilizing bi-modal fluidized bed technology is now presented. [10]. The propane dehydrogenation is the most economical route to propylene, but very complex due to endothermic reaction requirements, equilibrium limitations, stereo-chemistry and in engineering constraints. The state of the art idea of bi-model particle (Gas-Solid-Solid) fluidization was applied, in order to overcome alkane dehydrogenation reaction barriers in a fluidized bed technology. In this study, the propane dehydrogenation reaction was studied in an integrated fluidized bed reactor, using Pt-Sn/Al^2o_3-SAPO-34 novel catalyst (ZeeFLOTU) at 590 ºC. The results of fixed bed micro-reactor and integrated bi-model particle fluidized bed reactors were compared, and parametrically characterized. The results showed that the propylene selectivity is over 95 %, with conversion between 31-24 %. This significant enhancement is by using novel bi-model particle fluidization system, owing to uniform heat transfer throughout the reactor and transfer coke from principal catalyst to secondary catalyst, which increases principal catalyst's stability. Experimental investigation reveals that the novel Pt-Sn/Al^2o_3-SAPO-34 catalyst and proposed intensified design of fluidized bed reactor is a promising opportunity for direct propane dehydrogenation to propylene, with both economic and operational benefit. The experimental setup is shown in Figure 7 while catalyst and experimental design information can found elsewhere [10, 13-18].

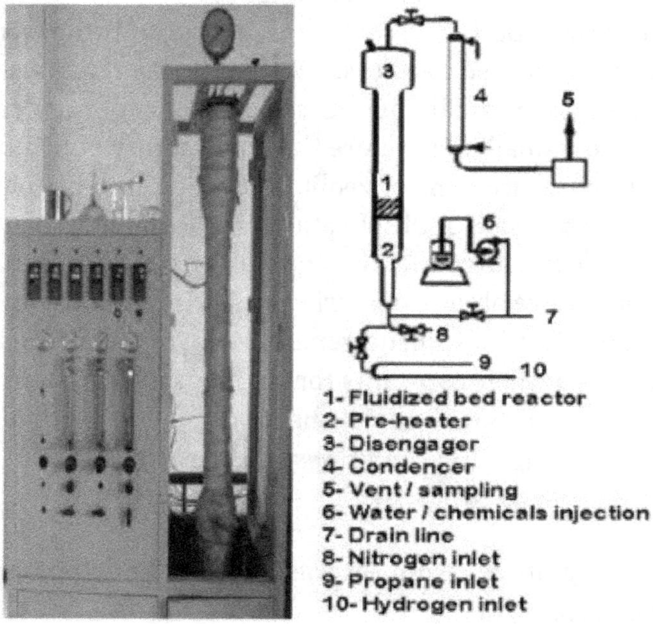

Fig. 7. Hot-model bi-model particle Fluidized Bed Reactor (FBR) apparatus.

The experimental results of propane dehydrogenation using novel catalyst Pt-Sn/Al^2o_3- SAPO-34 using bi-modal pilot scale fluidized bed reactor are shown in Figure 8. The comparison demonstrates that bi-model results are far better than fixed and single particle fluidized bed [10]. The influence of fluidization mode in an integrated bi-model particle fluidized bed was also investigated for propane dehydrogenation to propylene. It is observed that the propylene selectivity in a fluidized bed reactor was improved after 1 hr operation, when the reactor reaches steady state conditions. The steady and uniform conversion and yield is also achieved. Actually, in fixed bed reactor coke deposition is high as compared to fluidized bed reactor (see Table 1). In the two particles co-fluidized system it was observed that the coke deposited on SAPO-34 (fine catalyst particles) is higher than the metal incorporated SAPO-34 (principal catalyst). Therefore, it's easy for principal dehydrogenation catalyst (Pt-Sn/Al^2o_3-SAPO-34) to sustain its activity for longer duration.

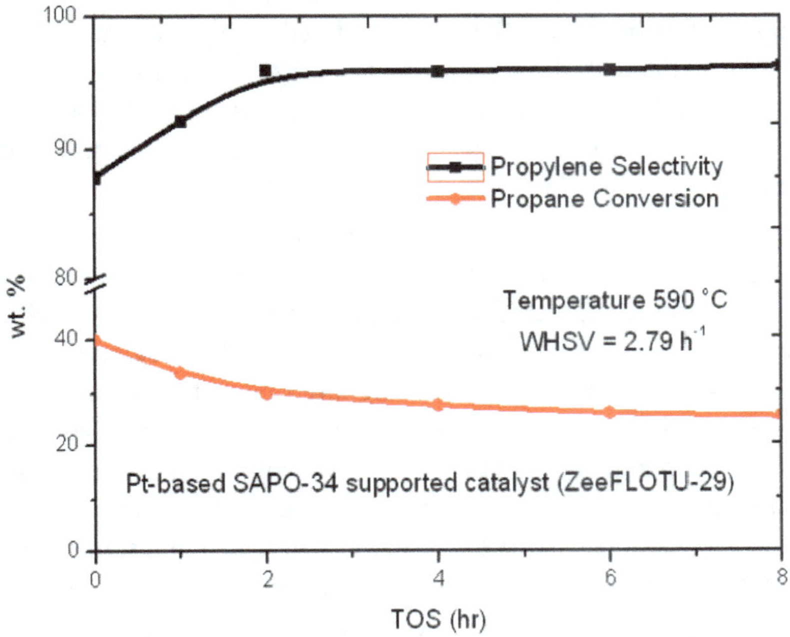

Fig. 8. Performance comparison of novel catalyst in fixed bed with proposed GSS-fluidized bed reactor

Table 1. Deactivation rate and amount of coke formed on principal catalyst (Pt-Sn/Al- SAPO-34)

Reactor	[a]Coke (wt.%)	[b]Deactivation (%)
Fixed bed	0.41	54
Fluidized Bed	0.24	45

[a]O_2-pulse coke analysis

[b]Deactivation = $[(X_0-X_f)/X_0 \times 100]$;

where, X_0 is the initial conversion at 5 min. and X_f is the final propane conversion

Moreover, in the continuous processing the small catalysts (those serve as heat carrier) were continuously regenerated and the process efficiency was improved. Above 96 % propylene selectivity was obtained at 8 hr time-on-stream. Sustainable conversion with lower deactivation rate (see table 1) was also observed. The lower propylene yield initially was due to lower conversion and selectivity, which increase gradually with time. Therefore we can say that the impressive results were obtained using this integrated fluidized bed reactor. It was interesting to find that the reaction stability and activity of catalyst will become superior, but also superior in coke management. Nevertheless, deactivation and/or activity loss of the bi-metallic catalysts is due to coke deposition and Pt sintering [13-15]. Therefore, in coke analysis of bi-model particle fluidized bed catalyst it was noted that large amount of coke is deposited over non-metallic SAPO-34, that is in a continuous recirculation through the regenerator, in continuous setup. It's an effective way to protect catalyst activity for longer time with stable activity, and so called coke management.

The overall picture of selective propane dehydrogenation to propylene over above said catalyst at 590 ºC in bi-modal fluidized bed reactor is shown in the OPE plot in Figure 9. The data was plotted with respect to yields and selectivity. The best propane conversion range to have high propylene yield and selectively is observed to be between 24-28% conversions. In designated operating range the propylene yield is above 25 % and selectivity is as high as 96 %. While at higher conversions both propylene yield and selectivity dropped sharply, with the increase in ethane formation. It is further noted that the higher conversion favours both cracking and hydride transfer reaction with the decrease in dehydrogenation rate. Moreover, the deactivation of catalyst may also lead to cracking.

The performance of the Pt-Sn/Al^2o_3-SAPO-34 is evaluated in a continuous mode of reaction-regeneration for three cycles. The results are shown in Table 2. The catalysts were regenerated with nitrogen mixed steam for 4 hr at 600 ºC. After regeneration, the Pt was re- dispersed using $C_2Cl_2H_4$ solution, injected with nitrogen at 500ºC. The detailed chlorination method can be finding elsewhere [13, 15]. After the regeneration and re-dispersion of Pt, catalyst was reduced in hydrogen environment, and reused for next reaction cycle at identical conditions. The results clearly demonstrate hydrothermal stability of the catalyst. Therefore, the robustness of proposed design of bi-model particle (Gas-Solid-Solid) fluidized bed reactor and Pt-Sn/Al^2o_3-SAPO-34 is successfully proved.

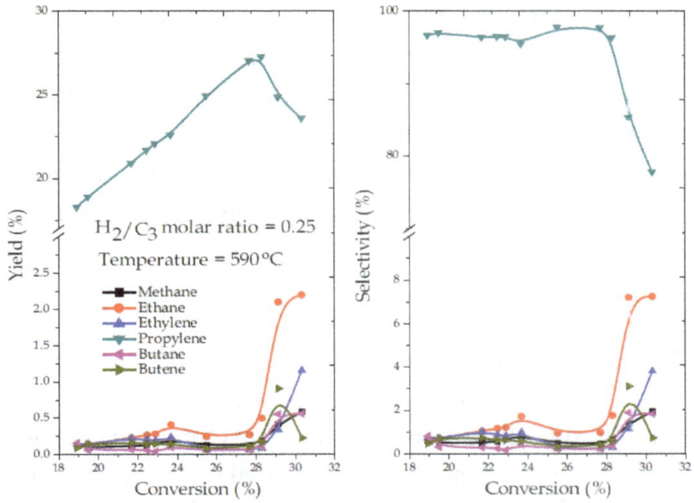

Fig. 9. OPE with respect to yield and selectivity

Table 2. Influence of hydrothermal treatment and catalysts performance in a continuous operation (Reaction conditions: T = 590 ºC, WHSV = 5.6 h-1, H2/ C_3H_8 molar ratio = 0.25)

Al-SAPO-34 supported	Cycle I				Cycle II				Cycle III			
	Conversion		Selectivity		Conversion		Selectivity		Conversion		Selectivity	
TOS	1 hr	8 hr	1 hr	8 hr	1 hr	8 hr	1 hr	8 hr	1 hr	8 hr	1 hr	8 hr
Pt-Sn-based	29.1	26.2	85.4	96.8	28.3	25.5	85.9	96.9	27.2	24.6	87.9	97.4

5. BI-MODAL (G-S-S) FLUIDIZATION SYSTEM: HEAT TRANSFER ANALYSIS

To assess the heat transfer efficiency of the bimodal fluidized bed reactor, the study is based on an energy balance and a mass balance, incorporating the following simplifying assumptions:

the powder particles are not considered individually but as a single phase, motion and temperature gradients are considered only in the direction of flow, the course particle bed is considered to be dilute packing as per total volume of the bed, then the heat transfer from the powder (fine particles) to packing (course particles) is considered as question. Under such assumptions, a plug flow model is established to study heat transfer. In the hot model experiments, since the fine particle mass flux could not be measured, the powder-to-gas mass flux ratio, $R_{sg}(G_s/G_g)$ is analyzed. The best value for R_{sg} has to be found by trial and error, by experiments. A suitable R_{sg} should satisfy the condition of higher conversion and selectivity, and is closed to axial distribution of temperature in experiments [19].

The experiments demonstrate that the temperature is almost uniform with +/- 10 ºC in the co-fluidized bed to keep propane conversion in reasonable range. The G_s/G_g values calculated were also calculated for the same superficial gas velocity. It is worthwhile to note that the temperature of the pre-heated propane injection (550 ºC) was taken to 590 ºC for a highly endothermic reaction and maintained because of the continuous supply of heat through fine particles. The dimensionless R_h is defined as follows:

$$R_h = \sum(\Phi_i Cp_{g,i}) + W_s Cp_s / \Phi_{propane} Cp_{g,propane} \quad (3)$$

where i stands for propane, propylene and hydrogen. The higher R_h is found to be beneficial for supplying reaction heat and its increases with the increase in fine particle mass flux. The substitution of tiny particles for a portion of input gases as a heat carrier boosts the heat transfer rate. The axial temperature distribution of the GSS-FBR far surpasses that of the adiabatic packed-bed reactor for endothermic reactions.

6. CONCISE SUMMARY

An integrated design of Fluidized Bed Reactors (FBR) including an advanced heat supply system without the inclusion of external or internal sources is presented for endothermic reactions. The advanced concept of bi-modal particle (Gas-Solid-Solid) fluidization is introduced and succinctly examined in all aspects. The enhanced operating advantages, particularly in heat transfer, were achieved by bi-modal fluidized bed technology. The method was examined using a case study of propane dehydrogenation with GSS-FRB. A comprehensive analysis of heat transfer in the bi-modal fluidization system according to the case study. The enhanced design of GSS-FBR presents a promising prospect for both endothermic and exothermic reactions, as the management of heat supply or removal remains a persistent challenge.

7. REFERENCES

K. Kato, T. Takarada, N. Matsuo, T. Suto, N. Nakagawa, Int. Chem. Engg. 34, 605 (1994).

T. Nakazato, J. Kawashima, T. Masagaki and K. Kato, Adv. Powder Technol. 17, 433 (2006).

G. Sun, J.R. Grace, Chem. Eng. Sci. 45, 2187 (1990).

H. Xue, W. Zhang, X. Yang, Eng. Chem. Metall. (China), 17, 235 (1996).

F. Wei, Y. Cheng, Y. Jin, Z.Q. Yu, Can. J. Chem. Eng. 76, 19 (1998).

B. Du, Master Thesis of Tsinghua University, Beijing, China, 1999.

B. Du, F. Wei, Chem. Eng. process. 41, 329 (2002).

Zeeshan Nawaz, Yujian Sun, Yue Chu and Fei Wei, "Mixing behavior and hydrodynamic study of GAS-SOLID-SOLID Fluidization system: Co-Fluidization of FCC and coarse particles", FLUIDIZATION XIII, May 16-21, 2010, Gyeong-ju, Korea.

Zeeshan Nawaz, PhD Thesis of Tsinghua University, Beijing, China, 2010.

Zeeshan Nawaz, Yue Chu, Wei Yang, Xiaoping Tang, Yao Wang and Fei Wei, Study of Propane Dehydrogenation to Propylene in an integrated Fluidized Bed Reactor using Pt-Sn/Al-SAPO-34 novel catalyst, Industrial & Engineering Chemistry Research, 49 (2010) 4614-4619.

ASTM Standard D-5757-Revised, ASTM, Philadelphia PA, (2006).

Zeeshan Nawaz*, Tang Xiaoping, Xiaobo Wei and Fei Wei, Attrition behavior of small particles in bimodal particles fluidized bed environment: Influence of density and size ratio, Korean J. Chem. Eng., 27(4), 1025-1031 (2010) (DOI: 10.1007/s11814-010- 0240-5)

Zeeshan Nawaz, Fei Wei and Shahid Naveed, Highly stable Pt-Sn-based, SAPO-34 supported, Al binded catalyst, and Integrated Fluidized Bed Reactor Design for Alkane Dehydrogenation, Pak. Patent Application No. 1127/2009; Chinese Patent Application No. 200910091226.6.

Zeeshan Nawaz, Tang Xiaoping, Qiang Zhang, Wang Dezheng and Fei Wei, A highly selective Pt-Sn/SAPO-34 catalyst for propane dehydrogenation to propylene, Catalysis Communications, 10 (2009) 1925-1930.

Zeeshan Nawaz and Wei Fei, Hydrothermal study of Pt-Sn-based SAPO-34 supported novel catalyst used for selective propane dehydrogenation to propylene, Journal of Industrial and Engineering Chemistry, 16 (2010) 774-784.

Zeeshan Nawaz, Xiaoping Tang, Yao Wang and Fei Wei, Parametric characterization and influence of Tin on the performance of Pt-Sn/SAPO-34 catalyst for selective propane dehydrogenation to propylene, Industrial & Engineering Chemistry Research, 49 (2010) 1274–1280.

Zeeshan Nawaz and Wei Fei, The Pt-Sn-based SAPO-34 supported novel catalyst for n- butane dehydrogenation, Industrial & Engineering Chemistry Research, 48 (2009) 7442-7447.

Zeeshan Nawaz, Tang Xiaoping, Shahid Naveed, Qing Shu and Fei Wei, Experimental Investigation of Attrition Resistance of Zeolite Catalysts in Two Particle Gas-Solid- Solid Fluidization System, Pakistan Journal of Scientific & Industrial Research, 53 (2010) 169-174.

Shuiyuan Huang, Zhanwen Wang, Yong Jin, Studies on gas-solid-solid circulating Fuidized-bed reactors, Chemical Engineering Science 54 (1999) 2067-2075.

2 Chapter

R134A CONDENSATION IN VERTICAL DOWNWARD FLOW WITH TWO-PHASE HEAT TRANSFER COEFFICIENTS AT HIGH MASS FLUX

1. INTRODUCTION: AN OVERVIEW

The transfer process of heat between two or more fluids of different temperatures in a wide variety of applications is usually performed by means of heat exchangers e.g. refrigeration and air-conditioning systems, power engineering and other thermal processing plants. In refrigeration equipment, condensers' duty is to cool and condense the refrigerant vapor discharged from a compressor by means of a secondary heat transfer fluid such as air and fluid. Examinations and improvements on the effectiveness of condenser have importance in case there is a design error which can cause the heat transfer failure occurrence in the condenser.

Industries have been trying to improve the chemical features of alternative refrigerants of CFCs due to the depletion of the earth's ozone layer. Because of the similar thermo-physical properties of HFC-134a to those of CFC-12, an intensive support from the refrigerant and air-conditioning industry has given to the refrigerant HFC-134a as a potential replacement for CFC-12. On the other hand, even though there isn't much difference in properties between these two refrigerants, the overall system performance may be affected significantly by these differences. For that reason, the detailed investigation of the properties of HFC-134a should be performed before it is applied.

A large number of researchers foCused on the heat transfer and pressure drop characteristics of refrigerants over the years, both experimentally and analytically, mostly in a horizontal straight tube. The CFCs' heat transfer and pressure drop studies inside small diameter vertical either smooth or micro-fin tubes for downward condensation has been concerned comparatively little in the literature Briggs et al. (1998) have studied on in-tube condensation using large diameter tubes of approximately 20.8 mm with CFC113. Shah (1979) used some smooth tubes up to 40 mm i.d. at horizontal, inclined and vertical positions to have a well- known wide-range applicable correlation. Finally, it has been compared by researchers commonly for turbulent condensation conditions and it is considered to be the most predictive condensation model for the annular flow regime in a tube. Oh and Revankar (2005) developed the Nusselt theory (1916) to investigate the PCCS condenser using 47.5 mm i.d. and 1.8 m long vertical tube in their experimental study for the validation of their theoretical calculations. Maheshwari et al. (2004) simulated PCCS condensers used in water-cooled reactors using a 42.77 mm i.d. vertical tube during downward condensation presence of non-condensable gas. The prediction of heat transfer considering mass transfer along the tube length was performed by means of a computer code in their study.

Convective condensation in annular flow occurs for many applications inside tubes such as film heating and cooling processes, particularly in power generation and especially in nuclear reactors. The one of the most important flow regimes is annular flow which is characterized by a phase interface separating a thin liquid film from the gas flow in the core region. This flow regime is the most investigated one either analytically or experimentally, because of its practical significance and common usage.

Akers et al. (1959) foCused on the similarity between two phase and single phase flows and developed a correlation using two-phase multiplier to predict frictional two phase pressure drop which is same rationale as the LoCkhart-Martinelli (1949) two-phase multiplier. His model is known as "equivalent Reynolds number model" in the literature. According to this model, an equivalent all liquid flow, which produces the same wall shear stress as that of the two phase flow, is replaced instead of annular flow inside a tube. Many researchers benefitted from his model such as Moser et al. (1998) and Ma and Rose (2004). Moser et al. (1998) predicted heat transfer coefficient in horizontal conventional tubes developing a model. Ma and Rose (2004) studied heat transfer and pressure drop characteristics of R113 in a 20.8 i.d. vertical smooth and enhanced tubes.

Dobson et al. (1998) used zeotropic refrigerants in their condensation tests including the wide range of mass flux in horizontal tubes and benefitted from the two-phase multiplier approach for annular flow. In their study, heat transfer coefficient increased with increasing mass flux and vapor quality in annular flow due to increased shear and thinner liquid film than other flow regimes. Sweeney (1996) modified their model their model for R407C considering the effect of mass flux.

Cavallini et al. (1974) have some significant theoretical analysis on the in-tube condensation process regarding the investigation of heat transfer and pressure drop characteristics of refrigerants condensing inside various commercially manufactured tubes with enhanced surfaces using a number of correlations in the literature. Lately, Cavallini et al. (2003) prepared a review paper on the most recent condensation works in open literature including the condensation inside and outside smooth and enhanced tubes.

Valladares (2003) presented a review paper on in-tube condensation heat transfer correlations for smooth and micro-fin tubes including the comparison of experimental data belong to various experimental conditions from different researchers. Wang and Honda (2003) made a comparison of well-known heat transfer models with experimental data belong to various refrigerants from literature and proposed some models, valid for the modified annular and stratified flow conditions, for micro-fin tubes. Bassi and Bansal (2003) compared various empirical correlations and proposed two new empirical models for the determination of condensation heat transfer coefficients in a smooth tube using R134a with lubricant oil. Jung et al. (2003, 2004) did condensation tests of many refrigerants such as R12, R22, R32, R123, R125, R134a, and R142b inside a smooth tube. They paid attention not only comparison of their experimental data with various well-known correlations but also proposition on a new correlation to predict condensation heat transfer coefficients.

Generally, the condensation heat transfer coefficients and pressure drops in tubes have been computed by empirical methods. The modifications of the Dittus-Boelter single-phase forced convection correlation (1930) are used in the literature, as in Akers et al. (1959), Cavallini and Zecchin (1974), and Shah (1979). Dalkilic et al. (2010a) made a comparison of thirteen well-known two-phase pressure drop models with the experimental results of a condensation pressure drop of R600a and R134a in horizontal and vertical smooth copper tubes respectively and revealed the main parameters of related models and correlations. Dalkilic et al. (2008a) used the equivalent Reynolds number model (1998) to propose a new correlation for the two-phase friction factor of R134a and also discussed the effect of main parameters such as heat flux, mass flux and condensation temperature on the pressure drop. Dalkilic et al. (2009b) made a comparison of eleven well-known correlations for annular flow using a large amount of data obtained under various experimental conditions to have a new correlation based on Bellinghausen and Renz's method (1992) for the condensation heat transfer coefficient of high mass flux flow of R134a. The effects of heat flux, mass flux and condensation temperature on the heat transfer coefficients also exist in their paper. Dalkilic et al. (2009c) showed the significance of the interfacial shear effect for the laminar condensation heat transfer of R134a using Carey's analysis (1992), which is the improved version of Nusselt's theory (1916), and proposed a new correlation based on Bellinghausen and Renz's method (1992) for the condensation heat transfer coefficient during annular flow of R134a at low mass flux in a vertical tube. Dalkilic et al. (2008b) investigated thirty-three void fraction models and correlations from the available literature and compared them each other using relevant data. The friction factors, based on the analysis of Ma et al. (2004), are obtained from various void fraction models and correlations and a comparison was made with each other and also with those determined from graphical information provided by Bergelin et al. (1946). The presentation for the effect of void fraction alteration on the momentum pressure drop was also shown in their paper. Dalkilic et al. (2009a) compared some simple void fraction models of the annular flow pattern for the forced convection condensation of pure R134a taking into account the effect of the different saturation temperatures in high mass flux conditions. The calculated film thickness from void fraction models and correlations and those from Whalley's annular flow model (1987) were compared each other using their experimental database. Dalkilic and Wongwises (2010b) used Barnea et al. (1982)'s mathematical model, based on the momentum balance of liquid and vapour phases, in order to determine the condensation film thickness of R134a in their paper. The discussions for the effects of heat flux, mass flux and condensation temperature on the film thickness and condensation heat transfer coefficient were also made for laminar and turbulent flow conditions. Six well-known flow regime maps from the literature were found to be in good agreement for the annular flow conditions in the test tube in spite of their different operating conditions. Dalkilic et al. (2010e) used Kosky and Staub's model (1971) to predict flow pattern transitions and validate the results of void fraction models and correlations proposed in their previous publications and also show the identification of flow regimes in data corresponding to annular flow downward condensation of R134a in a vertical smooth copper tube. Furthermore, investigation of twelve number of well-known flow regime correlations from the literature is performed to identify the flow regime occurring in the test tube. Dalkilic et al. (2010d) calculated the average predicted heat transfer coefficient of the refrigerant using Kosky and Staub's model (1971) and the Von Karman universal velocity distribution correlations by means of different interfacial shear stress equations valid for annular flow in horizontal and vertical tubes in order to validate Chen et al.'s annular flow theory (1987). The discussions for the effects of heat flux, mass flux and condensation temperature on the pressure drop were also made in their paper. A new correlation including dimensionless parameters such as the equivalent Reynolds number, Prandtl number, R number (ρ-μ ratio), LoCkhart and Martinelli parameter, Bond number and Froude number was proposed in their paper using a large number of data points for the determination of turbulent condensation heat transfer coefficient. Dalkilic et al. (2010c) proposed a new experimental approach on the determination of condensation heat transfer coefficient in a vertical tube by means of von Karman's universal velocity distribution and Kosky and Staub's annular flow film thickness model (1971). They benefitted from thirteen numbers of frictional pressure drop models and thirty five numbers of void fraction models in their model. Dalkilic and Wongwises (2009d) reported a detailed review of research on in-tube condensation by reason of its significance in refrigeration, air conditioning and heat pump applications. The investigations into heat transfer and pressure drop during in-tube condensation encompassed a comprehensive summary of pertinent research topics, including studies on condensation heat transfer and pressure drop relative to tube orientation (horizontal, vertical, inclined) and tube geometry (smooth and enhanced), flow pattern analyses of condensation, void fraction assessments, and the interaction of refrigerants with oil. In addition to the aforementioned, several more conference papers (2008, 2009, 2010) were utilized to substantiate and corroborate their offered models and connections in their publications. The aforementioned studies indicate that Dalkilic and Wongwises conducted a thorough investigation of the in-tube condensation process utilizing their experimental setup, which functions as a double-tube heat exchanger in the test portion. Despite the availability of certain information in the literature, there persists an opportunity for additional research. The primary objective of this chapter is to examine the application of established empirical annular flow correlations to annular flow condensation at elevated mass flux in a vertical double tube heat exchanger. This chapter further demonstrates the independence of annular flow heat transfer empirical correlations from tube orientation (1987) and their universal applicability to vertical short tubes.

2. DATA REDUCTION

Figure 1 illustrates the steady-state physical model of downward film condensation within a vertical tube. Nusselt-type analysis is applicable under specific assumptions, including: laminar film flow; saturated vapor state of R134a; condensed R134a film along the tube surface; consistent physical parameters matching to inlet pressure and temperature; and absence of entrainment. An interfacial shear effect arises at the interface due to the significantly higher vapor velocity compared to the film velocity. The solution of the problem can be started from the calculation of the force balance in Eq. (1) for the differential element in the control volume neglecting the inertia and downstream diffusion contributions as shown in Dalkilic et al.'s study (2009c).

The force balance for the differential element in the control volume can be expressed as follows:

$$\rho_l g dx dy dz + \tau_\delta(y + dy) dx dz + P(z) dx dy = \tau_\delta(y) dx dz + P(z + dz) dx dy \tag{1}$$

It should be noted that the solution of the investigated case in this chapter will be different from Nusselt's solution (1916) due to the high mass flux condition of the condensate flow in the tube.

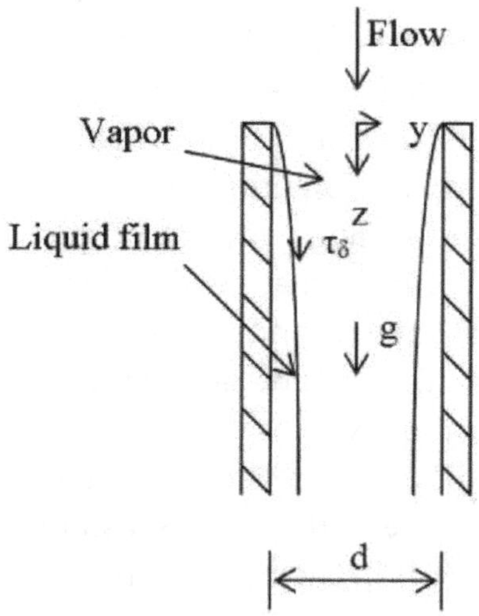

Fig. 1. System model for analysis of downward condensation

2.1 Experimental heat transfer coefficient

The details of determination of the experimental heat transfer coefficient by means of experimental setups can be seen in many papers given in references.

$$h_{\exp} = \frac{Q_{TS}}{A_i(T_{ref,sat} - T_{w,i})} \tag{2}$$

where h_{exp} is the experimental average heat transfer coefficient, QTS is the heat transfer rate in the test section, Twi is the average temperature of the inner wall, $T_{ref,sat}$ is the average temperature of the refrigerant at the test section inlet and outlet, and A_i is the inside surface area of the test section:

$$A_i = \pi dL \tag{3}$$

where d is the inside diameter of the test tube. L is the length of the test tube.

2.2 Uncertainties

The uncertainties of the Nusselt number and condensation heat transfer coefficient in the test tube, which belongs to a refrigerant loop consisting of an evaporator, test section and condenser, varied from ±7.64% to ±10.71%. The procedures of Kline and McClintoCk (1953) were used for the calculation of all uncertainties. Various uncertainty values of the study can be seen from Table 1 in detail.

Based on this usual method, suppose that a set of measurement is made and the uncertainty in each may be expressed with same odds. These measurements are then used to calculate some desired result of the experiments. The result P is a given function of the independent variables x_1, x_2, x_3, ...x_n. Thus:

$$P = P(x_1, x_2, x_3, \ldots\ldots, x_n) \tag{4}$$

Table 1. Uncertainty of experimental parameters

Parameters	Uncertainty
Tref, sat (°C)	0.19
xi	±6.96-8.24%
ΔT (K)	±0.191
$(T_{w,out}-T_{w,in})$TS (K)	±0.045
$(T_{w,in}-T_{w,out})$ph (K)	±0.13
m_{ref} (g s^{-1})	±0.023
$m_{w,TS}$ (g s^{-1})	±0.35
$m_{w,pre}$ (g s^{-1})	±0.38
q_{TS} (W m^{-2})	±6.55-8.93%
q_{pre} (W m^{-2})	±12-14.81%
h_{ref} (W $m^{-2}K^{-1}$)	±7.64-10.71%.
ΔP (kPa)	±0.15

Let w_p be the uncertainty in the result, and w_1, w_2, w_3, ...,w_n be the uncertainties in each independent variables. If the uncertainties in the independent variables are given with some odds, then the uncertainty in the result can be given as followed:

$$w_p = \pm\left[\left(\frac{\partial P}{\partial x_1}w_1\right)^2 + \left(\frac{\partial P}{\partial x_2}w_2\right)^2 + \left(\frac{\partial P}{\partial x_3}w_3\right)^2 + \ldots\ldots + \left(\frac{\partial P}{\partial x_n}w_n\right)\right]^{1/2} \tag{5}$$

2.2.1 Vapor quality (xin)

The vapor quality entering the test section is calculated from an energy balance, which gives the total heat transfer rate from hot water to liquid R134a in the evaporator as the sum of sensible and latent heat transfer rates, on the pre-heater:

$$x_{in} = \frac{1}{i_{ref,fg}}\left[\frac{Q}{m_{ref,T}} - C_{p_{ref}}\left(T_{ref,sat} - T_{ref,pre,in}\right)\right] \tag{6}$$

$$w_{x_{in}} = \pm\left[\left(\frac{\partial x_{in}}{\partial Q}w_Q\right)^2 + \left(\frac{\partial x_{in}}{\partial m_{ref,T}}w_m\right)^2 + \left(\frac{\partial x_{in}}{\partial T_{ref,sat}}w_{T_{ref,sat}}\right)^2 + \left(\frac{\partial x_{in}}{\partial T_{ref,pre,in}}w_{T_{ref,pre,in}}\right)^2\right]^{1/2} \tag{7}$$

$$w_{x_{in}} = \pm\left[\left(\frac{1}{m_{ref,T}\, i_{ref,fg}} w_Q\right)^2 + \left(\frac{Q}{i_{ref,fg}\, m_{ref}^{\ 2}} w_{m_{ref,T}}\right)^2 + \left(\frac{C_{p,ref}}{i_{ref,fg}} w_{T_{ref,sat}}\right)^2 + \left(\frac{Cp_{ref}}{i_{ref,fg}} w_{T_{ref,pre,in}}\right)^2\right]^{1/2} \quad (8)$$

2.2.2 Heat flux (Q)

The outlet quality of R134a from the evaporator is equal to the quality of R134a at the test section inlet. The total heat transferred in the test section is determined from an energy balance on the cold water flow in the annulus:

$$Q = m_w C_{p\ w}(T_{w,out} - T_{w,in}) \quad (9)$$

$$w_Q = \pm\left[\left(\frac{\partial Q}{\partial \dot{m}_w} w_{m_w}\right)^2 + \left(\frac{\partial Q}{\partial T_{w,in}} w_{T_{w,in}}\right)^2 + \left(\frac{\partial Q}{\partial T_{w,out}} w_{T_{w,out}}\right)^2\right]^{1/2} \quad (10)$$

2.2.3 Average heat transfer coefficient (h_i)

The refrigerant side heat transfer coefficient in Eq. (2) is determined from the total heat transferred in the test section which is a vertical counter-flow tube-in-tube heat exchanger with refrigerant flowing in the inner tube and cooling water flowing in the annulus:

$$w_{h_i} = \left[\left(\frac{\partial h_i}{\partial Q} w_Q + \frac{\partial h_i}{\partial\left(T_{ref,sat} - T_{ref,w,i}\right)} w_{(T_{ref,sat} - T_{ref,w,i})}\right)\right] \quad (11)$$

2.2.4 Nusselt number (Nu)

Nusselt number can be expressed as follows:

$$Nu = \frac{h_i L}{k_l} \quad (12)$$

$$w_{Nu} = \pm\left[\left(\frac{\partial Nu}{\partial h_i} w_{h_i}\right)^2\right]^{1/2} \quad (13)$$

2.2.5 Temperature difference

The temperature measurements for the determination of outer surface temperature of the test tube are evaluated as an average value for the uncertainty analysis and calculation procedure can be seen as follows:

$$T_0 = \frac{T_1 + T_2 + T_3 + \ldots.. + T_{10}}{10} \quad (14)$$

$$w_{T_0} = \pm\left[\left(\frac{\partial T_0}{\partial T_1} w_{T_1}\right)^2 + \left(\frac{\partial T_0}{\partial T_2} w_{T_2}\right)^2 + \dots + \left(\frac{\partial T_0}{\partial T_{10}} w_{T_{10}}\right)^2\right]^{1/2} \quad (15)$$

$$w_{T_0} = \pm\left[\left(\frac{1}{10} w_{T_1}\right)^2 + \left(\frac{1}{10} w_{T_2}\right)^2 + \dots + \left(\frac{1}{10} w_{T_{10}}\right)^2\right]^{1/2} \quad (16)$$

3. HEAT TRANSFER CORRELATIONS

The comparison of convective heat transfer coefficients with Shah correlation (1979) has been done by researchers commonly for turbulent condensation conditions especially in vertical tubes (2000) and it is found to be the most comparative condensation model during annular flow regime in a tube (2002). Shah correlation (1979) is based on the liquid heat transfer coefficient and is valid for Re1 ≥ 350. A two-phase multiplier is used for the annular flow regime of high pressure steam and refrigerants (2000, 2002) in the equation.

Dobson and Chato (1998) used a two-phase multiplier to develop a correlation for an annular flow regime. A correlation is also provided for a wavy flow regime. The researchers used commonly their correlations for zeotropic refrigerants. Its validated mass flux (G) covers the values more than 500 kg m-2 s-1 for all qualities in horizontal tubes.

Sweeney (1996) developed the Dobson and Chato model (1998) using zeotropic mixtures for annular flow.

Cavallini et al. (1974) used various organic refrigerants for the condensation inside tubes in both vertical and horizontal orientations and developed a semi empirical correlation as a result of their study.

Bivens and Yokozeki (1994) modified Shah correlation (1979) benefitted from various flow patterns of R22, R502, R32/R134a, R32/R125/R134a.

Tang et al. (2000) developed the Shah [(1979) equation for the annular flow condensation of R410A, R134a and R22 in i.d. 8.81 mm tube with Fr_{so}>7.

Fujii (1995) modified the correlation in Table 2 for shear-controlled regimes in smooth tubes. There is also another correlation belong to Fujii (1995) for gravity controlled regimes.

Chato (1961) used a two-phase multiplier to modifiy Dittus-Boelter's correlation (1930) for an annular flow regime.

Traviss et al. (1972) foCused on the flow regime maps for condensation inside tubes. Their correlation was suggested for the condensation of R134a inside tubes specifically. Their model takes into account of the variation in the quality of the refrigerant using LoCkhart- Martinelli parameter.

Akers and Rosson (1960) developed the Dittus–Boelter (1930)'s single-phase forced convection correlation. Their correlation's validity range covers the turbulent annular flow in small diameter circular tubes and rectangular channels.

Tandon et al. (1995) developed the Akers and Rosson (1960) correlation for shear controlled annular and semi-annular flows with $Re_g > 30000$. There is another correlation of Tandon et al. (1995) for gravity-controlled wavy flows with $Re_g < 30000$.

4. RESULTS AND DISCUSSION

It should be noted that it is possible for researcher to identify the experimental data of condensation by means of both flow regime maps and sight glass at the inlet and outlet of the test section. Dalkilic and Wongwises (2009b, 2010b, 2010e) checked the data shown in all figures and formulas that they were in an annular flow regime by Hewitt and Robertson's (1969) flow pattern map and also by sight glass in their experimental setup. The vapor quality range approximately between 0.7-0.95 in the 0.5 m long test tube was kept in order to obtain annular flow conditions at various high mass fluxes of R134a.

In Table 2, the list of correlations is evaluated to show the similarity of annular flow correlations which are independent of tube orientation (horizontal or vertical). Chen et al. (1987) also developed a general correlation to discuss this similarity in their article by relating the interfacial shear stress to flow conditions for annular film condensation inside tubes.

Table 2. Annular flow heat transfer correlations and models

Researcher	Model/Correlation
Shah (1979)	$\mathrm{Re}_l \geq 350$ $\quad \mathrm{Re}_l = \dfrac{Gd(1-x)}{\mu_l}$ $\quad h_{shah} = h_{sf}\left(\dfrac{1.8}{Co^{0.8}}\right)$ $\quad P_{red} = \dfrac{P_{sat}}{P_{critic}}$ $h_{sf} = h_l(1-x)^{0.8}$ $\quad Co = \left(\dfrac{1}{x}-1\right)^{0.8}\left(\dfrac{\rho_g}{\rho_l}\right)^{0.5}$ $\quad h_l = \dfrac{k_l}{d}\left(0.023\left(\dfrac{\mathrm{Re}_l}{1-x}\right)^{0.8}\mathrm{Pr}_l^{0.4}\right)$
Dobson and Chato (1998)	For $G > 500$ kg m^{-2} s^{-1} $\quad Nu = 0.023\mathrm{Re}_l^{0.8}\mathrm{Pr}_l^{0.4}\left[1+\dfrac{2.22}{X^{0.89}}\right]$ $\quad Ga = \dfrac{\rho_l(\rho_l-\rho_g)gd^3}{\mu_l^2}$ $X = \left(\dfrac{1-x}{x}\right)^{0.9}\left(\dfrac{\rho_g}{\rho_l}\right)^{0.5}\left(\dfrac{\mu_l}{\mu_g}\right)^{0.1}$ $\quad Fr_{so} = c_3\mathrm{Re}_l^{c_4}\left(\dfrac{1+1.09X^{0.039}}{X}\right)^{1.5}\dfrac{1}{Ga^{0.5}}$ $\mathrm{Re}_l = \dfrac{Gd(1-x)}{\mu_l}$ $\quad$ For $\mathrm{Re}_l > 1250$ and $Fr_{so} > 18$ $\quad Nu = 0.023\mathrm{Re}_l^{0.8}\mathrm{Pr}_l^{0.3}\dfrac{2.61}{X^{0.805}}$

Researcher	Model/Correlation
Sweeney (1996)	$Nu = 0.7\left(\frac{G}{300}\right)^{0.3} Nu_{Dobson-Chato}$
Cavallini et al. (1974)	$Nu_l = 0.05\mathrm{Re}_{eq}^{0.8}\mathrm{Pr}^{0.33}$ $\mathrm{Re}_{eq} = \mathrm{Re}_g(\mu_g/\mu_l)(\rho_l/\rho_g)^{0.5}\mathrm{Re}_l$ $\mathrm{Re}_g = Gdx/\mu_g$
,Bivens and Yokozeki (1994)	$Nu = Nu_{Shah}\left(0.78738 + \frac{6187.89}{G^2}\right)$ $P_{red} = P_{sat}/P_{critic}$ $Nu_{Shah} = 0.023\mathrm{Re}_l^{0.8}\mathrm{Pr}_l^{0.4}\left[1 + \frac{3.8}{P_{red}^{0.38}}\left(\frac{x}{1-x}\right)^{0.76}\right]$
Tang et al. (2000)	$Nu = 0.023\mathrm{Re}_l^{0.8}\mathrm{Pr}_l^{0.4}\left[1 + 4.863\left((-\ln(P_{red})\frac{x}{1-x}\right)^{0.836}\right]$ $\mathrm{Re}_l \leq 1250$ $c_3 = 0.025$ $c_4 = 1.59$ $Fr_{so} = c_3\mathrm{Re}_l^{c_4}\left(\frac{1+1.09X^{0.039}}{X}\right)^{1.5}\frac{1}{Ga^{0.5}}$ $\mathrm{Re}_l > 1250$ $c_3 = 1.26$ $c_4 = 1.04$
Fujii (1995)	$Nu_l = 0.0125\left(\mathrm{Re}_l\sqrt{\rho_l/\rho_g}\right)^{0,9}\left(\frac{x}{1-x}\right)^{0.1x+0.8}\mathrm{Pr}_l^{0.63}$
Chato (1961)	$Nu = 0.023\mathrm{Re}_l^{0.8}\mathrm{Pr}_l^{0.4}\left[\frac{2.47}{X^{1.96}}\right]$
Traviss et al. (1972)	$Nu = \mathrm{Re}_l^{0.9}\mathrm{Pr}_l\frac{F_1(X)}{F_2(\mathrm{Re}_l,\mathrm{Pr}_l)}$ $F_1(X) = 0.15\left[\frac{1}{X} + \frac{2.83}{X^{0.476}}\right]$ $\mathrm{Re}_l > 1125$ $F_2 = 5\mathrm{Pr}_l + 5\ln(1+5\mathrm{Pr}_l) + 2.5\ln(0.00313\mathrm{Re}_l^{0.812})$
Akers and Rosson (1960)	$Nu = 0.0265\mathrm{Re}_{eq}^{0.8}\mathrm{Pr}_l^{1/3}$ $\mathrm{Re}_{eq} = \frac{G_{eq}d}{\mu_l}$ $G_{eq} = G\left[(1-x) + x(\rho_l/\rho_g)^{0.5}\right]$
Tandon et al. (1995)	$\mathrm{Re}_g > 30000$ $Nu = 0.084\mathrm{Re}_g^{0.67}\mathrm{Pr}_l^{1/3}\left[\frac{1}{Ja_l}\right]^{1/6}$ $Ja_l = \frac{Cp_l\Delta T_{sat}}{i_{fg}}$

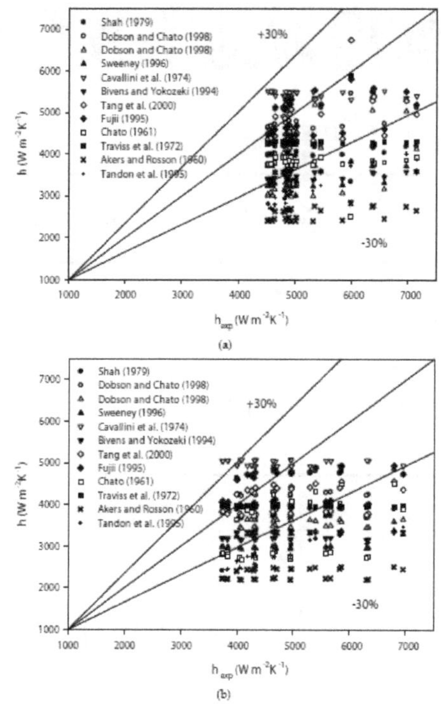

Fig. 2. Comparison of experimental condensation heat transfer coefficient vs. various correlations for G=300 kg m^{-2} s^{-1} (a) T_{sat}=40 °C and (b) T_{sat}=50 °C

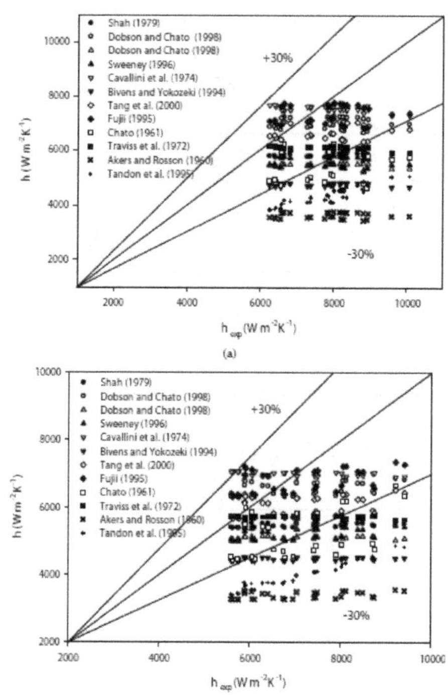

Fig. 3. Comparison of experimental condensation heat transfer coefficient vs. various correlations for G=456 kg m^{-2} s^{-1} (a) T_{sat}=40 °C and (b) T_{sat}=50 °C

The comparison of experimental heat transfer coefficients with various annular flow correlations are shown in Figs. 2-3 in a 30% deviation line for the condensation temperatures of 40 and 50 °C and mass fluxes of 300 and 456 kg $m^{-2}s^{-1}$ respectively. It can be clearly seen from these figures that the Dobson and Chato (1998) correlation, Cavallini et al. (1974) correlation, Fujii (1995) correlation are in good agreement with the experimental data. In addition to this, the majority of the data calculated by Shah (1979) correlation, Dobson and Chato (1998) correlation, Tang et al. (2000) correlation, Traviss et al. (1972) correlation fall within ±30%, whereas, Tandon et al. (1995) correlation and Akers and Rosson (1960) correlation are found to be incompatible with the experimental data. Nonetheless, the Chato (1961) correlation, the Sweeny (1996) correlation, and the Bivens and Yokozeki (1994) correlation are found to have poor agreement with experimental data.

In the literature, some correlations are developed for gravity-controlled regimes such as Fujii (1995) and Tandon et al. (1995) correlations. In this chapter, the correlations, proposed for gravity-controlled regimes, are not found to be in good agreement with the data as expected for annular flow regime. These kinds of correlations do not exist in this chapter due to their large deviations and validity for wavy flow. Chen et al. (1987) reported that the vapor shear stress acting on the interface of vapor-liquid phases affects the forced convective condensation inside tubes especially at high vapor flow rates. On that account, Fujii (1995) and Tandon et al. (1995)'s shear-controlled correlations were used to predict condensation heat transfer coefficient of R134a. Furthermore, Valladares (2003) obtained similar results on these explanations in this chapter for the condensation of various refrigerants in horizontal tubes by Valladares (2003).

5. CONCISE SUMMARY

This chapter proposes a method to ascertain the average heat transfer coefficient of R134a during condensation in vertical downhill flow at elevated mass flux in a smooth tube. The comparison between the various annular flow heat transfer correlations and experimental heat transfer coefficients is shown with ±30% deviation line. It can be noted that the Dobson and Chato (1998) correlation, the Cavallini et al. (1974) correlation, the Fujii (1995) correlation are found to have the most predictive results than others in an 8.1 mm i.d. copper tube for the mass fluxes of 300 and 456 kg m-2s-1 and condensation temperatures of 40 and 50 °C. This chapter's analysis demonstrates that annular flow models are orientation-independent, given the presence of an annular flow regime along the tube length, and can accurately predict condensation heat transfer coefficients within vertical test tubes, despite the majority of these correlations being formulated for annular flow condensation in horizontal tubes.

6. NOMENCLATURE

A surface area, m^2

Cp specific heat, J $kg^{-1}K^{-1}$

d internal tube diameter, m

Fr Froude number

G mass flux, kg m^{-2} s^{-1}

Ga Galileo number

g gravitational constant, m s^{-2}

h convective heat transfer coefficient, W $m^{-2}K^{-1}$

i entalphy, J kg^{-1}

i_{fg} latent heat of condensation, J kg^{-1}

Ja Jakob number

k thermal conductivity, W $m^{-1}K^{-1}$

L length of test tube, m

m mass flow rate, kg s^{-1}

Nu Nusselt number

P pressure, MPa

Pr Prandtl number

Re Reynolds number

T temperature

Q heat transfer rate, W

q mean heat flux, kW m^{-2}

y wall coordinate

x mean vapor quality

y radial coordinate

x axial coordinate

X LoCkhart-Martinelli parameter w uncertainty

Greek Symbols

ΔT vapor side temperature difference, $T_{ref,sat}$ -T_{wi}, °C

ρ density, kg m^{-3}

μ dynamic viscosity, kg $m^{-1}s^{-1}$

τ shear stress, N m^{-2}

Subscripts

eq equivalent

exp experimental

g gas/vapor

i inside

in inlet

l liquid

out outlet

pre preheater

red reduced

ref refrigerant

sat saturation

so Soliman

T total

TS test section

w water

wi inner wall

δ film thickness

7. REFERENCES

Akers, W.W., Rosson, H.F. (1960). Condensation inside a horizontal tube, *Chemical Engineering Progress Symposium Series*, Vol. 56, No. 30, 145-149.

Akers, W.W., Deans, A., Crosser, O.K. (1959). Condensing heat transfer within horizontal tubes. *Chemical Engineering Progress Symposium Series*, Vol. 55, No. 29, 171-176.

Barnea, D., Shoham, O., Taitel, Y. (1982). Flow pattern transition for vertical downward two phase flow. *Chemical Engineering Science*, Vol. 37, No. 5, 741-744.

Bassi, R., Bansal, P.K. (2003). In-tube condensation of mixture of R134a and ester oil: emprical correlations. *International Journal of Refrigeration*, Vol. 26, No. 4, 402-409.

Bellinghausen, R., Renz, U. (1992). Heat transfer and film thickness during condensation of steam flowing at high velocity in a vertical pipe. *International Journal of Heat and Mass Transfer*, Vol. 35, No. 3, 683-689.

Bivens, D.B., Yokozeki, A. (1994). Heat transfer coefficient and transport properties for alternative refrigerants. *ProC. 1994 Int. Refrigeration Conference*, 299-304, Indiana.

Bergelin, O.P., Kegel, P.K., Carpenter, F.G., Gazley, C. (1946). Co-current gas–liquid flow. II. Flow in vertical tubes. *ASME Heat Transfer and Fluid Mechanic Institute*, 19-28.

Briggs, A., Kelemenis, C., Rose, J.W. (2000). Heat transfer and pressure drop measurements for in-tube condensation of CFC-113 using microfin tubes and wire inserts. *Experimental Heat Transfer*, Vol. 13, No. 3, 163-181.

Briggs, A., Kelemenis, C., Rose, J.W. (1998). Condensation of CFC-113 with downflow in vertical, internally enhanced tubes. *ProCeedings of 11th IHTC*, August 23-28, Kyongju.

Carey, V.P. (1992). Liquid-Vapor Phase Change Phenomena, *Hemisphere Publishing*, 1992.

Cavallini, A., Zecchin, R. (1974). A dimensionless correlation for heat transfer in forced convection condensation. *6th International Heat Transfer Conference*, Vol . 3, 309- 313, Tokyo.

Cavallini, A., Censi, G., Del Col, D., Doretti, L., Longo, G.A., Rossetto, L., Zilio, C. (2003). Condensation inside and outside smooth and enhanced tubes - a review of recent research. *International Journal of Refrigeration*, Vol. 26, No. 4, 373-392.

Chato, J.C. (1961), Laminar Condensation inside horizontal and inclined tubes. *ASHRAE Journal*, Vol. 4, 52-60.

Chen, S.L., Gerner, F.M., Tien, C.L. (1987). General film condensation correlations. *Experimental Heat Transfer*, Vol. 1, No. 2, 93-107.

Dalkilic, A.S., Laohalertdecha, S., Wongwises, S. (2008). A comparison of the void fraction correlations of R134a during condensation in vertical downward laminar flow in a smooth and microfin tube. *ProCeedings of the Micro/Nanoscale Heat Transfer International Conference*, Parts A-B, 1029-1040, Tainan.

Dalkilic, A.S., Laohalertdecha, S., Wongwises, S. (2008). Two-phase friction factor obtained from various void fraction models of R-134a during condensation in vertical downward flow at high mass flux. *ProCeedings of the ASME Summer Heat Transfer Conference*, Vol. 2, 193-206, Jacksonville.

Dalkilic, A.S., Laohalertdecha, S., Wongwises, S. (2008). Two-phase friction factor in vertical downward flow in high mass flux region of refrigerant HFC-134a during condensation. *International Communications in Heat and Mass Transfer*, Vol. 35, No. 9, 1147-1152. (a)

Dalkilic, A.S., Laohalertdecha, S., Wongwises, S. (2008). Effect of void fraction models on the two-phase friction factor of R134a during condensation in vertical downward flow in a smooth tube. *International Communications in Heat and Mass Transfer*, Vol. 35, No. 8, 921-927. (b)

Dalkilic, A.S., Laohalertdecha, S., Wongwises, S. (2009). Experimental investigation on the condensation heat transfer and pressure drop characteristics of R134a at high mass flux conditions during annular flow regime inside a vertical smooth tube. *ASME Summer Heat Transfer Conference*, July 19-23, San Francisco.

Dalkilic, A.S., Agra, O. (2009). Experimental apparatus for the determination of condensation heat transfer coefficient for R134a and R600a flowing inside vertical and horizontal tubes. *ASME Summer Heat Transfer Conference*, July 19-23, San Francisco.

Dalkilic, A.S., Laohalertdecha, S., Wongwises, S. (2009). Experimental research on the similarity of annular flow models and correlations for the condensation of R134a at high mass flux inside vertical and horizontal tubes. *ASME International Mechanical Engineering Congress and Exposition*, November 13-19, Lake Buena Vista.

Dalkilic, A.S., Wongwises, S. (2009). A heat transfer model for co-current downward laminar film condensation of R134a in a vertical micro-fin tube during annular flow regime. *The Eleventh UK National Heat Transfer Conference, Queen Mary University of London,* September 6-8, London.

Dalkilic, A.S., Laohalertdecha, S., Wongwises, S. (2009). Effect of void fraction models on the film thickness of R134a during downward condensation in a vertical smooth tube. *International Communications in Heat and Mass Transfer*, Vol. 36, No. 2, 172-179. (a)

Dalkilic, A.S., Laohalertdecha, S., Wongwises, S. (2009). Experimental investigation of heat transfer coefficient of R134a during condensation in vertical downward flow at high mass flux in a smooth tube. *International Communications in Heat and Mass Transfer*, Vol. 36, No. 10, 1036-1043. (b)

Dalkilic, A.S., Yildiz, S., Wongwises, S. (2009). Experimental investigation of convective heat transfer coefficient during downward laminar flow condensation of R134a in a vertical smooth tube. *International Journal of Heat and Mass Transfer*, Vol. 52, No. 1-2,142-150. (c)

Dalkilic, A.S., Wongwises, S. (2009). Intensive literature review of condensation inside smooth and enhanced tubes. *International Journal of Heat and Mass Transfer*, Vol. 52, No. 15-16, 3409-3426. (d)

Dalkilic, A.S., Laohalertdecha, S., Wongwises, S. (2010). Comparison of condensation frictional pressure drop models and correlations during annular flow of R134a inside a vertical tube. *ASME-ATI-UTI Thermal and Environmental Issues in Energy Systems*, May 16-19, Sorrento.

Dalkilic, A.S., Wongwises, S. (2010). Experimental study on the flow regime identification in the case of co-current condensation of R134a in a vertical smooth tube. *ASME International Heat Transfer Conference,* August 8-13, Washington.

Dalkilic, A.S., Agra, O., Teke, I., Wongwises, S. (2010). Comparison of frictional pressure drop models during annular flow condensation of R600a in a horizontal tube at low mass flux and of R134a in a vertical tube at high mass flux, *International Journal of Heat and Mass Transfer,* Vol. 53, No. 9-10, 2052-2064. (a)

Dalkilic, A.S., Wongwises, S. (2010). An investigation of a model of the flow pattern transition mechanism in relation to the identification of annular flow of R134a in a vertical tube using various void fraction models and flow regime maps. *Experimental Thermal and Fluid Science*, Vol. 34, No. 6, 692-705. (b)

Dalkilic, A.S., Laohalertdecha, S., Wongwises, S. (2010). New experimental approach on the determination of condensation heat transfer coefficient using frictional pressure drop and void fraction models in a vertical tube. *Energy Conversion and Management*, Vol. 51, No. 12, 2535-2547. (c)

Dalkilic, A.S., Wongwises, S. (2011). Experimental study on the modeling of condensation heat transfer coefficients in high mass flux region of refrigerant HFC-134a inside the vertical smooth tube during annular flow regime. *Heat Transfer Engineering*, Vol. 32, No. 1, 1-12. (d)

Dalkilic, A.S., Wongwises, S. (2010). Validation of void fraction models and correlations using a flow pattern transition mechanism model in relation to the identification of annular vertical downflow in-tube condensation of R134a. International Communications in Heat and Mass Transfer Vol. 37, No. 7, 827-834. (e)

Dittus, F.W., Boelter, L.M.K. (1930). Heat transfer in automobile radiators of the tubular type, *University of California Publications on Engineering, Berkeley, CA*, Vol. 2, No. 13, 443-461.

Dobson, M.K., Chato, J.C. (1998). Condensation in smooth horizontal tubes. *Journal of Heat Transfer-Transactions of ASME*, Vol. 120, No. 1, 193-213.

Fujii, T. (1995). Enhancement to condensing heat transfer-new developments, *Journal of Enhanced Heat Transfer*, Vol. 2, No. 1-2, 127-137.

Hewitt, G.F., Robertson, D.N. (1969). Studies of two-phase flow patterns by simultaneous x-ray and flash photography. Rept AERE-M2159, UKAEA, Harwell.

Jung, D., Song, K., Cho, Y., Kim, S. (2003). Flow condensation heat transfer coefficients of pure refrigerants. *International Journal of Refrigeration*, Vol. 26, No. 1, 4-11.

Jung, D., Cho, Y., Park, K. (2004). Flow condensation heat transfer coefficients of R22, R134a, R407C and R41A inside plain and micro-fin tubes. *International Journal of Refrigeration*, Vol. 27, No. 5, 25-32.

Kim, S.J., No, H.C. (2000). Turbulent film condensation of high pressure steam in a vertical tube. *International Journal of Heat and Mass Transfer*, Vol. 43, No. 21, 4031- 4042.

Kline, S.J., McClintoCk, F.A. (1953). Describing uncertainties in single sample experiments.

Journal of the Japan SoCiety of Mechanical Engineers, Vol. 75, No. 1, 3-8.

Kosky, P.G., Staub, F.W. (1971). local condensing heat transfer coefficients in the annular flow regime, *AIChE Journal*, Vol. 17, No. 5, 1037-1043.

Liebenberg, L., Bukasa, J.P., Holm, M.F.K., Meyer, J.P., Bergles, A.E. (2002). Towards a unified approach for modelling of refrigerant condensation in smooth tubes. *ProCeedings of the International Symposium on Compact Heat Exchangers*, 457-462, Grenoble.

LoCkhart, R.W., Martinelli, R.C. (1949). Proposed correlation of data for isothermal two- phase, two-component flow in pipes. *Chemical Engineering Progress*, Vol. 45, No. 1, 39-48.

Ma, X., Briggs, A., Rose, J.W. (2004). Heat transfer and pressure drop characteristics for condensation of R113 in a vertical micro-finned tube with wire insert. *International Communications in Heat and Mass Transfer*, Vol. 31, No. 5, 619-627.

Maheshwari, N.K., Sinha, R.K., Saha, D., Aritomi, M. (2004). M., Investigation on condensation in presence of a noncondensible gas for a wide range of Reynolds number. *Nuclear Engineering and Design*, Vol. 227, No. 2, 219-238.

Moser, K., Webb, R.L., Na, B. (1998). A new equivalent Reynolds number model for condensation in smooth tubes. *International Journal of Heat Transfer*, Vol. 120, No. 2, 410-417.

Nusselt, W. (1916). Die oberflachen-kondensation des wasserdampfer. *Zeitschrift des Vereines Deutscher Ingenieure*, Vol. 60, No. 27, 541-569.

Oh, S., Revankar, A. (2005). Analysis of the complete condensation in a vertical tube passive condenser. *International Communications in Heat and Mass Transfer*, Vol. 32, No. 6, 716-722.

Shah, M.M. (1979). A general correlation for heat transfer during film condensation inside pipes. *International Journal of Heat and Mass Transfer*, Vol. 22, No. 4, 547-556. (3)

Sweeney, K.A. (1996). The heat transfer and pressure drop behavior of a zeotropic refrigerant mixture in a micro-finned tube. *M.S. thesis*, Dept. of Mechanical and Industrial Engineering, University of Illinois at Urbana-Champaign.

Tandon, T.N., Varma, H.K., Gupta, C.P. (1995). Heat transfer during forced convection condensation inside horizontal tube. *International Journal of Refrigeration*, Vol. 18, No. 3, 210-214.

Tang, L., Ohadi, M.M., Johnson, A.T. (2000). Flow condensation in smooth and microfin tubes with HCFC-22, HFC-134a, and HFC-410 refrigerants. part II: Design equations. *Journal of Enhanced Heat Transfer*, Vol. 7, No. 5, 311–325.

Travis, D.P., Rohsenov, W.M., Baron, A.B. (1972). Forced convection inside tubes: a heat transfer equation for condenser design. *ASHRAE Transactions*, Vol. 79, No. 1, 157- 165.

Valladeres, O.G. (2003). Review of in-tube condensation heat transfer coefficients for smooth and microfin tubes. *Heat Transfer Engineering*, Vol. 24, No. 4, 6-24.

Wang, H.S., Honda, H. (2003). Condensation of refrigerants in horizontal microfin tubes: comparison of prediction methods for heat transfer. *International Journal of Refrigeration*, Vol. 26, No. 4, 452-460.

Whalley, P.B. (1987). Boiling, condensation, and gas-liquid flow, *Oxford University Press*.

3 Chapter

IMPROVED BOILING HEAT TRANSFER FROM MICRO-PIN-FINNED SILICON CHIPS

1. INTRODUCTION: AN OVERVIEW

A computer mostly consists of chips that have several manufactured semiconductor switches. The demand for enhanced signal speed in computers has directed the electronics industry's efforts towards the creation of miniaturized electronic circuits and the development of high circuit density in chips. The integration technologies, having progressed to the very large scale integration (VLSI) level, result in an elevated power dissipation rate at the chip, module, and system levels. Sophisticated electronic cooling technology is needed to maintain relatively constant component temperature below the junction temperature, approximately 85°C for most mainframe memory and logic chips. Investigations have demonstrated that a single component operating 10°C beyond this temperature can reduce the reliability of some systems by as much as 50% (Nelson, 1978).

Traditionally, convection heat transfer from electronic hardware to the surroundings has been achieved through the natural, forced, or mixed convection of air; however, even with advances in air-cooling techniques, the improvements will not suffice to sustain the expected higher heat fluxes. As an effective and increasingly-popular alternative to air cooling, directly immersing the component in inert, dielectric liquid can remove a large amount of heat dissipation, of which pool and forced boiling possess the attractive attribute of large heat transfer coefficient due to phase change compared with single-phase.

An ideal boiling performance should provide adequate heat removal within acceptable chip temperatures. Direct liquid cooling, involving boiling heat transfer, by use of dielectric liquids has been considered as one of the promising cooling schemes. Primary issues related to liquid cooling of microelectronics components are mitigation of the incipience temperature overshoot, enhancement of established nucleate boiling and elevation of critical heat flux (CHF). Treated surface has been found to have great potential in enhancement of boiling heat transfer from electronic, significantly reducing the chip surface temperature and increasing CHF. Treated surfaces are used for nucleate boiling enhancement by applying some micro-structures on the chip surface to make the surface capable of trapping vapour and keeping the nucleation sites active or increasing effective heat transfer area. Since the 1970s, a number of active studies have dealt with the enhancement of boiling heat transfer from electronic components by use of surface microstructures that were fabricated directly on a silicon chip or on a simulated chip. These include a sand-blasted and KOH treated surface (Oktay 1982), a "dendritic heat sink" (brush-like structure) (Oktay and Schemekenbecher 1972), laser drilled cavities (3-15μm in mouth dia.) (Hwang and Moran 1981), re-entrant cavities (0.23-0.49 mm in mouth dia.) (Phadke et al. 1992).

Messina and Parks (1981) used flat plate copper surfaces sanded with 240 and 600 grit sandpaper to boil R-113 and found the sandpaper finished surfaces were very efficient in improving boiling heat transfer and elevating CHF as compared to a smooth surface, with 240 grit more efficient than 600 grit. Anderson and mudawar (1989) roughened a 12.7-mm square copper surface by longitudinal sanding with a 600 grit silicon wet/dry sand paper to examine the effect of roughness on boiling heat transfer of FC-72. The roughness was 0.6-1.0 μm and the roughened surface produced an earlier boiling incipience than smooth surface and shifted the boiling curve toward a reduced wall superheat. However, the CHF value was not affected by the roughness as compared with the results obtained by Messina and Parks (1981). Chowdhury and Winterton (1985) found nucleate boiling heat transfer improved steadily as the surface roughness level was increased. However, when they anodized a roughened surface covered with cavities of around 1 μm size, which had hardly any effect on the roughness, they observed that the nucleate boiling curves were virtually independent of roughness. They asserted that it was not roughness in itself but the number of active nucleation sites that influenced nucleate boiling heat transfer.

Oktay and Schmeckenbecher (1972) developed a brush-like structure called "dendritic heat sink" mounted on a silicon chip surface, and the thickness of the dendrite was 1 mm. The incipience boiling temperature in saturated FC-86 could be reduced to 60°C due to the high density of re-entrant and possibly doubly re-entrant cavities provided by the dendritic heat sinks, and an increase in CHF compared with a smooth surface was attributed, by the authors, to the deferred creation of Taylor instability on the dendritic surface.

Chu and Moran (1977) developed a laser-treated surface on a silicon chip, which consisted of drilling an array of cavities ranging in average mouth diameter from 3 to 15 μm staggered 0.25-mm centers. Boiling data in FC-86 revealed that the wall superheat at any particular heat flux decreased, and the critical heat flux was increased by 50%.

Phadke et al. (1992) used a re-entrant cavity surface enhancement for immersion cooling of silicon chip. The pool boiling heat transfer characteristics of the cavity enhanced surfaces were superior to those of a smooth surface, resulting in a substantial decrease in both the temperature overshoot and the incipient boiling heat flux.

Kubo et al. (1999) experimentally studied boiling heat transfer of FC-72 from micro- reentrant cavity surfaces of silicon chips. The effects of cavity mouth size (about 1.6μm and 3.1μm) and the cavity number density (811/cm2 and 9600/cm2) were also investigated. The heat transfer performance of the treated surface was considerably higher than that of the smooth surface. The highest performance was obtained by a treated surface with larger cavity mouth diameter and cavity number density.

Nakayama et al. (1982) developed a tunnel structure, in which parallel rectangular cross- sectional grooves with the dimensions of 0.25×0.4 mm2 (width×depth) were firstly gouged with a pitch of 0.55 mm on a copper surface (20×30 mm2), then covered by a thin copper plate having rows of 50-to-150 μm diameter pores. R-11 was boiled and the wall superheat was reduced as compared to a smooth surface. They attributed the boiling enhancement to the liquid suction and evaporation inside the grooves. Later, Nakayama et al. (1989) used a 5-mm high porous copper stud with micro-channels to enhance boiling heat transfer of dielectric fluid FC-72. The porous stud could reduce the threshold superheat for the boiling incipience and increasing CHF. The boiling heat transfer levelled off with further increasing stud height.

Anderson and Mudawar (1989) also attached mechanically manufactured cavities, micro- fins and micro-pin-fins to vertical 12.7 mm square copper chips immersed in a stagnant pool of FC-72. They found that large artificial cavities with the mouth diameter of 0.3 mm were incapable of maintaining a stable vapour embryo and had only a small effect on boiling heat transfer compared with a smooth surface, while micro-finned and micro-pin-finned surfaces significantly enhanced the nucleate boiling mainly due to a heat transfer area increase. The micro-pin-finned surface with the fin dimensions of 0.305×0.305×0.508 mm^3 (width×thickness×height) provided CHF values in excess of 50 W/cm^2 and 70 W/cm^2 for the liquid subcoolings of 0 and 35K, respectively.

In 1990's, You and his co-researchers made a noticeable progress in nucleate boiling enhancement by use of a series of micro-porous surfaces. You et al. (1992) applied a 0.3-3.0 μm alumina particle treatment on a simulated electronic chip surface with spraying method and tested in FC-72. Compared with a smooth reference surface, a reduction of 50% in incipient and nucleate boiling superheats and an increase of 32% in the CHF were realized. O'Connor and You (1995) further used the spraying application to apply the alumina particles (0.3-5.0 μm) on a simulated electronic chip surface. The enhancement of nucleate boiling heat transfer showed excellent agreement with those observed by You et al. (1992) with an exception of a much higher CHF increase (47% increase) due to the increased heater thickness (1 mm aluminium nitride) which provided CHF data free from thermal conductance/capacitance effects.

In their subsequent studies, O'Connor and You (1995), O'Connor et al. (1996) painted 3- 10μm silver flakes or 8-12 μm diamond particles on the copper surface. Chang and You (1996, 1997) used 1-50μm copper particles and 1-20 μm aluminium particles to form porous coatings. These micro-porous coating surfaces showed almost identical high boiling enhancement with a reduced incipient superheat, increased nucleate boiling heat transfer coefficient and CHF as compared to an unenhanced surface. These performance enhancements were due to the creation of micro-porous structures on the heater surfaces which significantly increased the number of active nucleation sites.

Bergles and Chyu (1982) reported a pool boiling from a commercial porous metallic matrix surface. Working fluids were R-113 and water. The excellent steady boiling characteristics of this type of surface were confirmed, however, high wall superheat were required in most cases to initiate the boiling. From the previously mentioned investigations, it is apparent that surface microstructure of the correct size plays an important role in the enhancement of boiling heat transfer. Most treated surfaces can reduce the boiling incipience temperature, improve the nucleate boiling heat transfer and increase CHF. However, the enhancement often deteriorated greatly in the high heat flux region, especially near CHF resulting in a too high wall temperature at the CHF point as compared to the maximum allowable temperature for the normal operation of LSI chips, making the enhancement not so sound in practical high-powered electronics cooling application.

Mudawar's group (Ujereh et al. 2007) studied the nucleate pool boiling enhancement by use of carbon nanotube (CNT) arrays, and found CNTs were quite effective in reducing incipience superheat and enhancing the boiling heat transfer coefficient.

Li et al. (2008) reported a well-ordered 3D nanostructured macroporous surfaces which was fabricated by elecrodeposition method for efficiently boiling heat transfer. Since the structure is built based on the dynamic bubbles, it is perfect for the bubble generation applications such as nucleate boiling. The result indicated that at heat flux of 1W/cm^2, the heat transfer coefficient is enhanced over 17 times compared to a plain reference surface.

El-Genk and Ali (2010) experimentally studied the enhancement of saturation boiling of degassed PF-5060 dielectric liquid on microporous copper dendrite surfaces. These surface layers were deposited by electroChemical technique.

The result showed that the thickest layer (145.6μm) of Cu nanodendrite surface is very promising for cooling electronic components, while keeping the junction temperature relatively low and no temperature excursion.

Nonetheless, it remains a problem for these treated surfaces to significantly enhance critical heat flux for the cooling of high-heat-flux chips. This study aims to provide a surface treatment that maintains a consistently high heat transfer rate across the whole nucleate boiling zone and significantly enhances critical heat flux while keeping the chip temperature within acceptable limits. The significant decline in heat transfer performance of the prior microstructured surfaces at elevated heat fluxes is attributed to the accumulation of vapors within the structures, which obstructs the bulk liquid from contacting the superheated wall for vaporization. The improvement, attributed to the heightened thermal resistance from the substantial quantity of vapors confined within the microstructures, diminishes significantly when the heat flux nears the critical heat flux (CHF) (See Fig. 1).

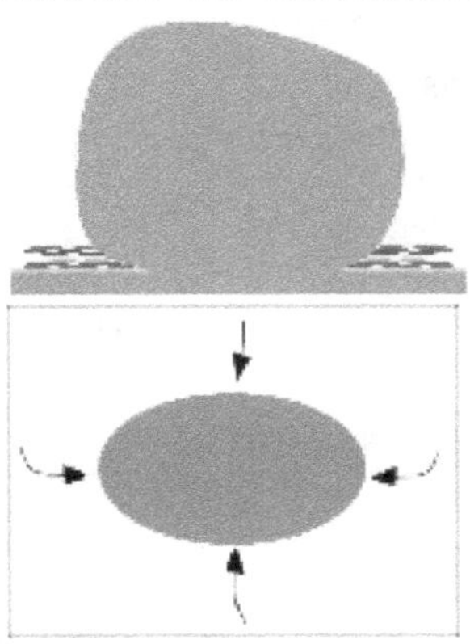

Fig. 1. Schematic of heat transfer phenomenon of smooth or previous porous structures

Consequently, the high-efficiency microstructures must offer a substantial driving force and a low-resistance pathway to facilitate the unobstructed entry of bulk liquid to the heater surface, even in the presence of big bubbles at elevated heat fluxes. Subsequently, Wei et al. (2003) formulated a novel model for heat transmission and fluid dynamics in the vapor mushroom zone of saturated nucleate pool boiling. The vapor mushroom region is defined by the development of a liquid layer interspersed with multiple, continuous columnar vapour stems beneath an expanding mushroom-shaped bubble illustrated in Fig. 2. The liquid layer between the vapor mushroom and the heater surface is referred to as the macrolayer, while the thin liquid film created beneath the vapor stems is called the microlayer. Consequently, Wei et al. (2003) proposed three highly efficient heat transfer mechanisms, which include conduction and evaporation in the microlayer region, conduction and evaporation in the macrolayer region, and Marangoni convection in the macrolayer region. Moreover, the computational findings of Wei et al. (2003) shown that heat transfer can be effectively conveyed to the vapor-liquid interface via Marangoni convection. The evaporation at the triple point (the contact point of liquid, vapor, and solid) significantly influences heat transfer, accounting for approximately 60% of the heat flux across the examined ranges, while the relative evaporations at the bubble-liquid interface and the stem-liquid interface are approximately 30% and 10%, respectively. Nonetheless, the vapor stem will ultimately fail, leading to the cessation of Marangoni convection and microlayer evaporation in the vapor mushroom zone of saturated pool nucleate boiling heat transfer. In the aforementioned circumstance, additional investigations were conducted by Wei et al. (2003) for cases where Marangoni convection and/or microlayer evaporation were excluded. The findings demonstrated that the maximum wall temperature occurs in scenarios devoid of Marangoni convection and microlayer evaporation. This suggests that both Marangoni convection and microlayer evaporation are significant factors in the mushroom region of saturated pool nucleate boiling heat transfer.

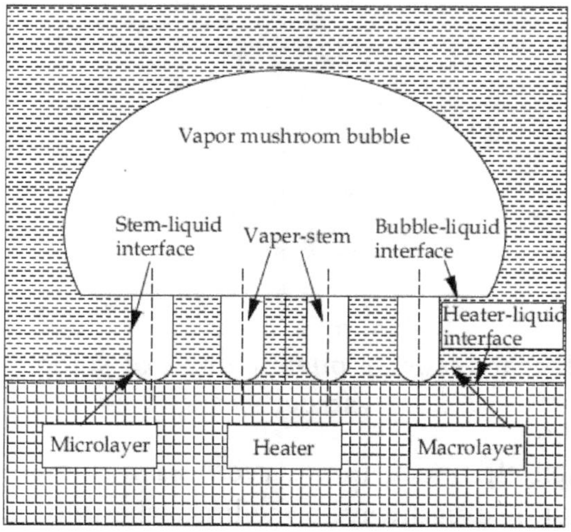

Fig. 2. Schematic of vapour mushroom structure near heated surface

Therefore, to overcome the above problems occur, we developed a micro-pin-finned surface with the fin thickness of 10-50 μm and the fin height of 60-120 μm. The fin gap was twice the fin thickness. The generated bubbles staying on the top of the micro-pin-fins can provide a capillary force to drive plenty of fresh liquid into contact with the superheated wall for vaporization through the regular interconnected structures formed by micro-pin-fins, as well as improve the microlayer evaporation and the Marangoni convection heat transfer by the motion of liquid around the micro-pin-fins (See Fig. 3).

So, the boiling heat transfer performance of FC-72 for the micro-pin-finned surfaces was firstly carried out in pool boiling test system. The maximum cooling capacity of this type of cooling module is determined by either the occurrence of CHF or complete vapor-space condensation (Kitching et al.1995). Then, some researchers such as Mudawar and Maddox (1989), Kutateladze and Burakov (1989), Samant and Simon (1989), and Rainey et al. (2001), have found that both of fluid velocity and subcooling had significant effects on the nucleate boiling curve and the critical heat flux of their thin film heater. Therefore, the combined effects of fluid velocity and subcooling on the flow boiling heat transfer of FC-72 over micro- pin-fined surfaces were investigated for further enhancement of boiling heat transfer to cool high-heat-flux devices.

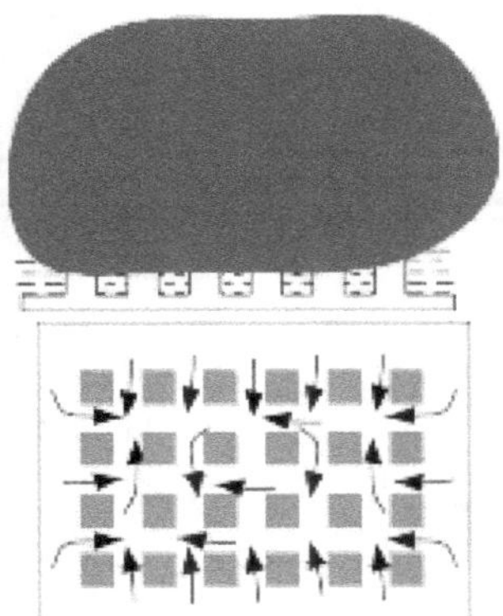

Fig. 3. Schematic of boiling heat transfer phenomena of micro-pin-fins at high heat flux near CHF

2. EXPERIMENTAL APPARATUS AND PROCEDURE

2.1 Test facility of pool boiling

The first test facility for pool boiling heat transfer is shown schematically in Fig. 4. The test liquid FC-72 was contained within a rectangular stainless vessel with an internal length of 120 mm, width of 80mm, and height of 135 mm (1.3L), which was submerged in a thermostatic water bath (42L) with a temperature adjusted range of 5-80°C. The bulk temperature of FC-72 within the test chamber was maintained at a prescribed temperature by controlling the water temperature inside the water bath. Additional liquid temperature control was provided by an internal condenser, which was attached at the ceiling of the test vessel and through which water was circulated from cooling unit. The pressure inside the test vessel was measured by pressure gauge and a nearly atmospheric pressure was maintained by attaching a rubber bag to the test vessel. For the visual observation of boiling phenomena, the test chamber was fitted with glass windows in both the front and back. The test heater assembly consisting of a test chip bonded on a pyrex glass plate and a vacuum chuck made of brass was immersed horizontally in the test chamber with the test chip facing upward. The local temperatures of the test liquid at the chip level, and 40mm and 80mm above the chip level were measured by T-type thermoCouples the hot junctions of which were loCated on a vertical line 25mm apart from the edge of test chip.

Details of the test section are shown in Fig. 5. The test chip was a P doped N-type silicon chip with the dimensions of $10\times10\times0.5mm^3$. The specific resistance of the test chip was 1- 2Ωcm, and the thermal conductivity was about 156W/m. K at room temperature.

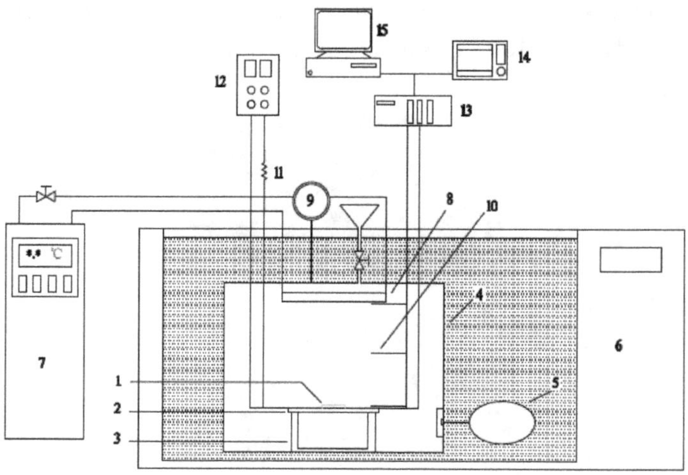

1, Test chip; 2,Glass plate; 3,Vacuum chuck; 4, Test vessel; 5,Rubber bag; 6,Water bath; 7,Cooling unit; 8, Condenser; 9,Pressure gauge; 10,ThermoCouples; 11, Standard resistance; 12, DC power supply; 13, Digital multimeter; 14,Imange acquisition System; 15, Computer

Fig. 4. Schematic diagram of experimental apparatus for pool boiling

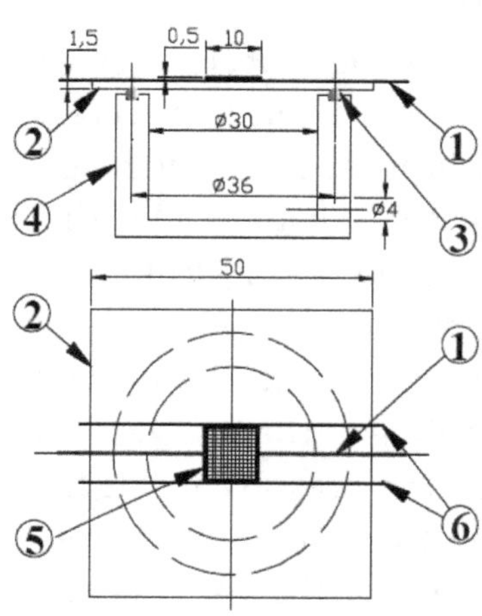

1, ThermoCouple; 2, Pyrex glass plate; 3, O-ring; 4, Vacuum chuck; 5, Silicon chip; 6, Copper lead wire

Fig. 5. Test section for pool boiling

The chip was Joule heated by a direct current. Two 0.25-mm diameter copper wires for power supply and voltage drop measurement were soldered with a low temperature solder to the side surfaces at opposite ends. In order to secure the Ohmic contact between the test chip and the copper wire, a special solder with the melting point of 300°C was applied to the chip with ultrasonic bonding method before soldering the copper wires.

The local wall temperatures at the center and about 1.5mm from the edge on the centreline of the chip were measured by two 0.12-mm diameter T-type thermoCouples bonded under the test chip. The test chip was bonded on the top of a 50×50×1.2 mm^3 pyrex glass plate using epoxy. Then the glass plate on which the test chip was bonded was pressed firmly to the O-ring on a brass vacuum chuck when the inside of the chuck was evacuated by using a vacuum pump. This enabled a seamless swap of the test chip and reduced thermal losses from conduction and convection at the chip's rear surface. The lateral surfaces of the chip were coated with an adhesive to reduce thermal dissipation. Consequently, only the superior surface of the chip facilitated heat transfer.

2.2 Test facility of forced flow boiling

Figure 6 roughly depicts the second test facility for forced flow boiling heat transfer. The system comprises a closed-loop circuit that includes a tank, a scroll pump, a test section, two heat exchangers, and a turbine flowmeter. The tank functioned as a fluid storage and pressure regulator throughout the testing phase.

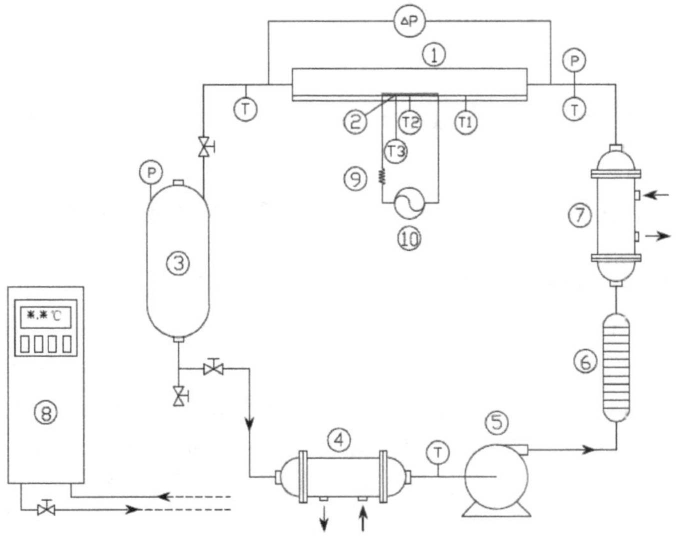

1, Test Section; 2, Test chip; 3, Tank; 4, Condenser; 5, Pump; 6, Flowmeter; 7, Pre-heater; 8, Cooling unit; 9, Standard resistance; 10, Direct current

Fig. 6. Forced flow boiling test loop

The condenser preceding the pump was employed to chill the fluid and avert cavitation in the pump. The pre-heater before the test section was utilized to regulate the inlet temperature of the test section. The pump was integrated with a converter to regulate the mass flow rate. A pressure transducer was put at the intake of the test section to guarantee accurate inlet pressure regulation. The pressure differential throughout the test segment was measured using a pressure difference transducer. The flowmeter and the sensors for pressure and pressure difference have the function of outputting 4~20mA current signals and were measured directly by a data acquisition system.

The test silicon chip is bonded to a substrate made of polycarbonate using epoxy adhesive and fixed in the horizontal, upward facing orientation on the bottom surface of a 5mm high and 30mm wide horizontal channel as shown in Fig. 7. The chip is loCated 300mm from the inlet of the test section so that the fluid flow on it is estimated to be fully developed turbulent flow for the present fluid velocity range.

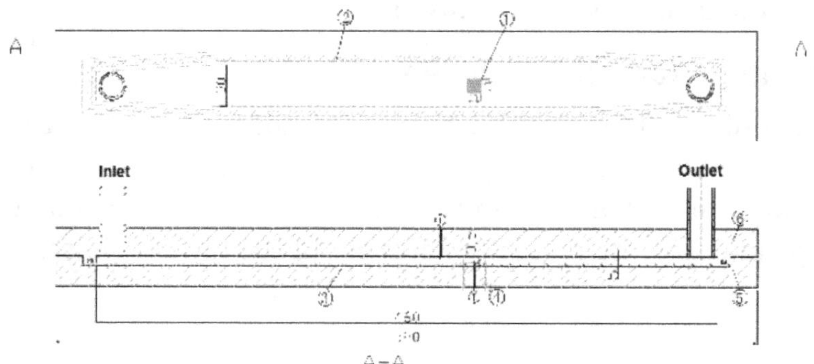

1, Test chip; 2, O-ring; 3, Polycarbonate plate; 4, Lead wires; 5, Lower cover; 6, Upper cover

Fig. 7. Test section for forced flow boiling

2.3 Experimental conditions and test procedure

Experiments were performed at three fluid veloCities of 0.5, 1 and 2m/s for the second test system and the liquid subcoolings of 3, 15, 25, 35 and/or 45K for both of test systems. Six kinds of silicon chips, one with a smooth surface and five with square micro-pin-fins having fin dimensions of 30×60, 30×120, 30×200, 50×60 and 50×120 μm^2 (thickness×height) are tested. The fin pitch *p* was twice the fin thickness. The micro-pin-finned chip was fabricated by the dry etching technique. These chips were named chips S, PF30-60, PF30-120, PF30-200, PF50-60 and PF50-120, respectively. The scanning electron microscope (SEM) images of the five micro-pin-finned chips are shown in Fig. 8a-e, respectively. A smooth chip was also tested for comparison.

In the above two test systems, the test chips were Joule heated by using a programmable d.c. power supply. The power supply was connected in series to a standard resistance (1Ω) and the test chip. The standard resistance was used to measure the electric current in the circuit. Power input to the test chip was increased in small steps up to the high heat flux region of nucleate boiling. The heat flux q was obtained from the voltage drop of the test chip and the electric current. In order to prevent real heater burnout, an overheating protection system was incorporated in the power circuit. If the wall temperature sharply increases by more than 20K in a short time, the data acquisition algorithm assumed the occurrence of CHF condition and the power supply was immediately shut down. The CHF value was computed as the steady state heat flux value just prior to the shut down of the power supply. The uncertainties in the chip and bulk liquid temperature measurements by the thermoCouple and the resistance thermometry for the pool boiling is estimated to be less than 0.1K and that to be less than 0.3K for the forced flow boiling. The uncertainty in the calculated value of q for the pool boiling is mostly due to the heat loss and estimated to be less than 15 and 5.0 percents for natural convection and nucleate boiling regions, respectively. For the forced flow boiling, the uncertainty in the calculated value of q is estimated to be less than 16 and 6.0 percents for the forced convection and nucleate boiling regions respectively. It is relevant to note here that q includes the heat transferred to the bulk liquid by conduction through the copper lead wires and the polycarbonate substrate.

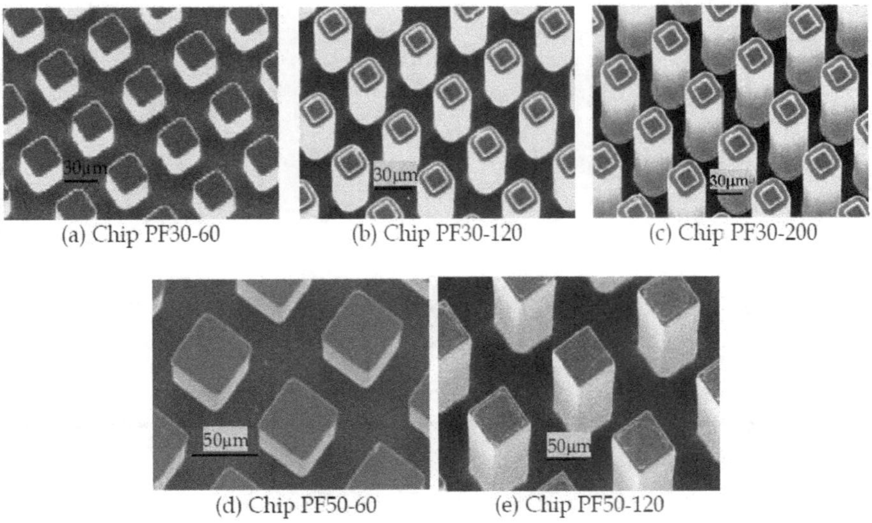

(a) Chip PF30-60 (b) Chip PF30-120 (c) Chip PF30-200

(d) Chip PF50-60 (e) Chip PF50-120

Fig. 8. SEM Images of micro-pin-fins

The experiment was repeated two or three times for all chips. The time interval between the subsequent runs was greater than 0.5 h. The boiling curves showed a good repeatability for all cases except for the boiling incipience point. Thus, only the results for the third runs are presented in the next section.

3. RESULTS AND DISCUSSION

3.1 Pool boiling performance of micro-pin-finned surfaces

Figure 9 shows the boiling curves of micro-pin-finned surfaces of PF30-60 (fin thickness of 30μm and height of 60μm) and PF30-200 (fin thickness of 30μm and height of 200μm) at ΔT_{sub}=3K. The boiling curve of the smooth surface chip S is also shown for comparison. All chips follow almost the same q versus ΔT_{sat} relation in the non-boiling region despite that chip PF30-200 has a large fin height of 200 μm. This indicates that the fin height up to 200 μm is not effective in the enhancement of natural convection heat transfer. However, at the nucleate boiling region, the micro-pin-finned surfaces show considerable heat transfer enhancement compared to chip S. Furthermore, the boiling curves of chips PF30-60 and PF30-200 are very steep and the wall superheats show a very small change with increasing heat flux q up to the critical heat flux (CHF) point. It is supposed that the increased surface activated in the nucleate region to be much larger for chips PF30-60 and PF30-200, hence the boiling heat transfer is enhanced more greatly. The q_{CHF} increases in the order of chips S, PF30-60 and PF30-200. For the micro-pin-finned chips in the present study, the wall temperature at the CHF point is lower than 85°C.

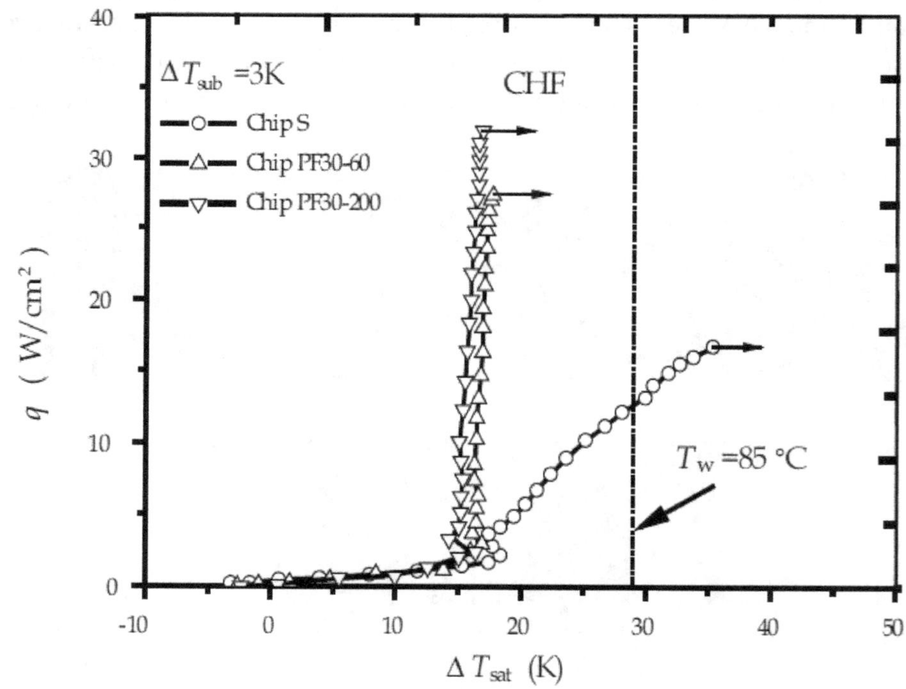

Fig. 9. Boiling curves of chips PF30-60, PF30-200 and S, ΔT_{sub}=3K

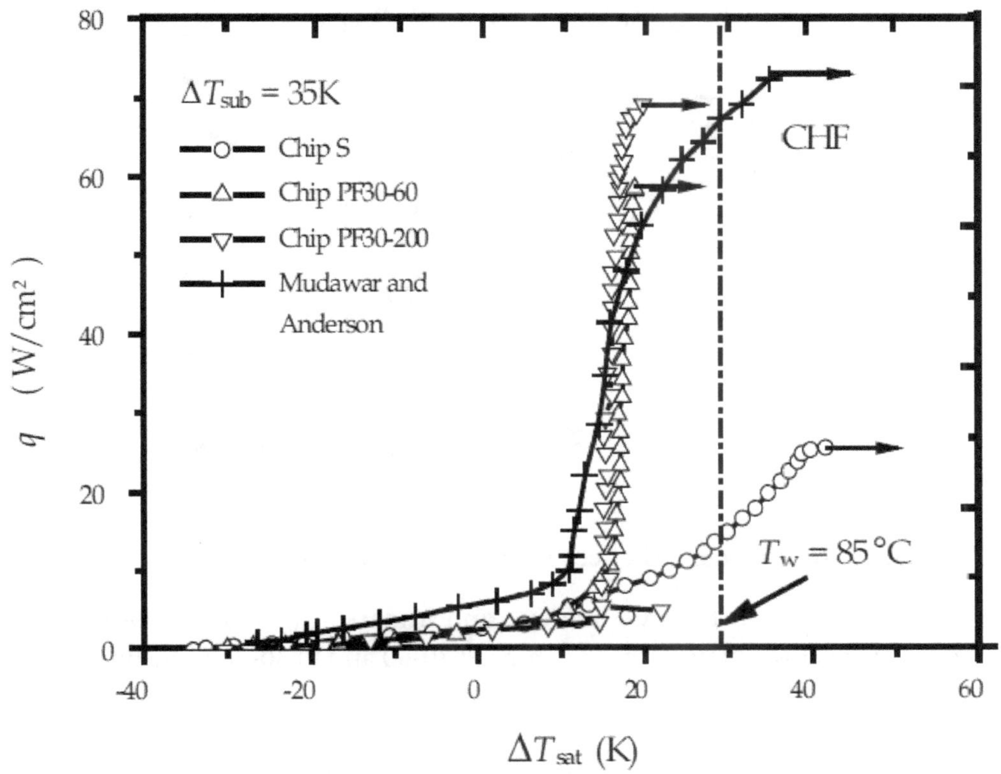

Fig. 10. Boiling curves of chips PF30-60, PF30-200 and S, ΔT_{sub}=35K

Figures 10 and 11, respectively, show the boiling curves of micro-pin-finned chip PF30-60 and PF30-200 at ΔT_{sub}=35K and ΔT_{sub}=45K. In Fig. 10, the boiling curves of the micro-pin- finned surface reported by Mudawar and Anderson (1989) and a smooth surface are also shown for comparison. The test surface of Mudawar and Anderson (1989) had square micro- pin-fins with the dimensions of 0.305×0.508 mm^2 (thickness×height). These fin dimensions are one order of magnitude greater than those of chip PF30-60 in the present experiments. The ΔT_{sat} values in the low-heat-flux region are smaller than that for chips PF30-60 and PF30- 200, whereas the wall superheats at the CHF point $\Delta T_{sat,CHF}$ are greater than that for the latter. While the measured heat flux in the free convection region for chip S and the micro- pin-finned chips in this study is almost the same, the heat flux for the micro-pin-finned chip reported by Mudawar and Anderson (1989) is about 40% higher than that for chip S, indicating that the increased area of their micro-pin-finned surface over a smooth surface is effective in natural convection heat transfer.

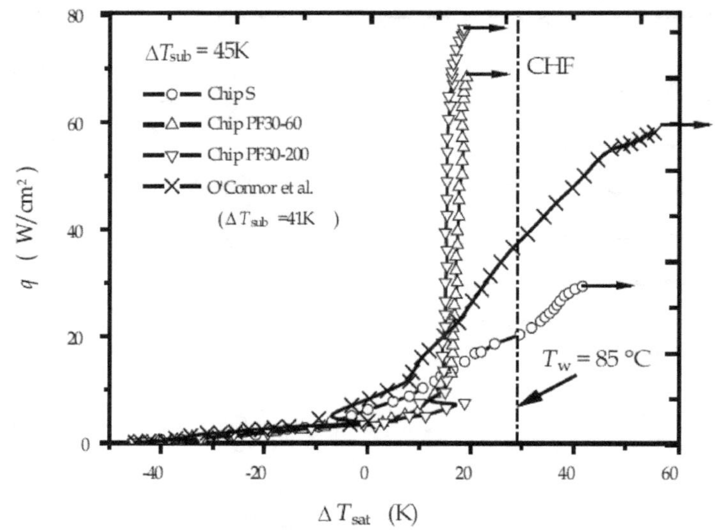

Fig. 11. Boiling curves of chips PF30-60, PF30-200 and S, ΔT_{sub}=45K

Although the q_{CHF} reported by Mudawar and Anderson (1989) for the micro-pin-finned surface is higher than that of chips PF30-60 and PF30-200, the boiling curve near the CHF point shows a much smaller slop than the micro-pin-finned chips in the present study.

In Fig. 11 boiling occurs at a much smaller value of ΔT_{sat} and the temperature overshoots at the boiling incipience are small, compared to the case of Fig. 10. The boiling curves of the microporous surface developed by O'Connor et al. (1996) and a smooth surface are also shown for comparison. The test surface of O'Connor et al. (1996) had a porous layer consisting of 8-12 μm diamond particles produced by painting technique. The porous surface shows a severe deterioration of boiling heat transfer performance at high heat flux region and the value of q_{CHF} is smaller than that of the micro-pin-finned chips. However, the micro-pin-finned surfaces in the present study show a sharp increase in the heat flux with increasing wall superheat from the boiling incipience to the critical heat flux. The increase of CHF for micro-pin-finned surfaces could reach more than twice that of a smooth chip. Comparison of the experimental results reveals that the wall temperature at the CHF point $T_{w,CHF}$ is higher than 85°C for chip S and the previous results reported by Mudawar and Anderson (1989) and O'Connor et al. (1996). On the other hand, the micro-pin-finned chips PF30-60 and PF30-200 show $T_{w,CHF}$ smaller than 85°C.

As stated previously, the upper limit of temperature for a reliable operation of electronic chips is given by 85°C. Thus the maximum allowable heat flux q_{max} is given by the CHF if $T_{w,CHF}$<85°C and by q at T_w=85°C if $T_{w,CHF}$>85°C. Figure 12 shows the variation of maximum heat flux of chips PF30-60 and PF30-200 with liquid subcooling ΔT_{sub}. The experimental data by Mudawar and Anderson (1989) and

O'Connor et al (1996) are also shown for comparison. We can see that the maximum heat flux of the micro-pin-finned surfaces in the present study is much higher than that of the porous and other large scale micro-pin-finned and smooth surfaces, and increases greatly with the liquid subcooling. The micro-pin- finned surface of Mudawar and Anderson (1989) shows almost the same q_{max} with chip PF30-200. The porous surface of O'Connor et al (1996) shows the q_{max} value in between those for micro-pin-finned chips and smooth chip. The difference of q_{max} between the porous surface and micro-pin-finned surfaces increases greatly with subcooling, indicating that the subcooling effect is larger for the micro-pin-finned surfaces.

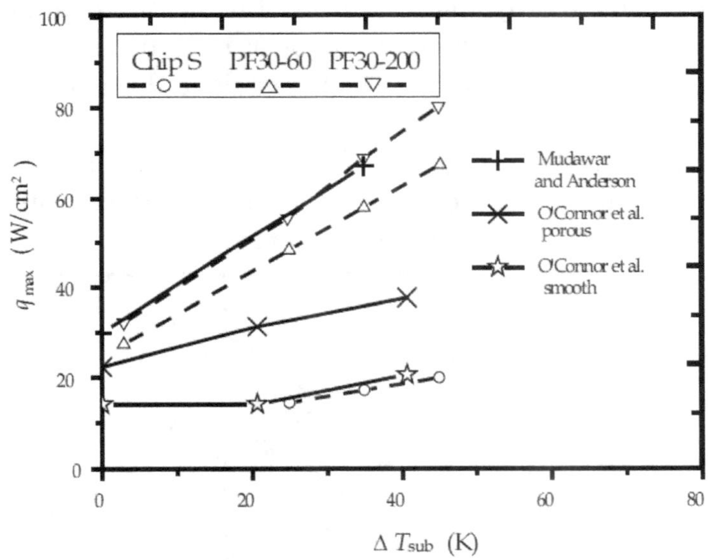

Fig. 12. Variation of q_{max} with ΔT_{sub}

3.2 Flow boiling performance of micro-pin-finned surfaces

Figure 13 shows the comparison of boiling curves for all surfaces with ΔT_{sub}=15K. For the fluid velocity V<2m/s, the wall superheat in the nucleate boiling region decreases in the order of chip S, PF50-60, PF30-60, PF30-120 for the same fluid velocity, again showing that the boiling heat transfer can be enhanced by increasing total surface area. However, the boiling curve of chip PF50-120 shows the smallest wall superheat despite of the surface area ratio of 3.4. Although the condition for the etching process is set to the same, the etching size is different for different micro-pin-finned chips, which may affect the etching results. From the scanning-electron-micrograph (SEM) image of micro-pin-fin (Fig. 8), we found that the roughness on the fin flank near the fin root usually increases with increasing etching depth. Chip PF50-120 with the largest fin thickness and height is observed to have a large roughness scale on the fin flank. It is considered that the roughness causes the earlier boiling incipience and thus smaller wall superheat in the nucleate boiling region. For the fluid velocity of 2m/s, the nucleate boiling curves of all micro-pin-finned chips are more or less affected by forced convection heat transfer since the slopes of boiling curves become smaller than those at lower veloCities, but the nucleate boiling curves of 50-μm thick micro-pin-fins show a much larger slope than those of 30-μm thick micro-pin-fins. According to enhanced boiling heat transfer mechanism for micro-pin-fins developed by Wei et al. (2003), the microlayer evaporation and the Marangoni convection caused by thermal capillary force in the micro-pin-fin formed interconnect tunnel play an important role for boiling heat transfer. It is considered that bulk fluid flow may affect the Marangoni convection greatly and the smaller fin gap of 30 μm may generate a larger flow resistance for Marangoni convection around the fin sidewalls, resulting in a lower heat transfer performance. The larger slope shifts the boiling curve of chips PF50-60 and PF50-120 to a smaller wall superheat than that of chips PF30-60 and PF30-120, respectively, at high heat flux for V=2m/s. For comparison, Lie et al. (2007)'s saturated boiling curves for two pin-finned surfaces with the larger fin thicknesses of 100 and 200 μm are also shown in Fig. 13.

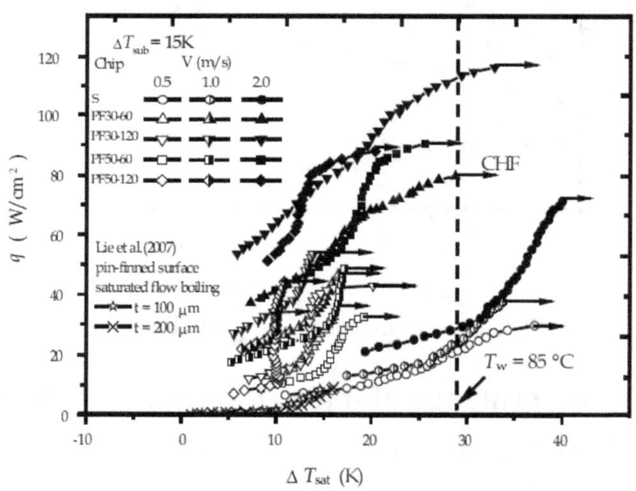

Fig. 13. Comparison of flow boiling curves for all chips, ΔT_{sub}=15K

The boiling curves of Lie et al. (2007) show a much smaller slope and a larger wall superheat compared with that of the present micro-pin-finned surfaces. Generally, the wall superheat deceases in the order of 200, 100, 50 and 30-μm micro-pin-fins for the fin height of about 70μm. We have found that the boiling heat transfer performance for the micro-pin-finned surface with the fin thickness of 10-50 μm is much better than the pin-finned surface with the larger fin thickness of about 300 μm used by Anderson and Mudawar (1989), and the fin thickness of 30-50μm is a suitable range of effectively enhancing boiling heat transfer. The optimum fin gap size is considered to be determined by the balance of the capillary force caused by evaporation of bubbles for driving the micro-flow in the gap of micro-pin-fins and the flow resistance. Large fin gap usually has small flow resistance but generates small microConvection heat transfer proportion, and *vice versa*. The present experimental study again shows the larger fin thickness above 100 μm is not so remarkably effective compared with the fin thickness of 30-50 μm.

Figures 14 and 15 show the comparison of boiling curves for all surfaces with ΔT_{sub}=25 and 35K, respectively. The trend of experimental data is basically the same as the case of ΔT_{sub}=15K shown in Fig. 13 except that the values of the CHF increase considerably in the order of 25 and 35 K.

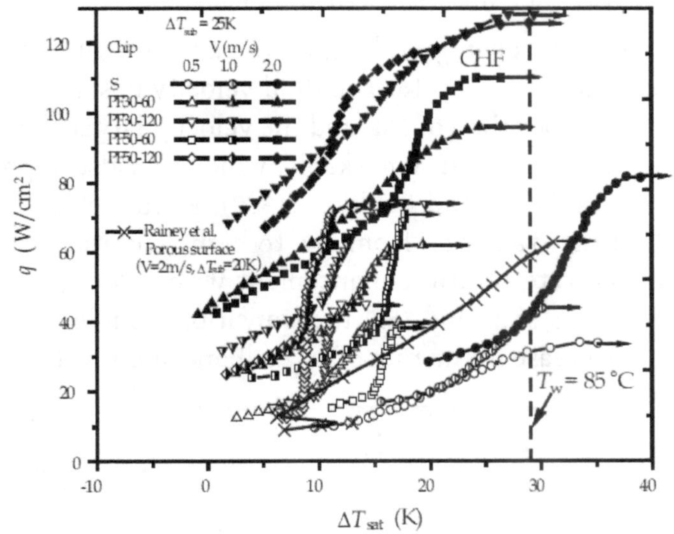

Fig. 14. Comparison of flow boiling curves for all chips, ΔT_{sub}=25K

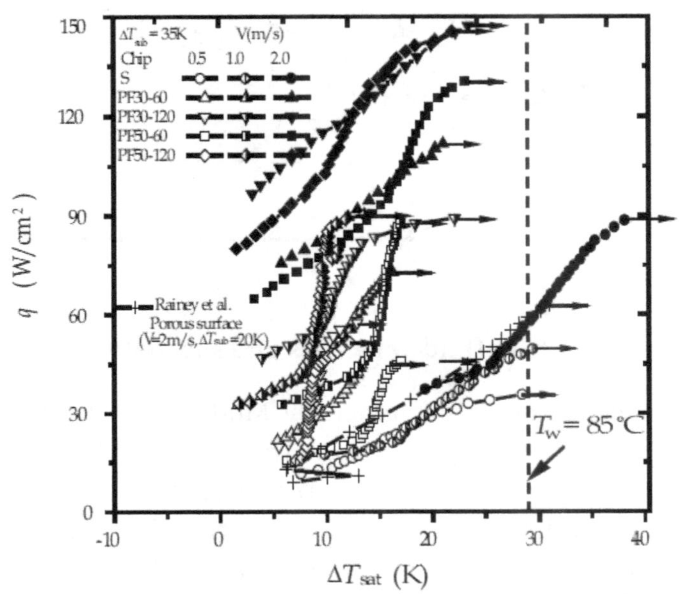

Fig. 15. Comparison of flow boiling curves for all chips, ΔT_{sub}=35K

The highest value of CHF (=148 W/cm^2), about 1.5 times as large as that for the smooth surface, was obtained by chips PF30-120 (fin thickness of 30 μm and height of 120 μm) and PF50-120 (fin thickness of 50 μm and height of 120 μm) at liquid subcooling ΔT_{sub} = 35 K and flow velocity V = 2 m/s. For comparison, Rainey et al.(2001) 's boiling curve with the liquid subcooling of 20 K for the microporous surface at V = 2 m/s is also shown in Figs. 14 and 15. Although having an earlier boiling incipience, the microporous surface shows a larger wall superheat than the micro-pin-finned surfaces in the nucleate boiling region, and the heat flux at 85°C is less than half the CHF of chips PF30-120 and PF50-120.

We plot the relationship of q_{max} with fluid velocity for all surfaces with fluid subcooling as a parameter in Fig. 16. The fluid velocity has a very large effect on q_{max}. For the low fluid subcooling of 15K and the velocity larger than 1m/s, the rate of q_{max} enhancement is increased remarkably, which was also supported by many researches such as Mudawar and Maddox (1989), Rainey et al. (2001), and etc., who had noted that the transition from low to high veloCities was characterized by an increase in the rate of CHF enhancement with velocity. However, for the larger liquid subcoolings of 25 and 35K, the transition from low to high velocity is characterized by a decrease in the rate of q_{max} enhancement with velocity.

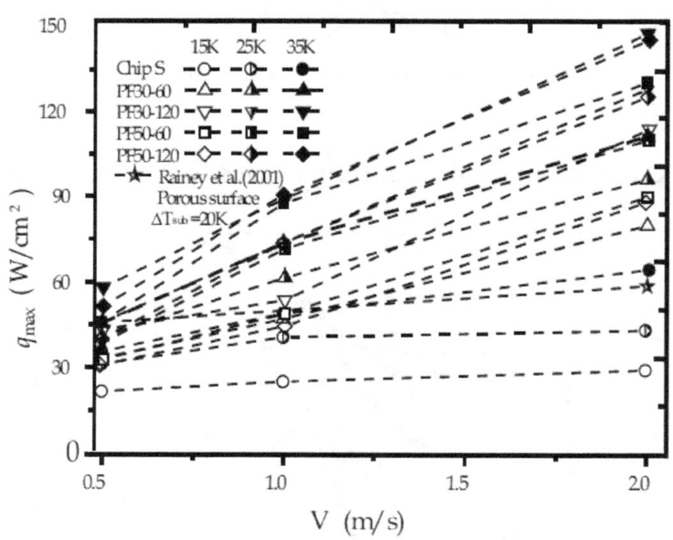

Fig. 16. Effects of fluid veloCities and subcoolings on CHF

For a low fluid subcooling, as explained by Mudawar and Maddox (1989), the low velocity q_{max} was caused by dryout of the liquid sublayer beneath a large continuous vapor blanket near the downstream edge of the heater; however in the high velocity q_{max} regime, the thin vapor layer covering the surface was broken into continuous vapor blankets much smaller than the heater surface, decreasing the resistance of fluid flow to rewet the liquid sublayer and thus providing an additional enhancement to q_{max} and subsequent increase in slope. For a large fluid subcooling, the bubble size becomes small and the heater surface was not fully oCcupied with vapor layer for the fluid velocity range in this study. Therefore, there is no obvious slope change as seen in Fig. 16. Moreover, from the slope of boiling curves we can see that the effect of the fluid velocity on micro-pin-finned surfaces is more noticeably compared with the porous surface (Rainey et al. 2001) and the smooth surface, and the enhancement of q_{max} sharply increases with fluid velocity. For chips PF30-120 and PF50-120, the q_{max} reaches nearly 148W/cm^2 at V=2m/s and ΔT_{sub}=35K.

4. CONCISE SUMMARY

All micro-pin-fined surfaces have considerable heat transfer enhancement compared with a smooth surface and other microstructured surfaces, and the maximum CHF can reach nearly 148W/cm^2 by chips PF30-120 and PF50-120 at V= 2 m/s and ΔT_{sub} = 35 K. The wall temperature for the micro-pin-finned surfaces is less than the upper temperature limit for the normal operation of LSI chip, 85°C. Consequently, micro-pin-finned surfaces have significant promise for high-efficiency electronic cooling. The utilization of boiling heat transfer for electronics cooling in space and on planetary bodies has emerged as a notably important topic. In microgravity, the buoyancy effect diminishes during boiling heat transfer, leading to an extended duration for bubble departure. These factors may impede the timely entry of fresh bulk liquid to the heater surface, leading to diminished boiling heat transfer efficiency at elevated heat flux (Wan and Zhao 2008). Enhancing boiling heat transmission in microgravity is a significant concern. Given the superior boiling heat transfer characteristics of micro-pin-finned surfaces and the augmented boiling heat transfer mechanism, it is posited that although bubbles adhering to the apex of the micro-pin-fins may not detach promptly in microgravity, the fresh bulk liquid can still reach the heater surface via interconnecting tunnels created by the micro-pin-fins, facilitated by capillary forces, irrespective of the gravitational context. Consequently, we are keen to investigate the boiling heat transfer efficacy of micro-pin-finned surfaces in microgravity in the future.

5. REFERENCES

Anderson, T.M. & Mudawar, I. (1989). Microelectronic cooling by enhanced pool boiling of a dielectric fluoroCarbon liquid. ASME J. Heat Transfer, 111:752-759

Bergles, A.E. & Chyu, M.C. (1982). Characteristics of nucleate pool boiling from porous metallic coatings. ASME J. Heat Transfer, 104:279-285

Chang, J.Y. & You, S.M. (1996). Heat orientation effects on pool boiling of micro-porous- enhanced surfaces in saturated FC-72. ASME J. Heat Transfer, 118:937-943

Chang, J.Y. & You, S.M. (1997). Enhanced boiling heat transfer from micro-porous surfaces: effects of a coating composition and method. Int. J. Heat Mass Transfer, 40:4449-4460

Chowdhury, S.K.R. & Winterton, R.H.S. (1985). Surface effects in pool boiling. Int. J. Heat Mass Transfer, 28:1881-1889

Chu, R.C. & Moran, K.P. (1977). Method for customizing nucleate boiling heat transfer from electronic units immersed in dielectric coolant. US Patent 4,650,507

El-Genk, M.S. & Ali, A.F. (2010). Enhancement of saturation boiling of PF-5060 on microporous copper dendrite surfaces. ASME J. Heat Transfer, 132:071501(1-9)

Hwang, U.P. & Moran, K.F. (1981). Boiling heat transfer of silicon integrated circuits chip mounted on a substrate. In: ASME HTD, 20:53-59

Kitching, D.; Ogata, T. & Bar-cohen, A. (1995). Thermal performance of a passive immersion cooling multichip module. J. Enhanced Heat Transfer, 2:95-103

Kubo, H.; Takamatsu, H. & Honda, H. (1999). Effects of size and number density of micro- reentrant cavities on boiling heat transfer from a silicon chip immersed in degassed and gas-dissolved FC-72. Enhanced Heat Transfer, 6:151-160

Kutateladze,S.S. & Burakov, B.A. (1989).The critical heat flux for natural convection and forced flow of boiling and subcooled dowtherm. Problems of Heat Transfer and Hydraulics of Two-Phase Media, Pergamon,Oxford, pp. 63–70

Li, S.; Furberg, R.; Toprak, M.S.; Palm, B. & Muhammed, M. (2008). Nature-inspired boiling enhancement by novel nanostructure macroporous surfaces. Adv. Funct. Mater., 18:2215-2220

Lie, Y.M.; Ke, J.H.; Chang, W.R.; Cheng, T.C. & Lin, T.F. (2007). Saturated flow boiling heat transfer and assoCiated bubble characteristics of FC-72 on a heated micro-pin- finned silicon chip. Int. J. Heat Mass Transfer, 50:3862-3876

Messina, A.D.; & Park, E.L. (1981). Effects of precise arrays of pits on nucleate boiling. Int. J. Heat Mass Transfer, 24:141-145

Mudawar, I. & Anderson, T.M. (1989). High flux electronic cooling by means of pool boiling – Part I; parametric investigation of the effects of coolant variation, pressurization, subcooling and surface sugmentation. Heat Transfer Electron ASME-HTD, 111:25-34

Mudawar,I. & Maddox,D.E. (1989). Critical heat flux in subcooled flow boiling of fluoroCarbon liquid on a simulated electronic chip in a vertical rectangular channel. Int. J. Heat and Mass Transfer, 32: 379–394.

Nakayama, W.; Nakajima, T.; Ohashi, S. & Kuwahara, H. (1989). Modeling of temperature transient of micro-porous studs in boiling dielectric fluid after stepwise power application. ASME HTD, 111:17-23

Nelson, L.A.; Sekhon, K.S.; & Fritz, J.E. (1978). Direct heat pipe cooling of semiconductor devices. In: ProCeedings of the 3rd international heat pipe conference, CA, USA, pp.373-376

Nakayama, W.; Daikoku, T. & Nakajima, T. (1982). Effects of pore diameters and system pressure on saturated pool nucleate boiling heat transfer from porous surfaces. ASME J. Heat Transfer, 104:286-291

O'Connor, J.P. & You, S.M. (1995). A painting technique to enhance pool boiling heat transfer in saturated FC-72. ASME J. Heat Transfer, 117:387-393

O'Connor, J.P.; You, S.M. & Chang, J.Y. (1996). Gas saturated pool boiling heat transfer from smooth and microporous surfaces in FC-72. ASME J. Heat Transfer, 118:662-667

Oktay, S. (1982). Departure from natural convection (DNC) in low-temperature boiling heat transfer encountered in cooling microelectronic LSI devices. In: ProC. 7th Int. Heat transfer Conf., 4:113-118

Oktay, S. & Schmeckenbecher, A. (1972). A method for forming heat sinks on semiconductor device chips. US Patent 3,706,127

Phadke, N.K.; Bhavnani, S.H.; Goyal, A.; Jaeger, R.C. & Goodling, J.S. (1992). Re-entrant caviy surface enhancements for immersion cooling of silicon multichip packages. IEEE Trans. Comp., Hybrids, Manuf. Technol.,15:815-822

Rainey, K.N.; Li, G. & You, S.M. (2001). Flow Boiling Heat Transfer From Plain and Microporous Coated Surfaces in Subcooled FC-72. ASME J. Heat Transfer, 123(5): 918-925

Samant, K.R. & Simon, T.W. (1989). Heat transfer from a small heatedregion to R-113 and FC-72. ASME J. Heat Transfer, 111:1053–1059

Ujereh, S.; Fisher, T.; & Mudawar, I. (2007). Effects of carbon nanotube arrays on nucleate pool boiling. Int. J. Heat Mass Transfer, 50:4023-4038

Wan, S.X. & Zhao, J.F. (2008). Pool boiling in microgravity: recent results and perspectives for the project DEPA-SJ10. Microgravity Sci. Technol., 20:219-224

Wei, J.J.; Yu, B. & Wang, H.S. (2003). Heat transfer mechanisms in vapor mushroom region of saturated nucleate pool boiling. Int. J. Heat and Fluid Flow, 24:210-222

You, S.M.; Simon, T.W. & Bar-Cohen, A. (1992). A technique for enhancing boiling heat transfer with application to cooling of electron equipment. IEEE Trans CHMT, 15:823-831

4 Chapter

THERMAL TRANSFER IN MINICHANNELS AND MICROCHANNELS FOR CPU COOLING SYSTEMS

1. INTRODUCTION: AN OVERVIEW

The 1970s of the previous century witnessed the shrinking of electronic components and the advancement of micro and nano production processes, leading to an increase in various applications in fields like as biomedical devices, MEMS, and cooling nanotechnologies. The overheating of these micro components and devices necessitated the implementation of mini and microchannels in the aforementioned technologies. The objective is to rapidly dissipate the maximum amount of heat from these systems to enhance dependability and functional stability (Kim & Kim, 2007). Operating CPUs at elevated temperatures can lead to immediate system breakdowns and significantly diminish the lifespan of the CPU over time. In severe instances, your CPU may overheat or fuse with the motherboard. The evacuation of substantial heat flow via conduction and forced convection of air cannot be sufficiently accomplished through traditional means. Therefore, it is imperative that CPU cooling systems provide adequate cooling for the CPU. A comparison analysis of the peak temperatures attained within CPU cores would be intriguing. Figures 1 and 2 illustrate the CPU's specified maximum temperature, sometimes referred to as critical temperature.

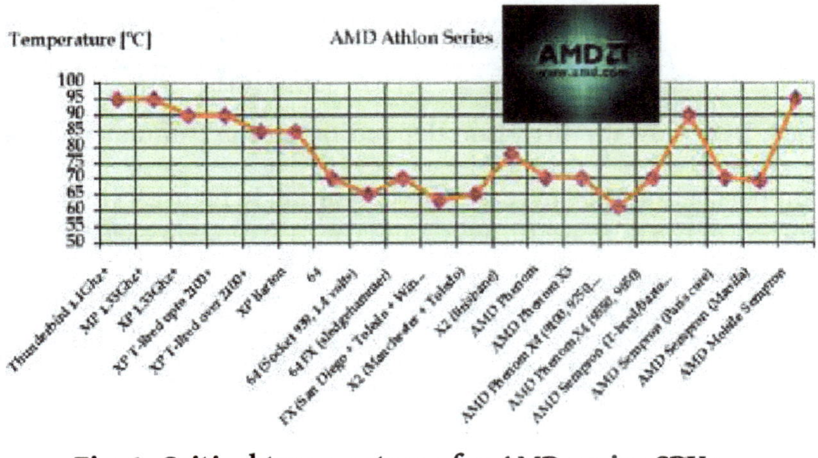

Fig. 1. Critical temperatures for AMD series CPUs

For practical reasons, we determined that it is advisable to sustain the functional values (as indicated by the computer's software) at around 20 °C below the maximum values depicted in the two diagrams.

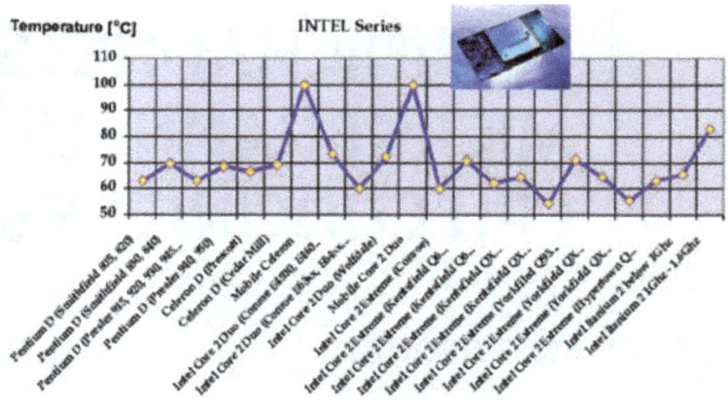

Fig. 2. Maximal values of the temperature for INTEL series CPUs

In both cases we notice that the maximal values of the temperature vary from a maximum of 100°C, for an Intel Mobile Celeron and Core 2 Duo, to a minimum of 54,8 °C for an Intel Core 2 Extreme (Kentsfield QX6800), whereas for an AMD Athlon CPU we have a maximum of 95 °C and a minimum of 61 °C, for the Phenom X4 models (9100, 9750, 9850).

Together with CPU development, as it results from Figure 3, the heat flow increases significantly, depending on the frequency variation.

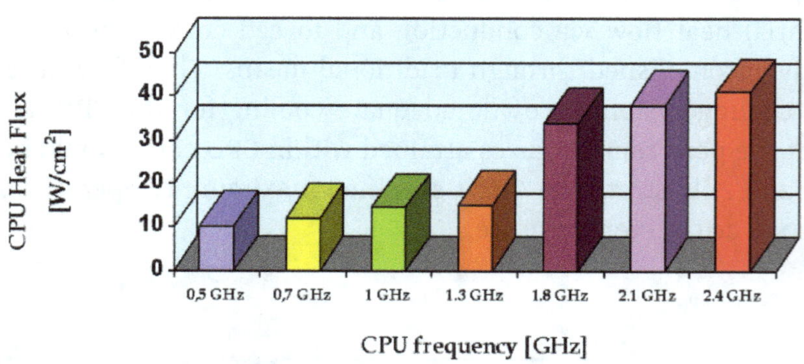

Fig. 3. Heat flow modification depending on the frequency variation

We can emphasize that this increase is exponential. In Figure 4 we show a graphical representation of the temperature flow and the CPU die size variation starting 1975 until the present day.

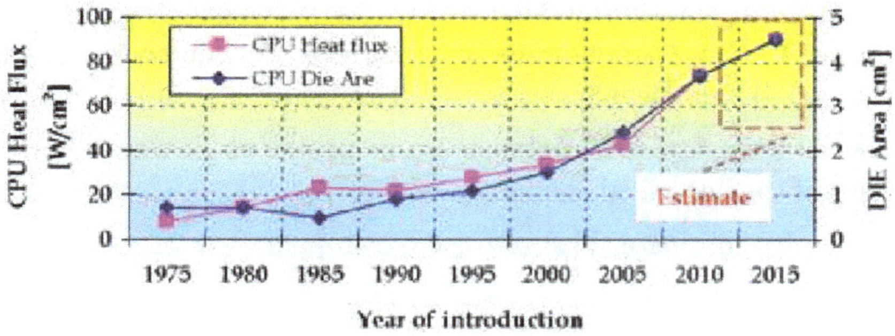

Fig. 4. Variation of the thermal flow and the CPU die size depending on the production year

By analyzing the information that was described above, one can observe that, no matter what is the manufactured CPU, an important density of thermal flow develops on the CPU die. High values of the thermal flow imply taking heat dissipation measures. If this requirement is not met then the risk of CPU thermal inducted damage might occur. Therefore, different cooling methods are needed which, under the circumstances of reaching extremely low temperatures, can lead to a significant increase in the CPU's processing speed.

Most recently, the attention was foCused on the study of flow processes and heat transfer in microdevices. In these systems (Hadjiconstantinou & Simek, 2002), the flow and heat transfer processes are of nano and microscopic type and differ as basic mechanism from the macroscopic ones due to dimensional characteristics and molecular type phenomena.

2. CPU COOLING METHODS AND THEIR LIMITS

Until now (Viswanath et al., 2000, Meijer et al., 2009) indicate two CPU cooling systems, as they are described in Figure 5.

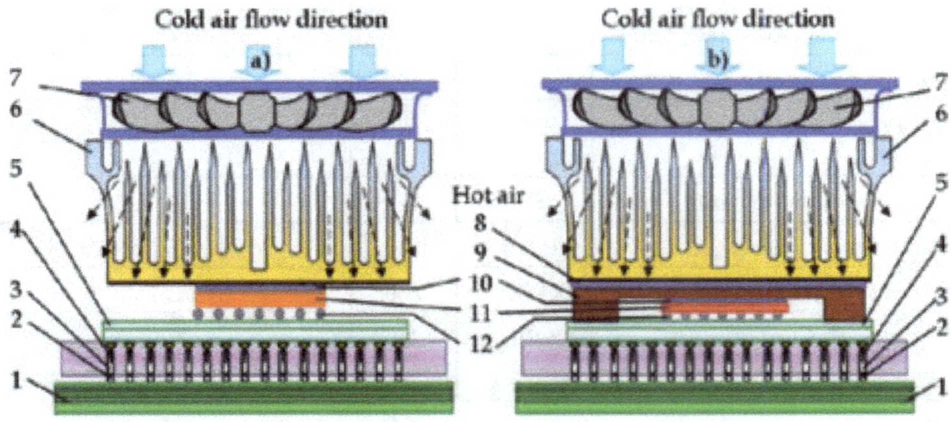

Fig. 5. Types of CPU cooling systems, where: (a) Package with Exposed Die and Heat Sink and (b) Lidded package and Heat Sink

Description of the annotations in Figure 5: 1 – Motherboard planes, 2 – solder ball, 3 – Receptacle, 4 – Pin, 5 – Substrate, 6 – Heat Sink, 7 – cooler, 8 – TIM (Thermal Interface Material) 2, 9 – Heat Spreader, 10 – TIM 1, 11 – CPU die, 12 – C4 Solder ball.

Strictly from a thermal transfer perspective, as (Meijer et al., 2009) also points out, the functions of the described elements are:

- Heat Sink – Dissipates heat to the environment;
- TIM 2 – Transmits heat to the Heat Sink;
- Heat Spreader:
  - Evens out heat spot;
  - Spreads heat horizontally;
  - Protects the die;
- TIM 1 – Transmit heat to the spreader;
- Die:
  - Heat generation;
  - Hot spot;

The Heat Sink place can be replaced by other cooling systems. A description of these systems (Pautsch, 2005) is represented in Table 1. Also, we mention each one's performance.

Table 1. CPU cooling systems

Cooling Technology	Cray Product	Power [W]	Power Density Removal [W/cm^3]
Air Cooling – Open loop	SV1	40	127,0
Air Cooling – Closed loop	HPAC	75	203,2
Conduction to Liquid – Cold plate	T3E	80	457,2
Conduction to Liquid – Top Hat	MTA	100	508,0
Single Phase Forced convection (immersion)	T90	75	381,0
Single Phase Impingement	SS1	120	762,0
Heat Pipe Assisted Heat Sink	Rainier	140	66,04
Spray Evaporative Cooling	X1/X1E	200	1016,0
Film Evaporative Cooling (Spray cooling with Phase Change)	Future	275	1397,0 Projected

In addition to the methods described in Table 1 (Banton & Blanchet, 2004) indicates other possibilities regarding major cooling technology trends, such as:

- Spray cooling with mercury inside – first circa 1999;
- Hollow core liquid cooled electronic modules – Mercury circa 1998;
- Microchannel cooling;
- Refrigeration, air chillers;
- Heat pipes;
- Thermo-electric coolers (TEC).

Recently, development boosted in the area concerning cooling methods that ensure dissipation of increasingly larger power flows. In this respect, (Meijer et al., 2009) indicate various possibilities to increase the cooling process efficiency by either operating on TIM or using one of the following: Ultra Thin High Efficiency Heat Sinks and Manifold Micro- channel Heat Sinks, Radially Oscillating Flow Hybrid Cooling Systems, Oscillating Flow Liquid Cooling and the Phonon Transport Engineering method.

A representation of the limits that can be reached by dissipating the heat flow to the maximum is described in Figure 6. In the figure one can observe the CPU cooling methods, the data belonging to (Pautsch, 2005) for the CRAY technologies.

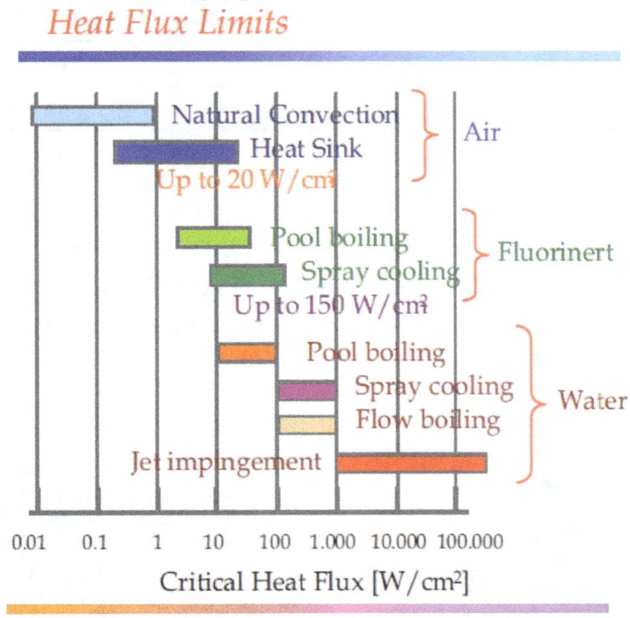

Fig. 6. Maximal values of the Critical Heat Flow (Pautsch) depending on the cooling methods that were used

Regarding the aforementioned information, it can be concluded that the operation of a CPU system is highly intricate, and a direct correlation between operational speed and cooling efficiency has been experimentally demonstrated.

3. CONSIDERATIONS PERTAINING TO SMALL AND MICRO CHANNELS

3.1 Presence of the mini and micro channels within CPU cooling systems

Heat and mass transfer is accomplished across the channel walls in many man-made systems, such as micro heat exchangers, desalination units, air separation units, CPU cooling etc.A channel serves (Kandlikar et al., 2005) to bring a fluid into intimate contact with the channel walls and to bring fresh fluid to the walls and remove fluid away from the walls as the transport process is accomplished. A classification of the channels conducted by (Kandlikar et al., 2005, Kandlikar & Grande, 2003) is described in Table 2.

Table 2. Channel classification scheme (Kandlikar et al., 2005); D value represents the smallest channel dimension.

Type of channel	Overall dimensions
Conventional channels	>3mm
Minichannels	3mm≥D>200μm
Microchannels	200μm≥D>10μm
Transitional Microchannels	10μm≥D>1μm
Transitional NanoChannels	1μm≥D>0.1μm
NanoChannels	0.1μm≥D

Advanced solutions for CPU cooling increasingly utilize small or micro channels of various types. Consequently, Escher et al. (2009) recommend the utilization of the Ultra-Thin Manifold Micro-Channel Heat Sink; Figure 7a illustrates the operational criteria, while Figure 7b presents a scanning electron microscope view of the microchannels.

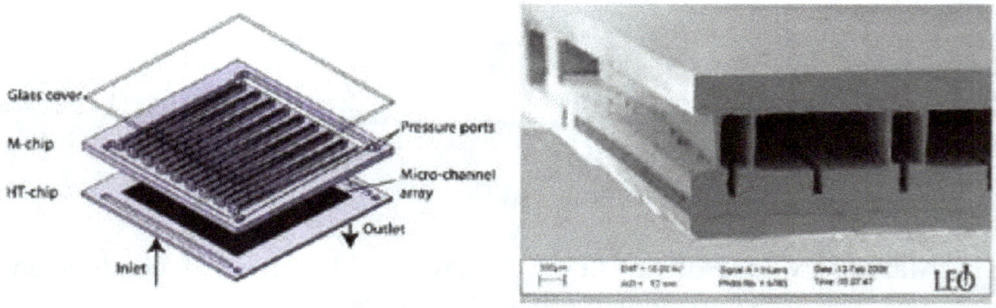

(a) (b)

Fig. 7. Micro channels used within CPU systems (Escher et al., 2009): (a) basic schematics; (b) SEM image of the micro channels

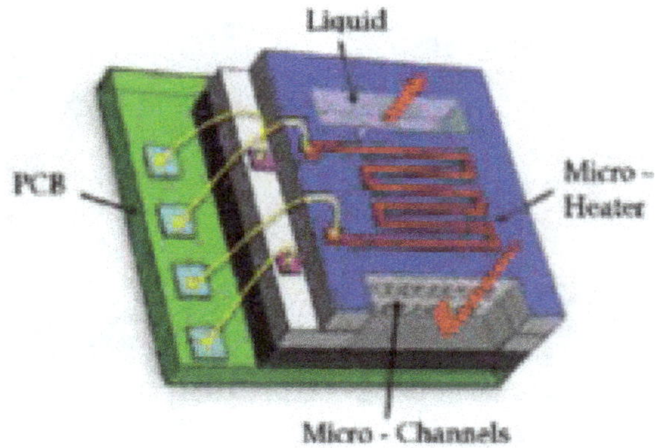

Fig. 8. Pyramid chip stack

The usage of the micro channels in the CPU systems is reminded by (Meijer et al., 2009) for the structures type Pyramid chip stack.

In Figure 8 we present such a structure, emphasizing on the micro channels that are present in the lower part.

IBM Zurich Research Laboratory in 2009 in its research report (Meijer et al., 2009) presents images of the micro channels that can be found in Figure 9.

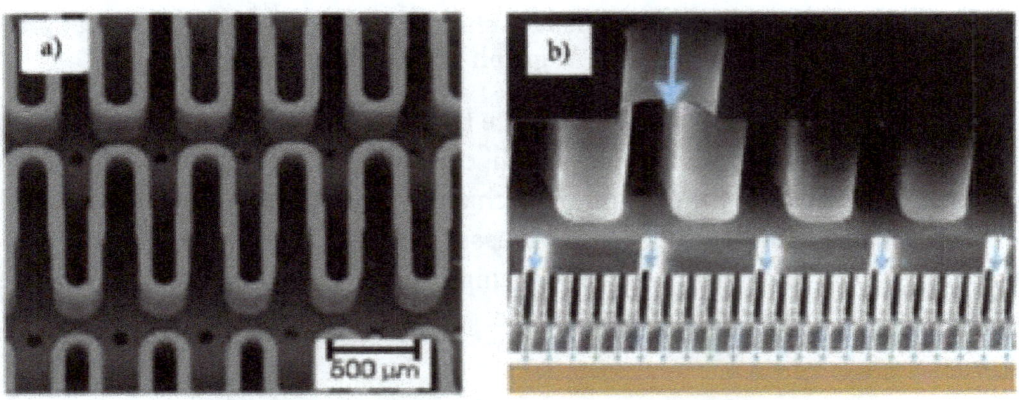

Fig. 9. Micro channels: (a) overview of the micro channels, (b) SEM cross-section of two-level jet plate with diameter of 35µm

Inside the *Heat transfer* laboratory of the “Stefan cel Mare” University of Suceava, Romania, we developed mini heat exchangers that integrate micro channels. Image 10a shows a plate from a mini heat exchanger and image 10b shows an AFM image (19.7 x 80.0 µm) of a micro channel. Also, image 10b shows, on the right size, the threshold corresponding to the copper material and, on the left side, the threshold corresponding to the composite material. In figure 10c the graph shows the values measured by the AFM.

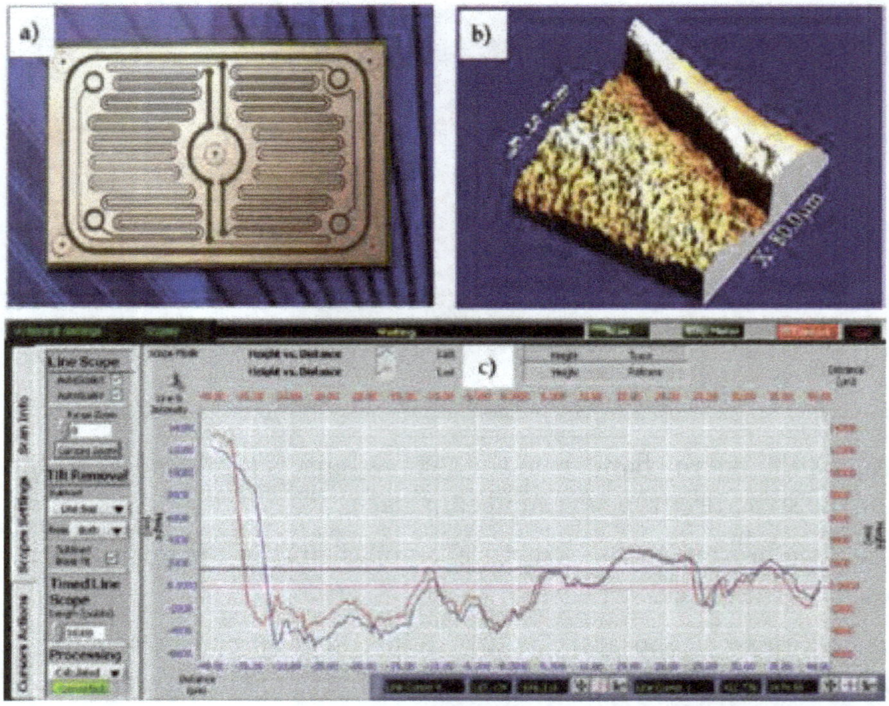

Fig. 10. Images specific to a mini heat exchanger with micro channels: (a) plate, (b) AFM image, (c) AFM graph with measurement data

The data in diagram 10.c show a difference between the lower part of the 1250nm channel and the maximum limit which corresponds to the 13600nm copper surface. This indicates a medium micro channel height of 14850 nm. In order to perform accurate surface topography measurements under various conditions, a NanofoCus µscan laser profilometer was employed. The images that were obtained for the micro channels, and a detail image, can be seen in Figure 11.

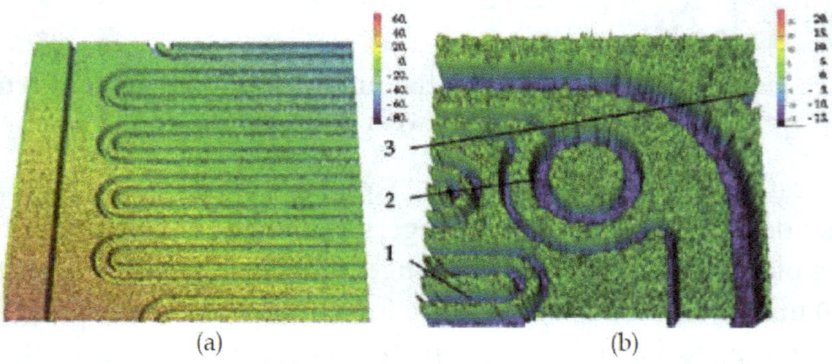

Fig. 11. Images obtained with the NanofoCus µscan profilometer: (a) micro channels, (b) micro channels detail image, were: 1 - microchannels, 2 - admission channelling, 3 - centre holes

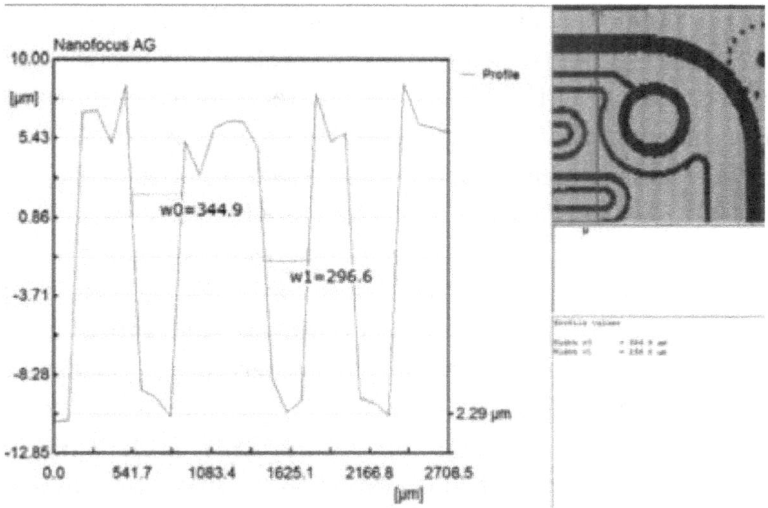

Fig. 12. Heat exchanger micro channels parameters that were obtained with the help of a NanofoCus µscan profilometer

The values obtained with the help of a profilometer, as they are described in figure 12, indicate a height approximately equal to the one indicated by the AFM, resulting in an average of 14,85 µm which is practically doubled by overlapping two plates. The width of a micro channel measured at the middle point of its height is 296.6 µm.

From what was described above we can ascertain that mini and micro channels are often used in various areas, one of them being CPU cooling.

3.2 Experimental evidence of defects occurred in interface layer of cooler-CPU

When the contact between two surfaces is imperfect, the specific thermal resistance of interface layer suddenly increases, so it became of frequent use to apply diverse materials between the CPU and radiator. Previous research showed that when applied in thin layers, CPU thermal grease may form micro and nanoChannels on the surface. These can have a significant impact on the heat transfer between CPU and the cooling system radiator. These materials should both fill the gaps occurred due to surfaces roughness, material's fatigue, loading pressure etc. and transfer as much heat as possible during a short period of time (Mihai et al., 2010).

Inside the *Heat transfer* laboratory of the "Stefan cel Mare" University of Suceava, using an atomic force microscope AFM-universal SPM, Park Scientific instruments firm, processing unit module SPC 400, electronic control unit SFM 220A, we scanned the interface layer of different processors. The images obtained for an AMD processor are presented in Figures 13a (scale 1x1 µm) and 13b (scale 1.6x1.6 µm) respectively.

By analyzing the obtained images, with the help of the AFM, two aspects can be ascertained. Thus, from Figure 13a one can observe that flowing channels may occur in interface layer, having in this case, an estimated width of only 2µm for a height of 2000 Å. The second assertion can be deducted from Figure 13b, the appearance of the steady flow zones being clearly visible. Whatever the situation is, a change in the CPU-TIM-cooler contact appears, change which leads to the alteration of the heat exchange transfer. First, a special attention is paid to assess the effect of mechanical and thermal properties of the contacting bodies, applied contact pressures and surface roughness characteristics as well as the use of different thermal interface materials on the maximum temperature experienced by the CPU. Second, it can be appreciated that good wetting of the mating surfaces and the retention of asperity micro-contacts can become critical elements in effectively removing the heat generated by the CPU (Mihai et al., 2010). The roughness effect is well acknowledged to influence the performance of microchannels and microtubes, affecting both pressure drop and heat transmission. From the aforementioned factors, we can infer that modifications in the TIM layer resulting from improper installation, insufficient push pressure values, material degradation, erroneous TIM application techniques, excessive operational temperatures, etc., alter the thermal conduction coefficient and, consequently, the thermal resistivity, which may result in CPU damage.

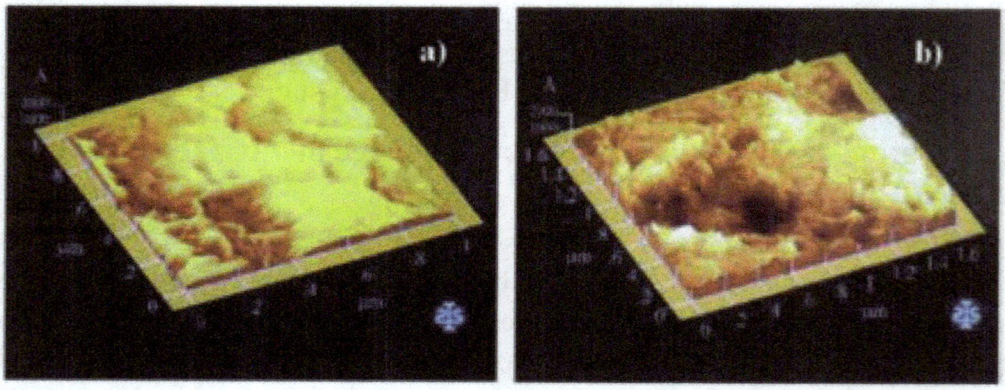

Fig. 13. (a) (b) AFM images of interface layer CPU – Heat Sink: (a) - with the micro channels highlighted, (b) – with the concavities and protuberances emphasized

3.3 Aspects regarding TIM dilatation

A test rig was designed and constructed in the Heat Transfer Laboratory at Suceava University to investigate the thermal grease behavior of CPUs under elevated temperatures, approaching those that may cause CPU failure.

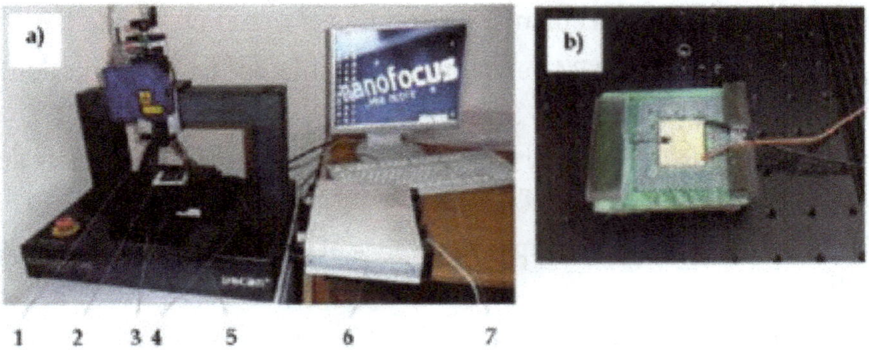

Fig. 14. (a) Test rig Image and (b) CPU and heating system

The description of the annotations in Image 14a is: 1 – laser head, 2 – CPU, 3 – work bench with coordinates, 4 – joystick, 5 - stand, 6 – DC Power supply for the resistors, 7 – nanofoCus software. In order to simulate the CPU heating process, under the CPU were mounted 3 electric resistors. The assembly allows reaching a maximum temperature of 110 ºC. A thin layer of thermal grease Keratherm Thermal Grease KP97 is applied on the CPU surface and then the temperature is gradually elevated. After the thermal condition is stabilized, the layer is subjected to profilometer scanning. The surface micro-topography is then analyzed for thermal grease volume and surface variations shown.

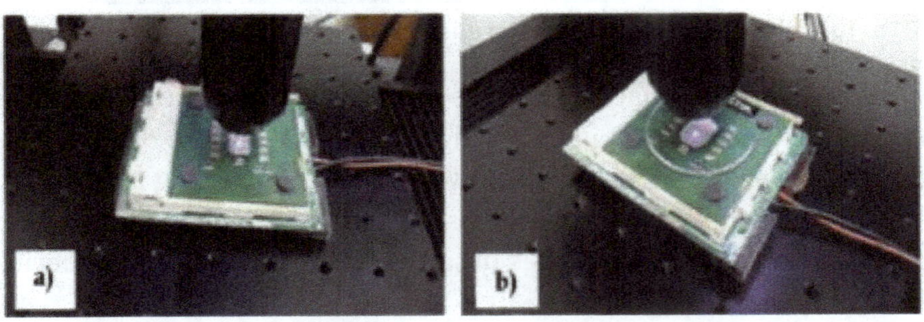

Fig. 15. Experimental investigations of thermal grease layer. (a) without crystal glass and (b) with crystal glass

The experimental determinations were conducted for three hypotheses:

- CPU without a thermal grease layer applied;
- CPU with a thermal grease layer applied (Figure 15a);
- Having a 2,7 mm crystal glass plate placed on the thermal grease covered CPU (Figure 15b), and a 12,5 N/cm2 pressure applied in order to simulate cooler interface.

For the last two hypotheses, changing the voltage of the power supply led to the three resistors heating up to various predetermined values. The temperature was measured at the surface of the TIM material and CPU with the help of a laser thermometer.

During experimental measurements, several issues were investigated:

- In order to analyze TIM behaviour we simulated the CPU functioning with freshly deposited thermal grease on the surface;
- investigations on whether or not micro or nanometer channels appear in an incipient phase at the TIM surface during its heating stage;
- emphasizing the "pump-up" occurrence of the TIM caused by the expansion effect;
- detection of surface discontinuities appearing during heating (localized lack of material);
- monitoring thermal grease layer profile shape, roughness and waviness evolution.

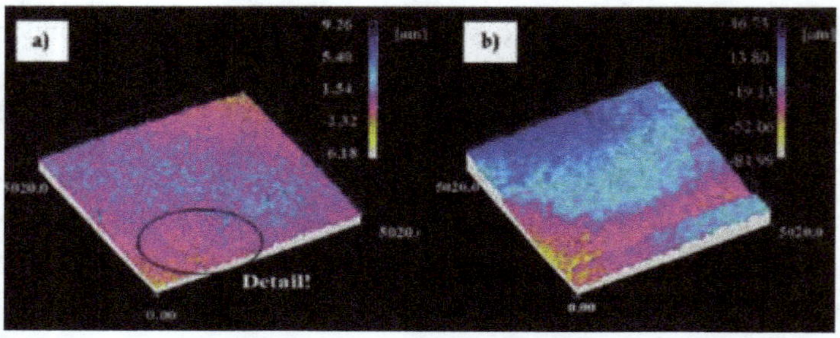

Fig. 16. (a) CPU without thermal grease; (b) Thermal grease layer at 20 °C

Next we will describe the images that were obtained by using the assembly specified in Figure 14a. On the left side image of figure 16a, a 3D image of the CPU surface obtained by laser profilometry is shown without the TIM being applied. In Figure 16b, on the CPU surface, a thermal grease layer was applied. It can be observed that, in contrast with the hypothesis described in Figure 14a, the asperity of the surface is modified.

The detail from Figure 16a allows observing the AMD sign. In Figure 17 the profile shape alteration is shown together with the heating of the CPU and, subsequently, the heating of the thermal grease surface.

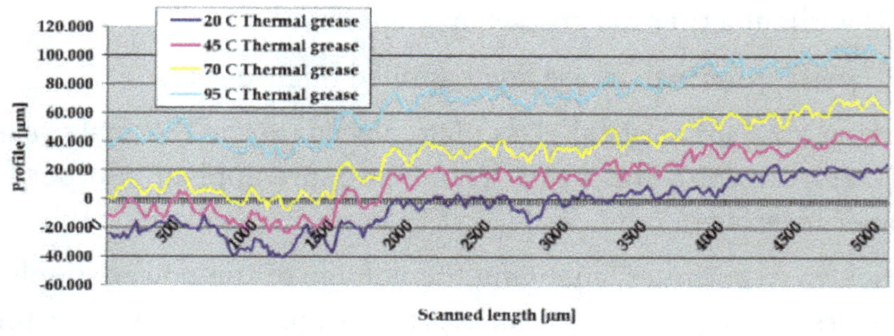

Fig. 17. Profile Shape variation depending on the temperature that was obtained by laser profilometry

The results obtained by profilometry of the applied thermal grease paste are shown in Figures 18a,b and for the roughness and waviness, for temperatures of 20, 45, 70 and 95 ºC, in Figures 19a,b.

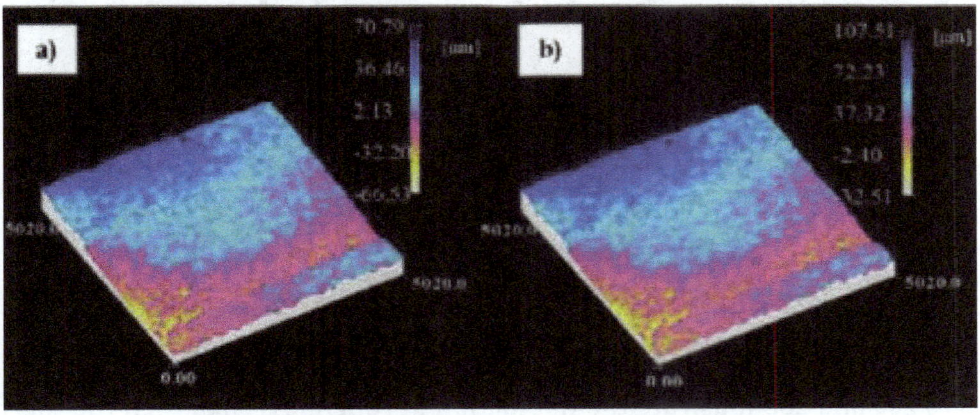

Fig. 18. Thermal grease layer at: (a) 45 ºC and (b) 95 ºC

In the second instance, we placed a crystal glass over the TIM layer, applying downward pressure equivalent to the previously specified value. The utilization of this crystal glass is motivated by our intention to replicate the TIM behavior under thermal conditions while rectifying the imperfections of the heat sink in a practical setting. Figure 20 displays the images acquired using profilometry.

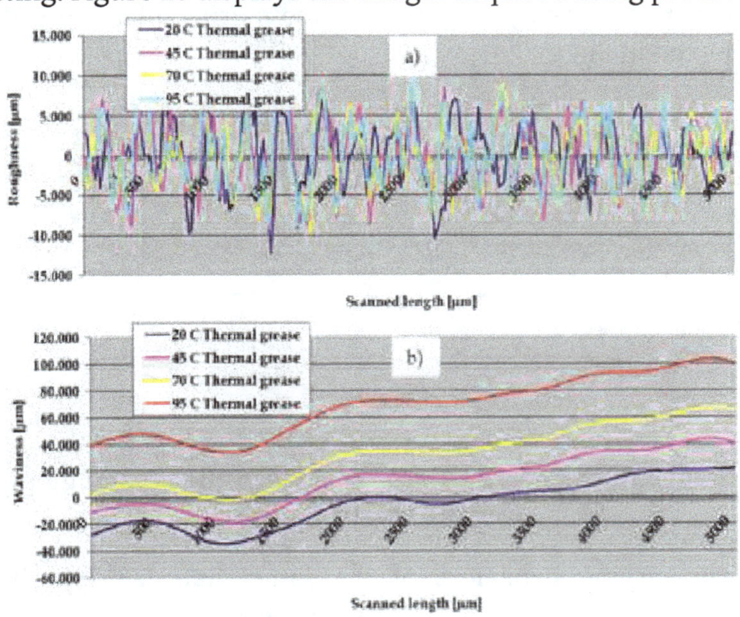

Fig. 19. The results obtained by profilometry for TIM. (a) roughness and (b) waviness

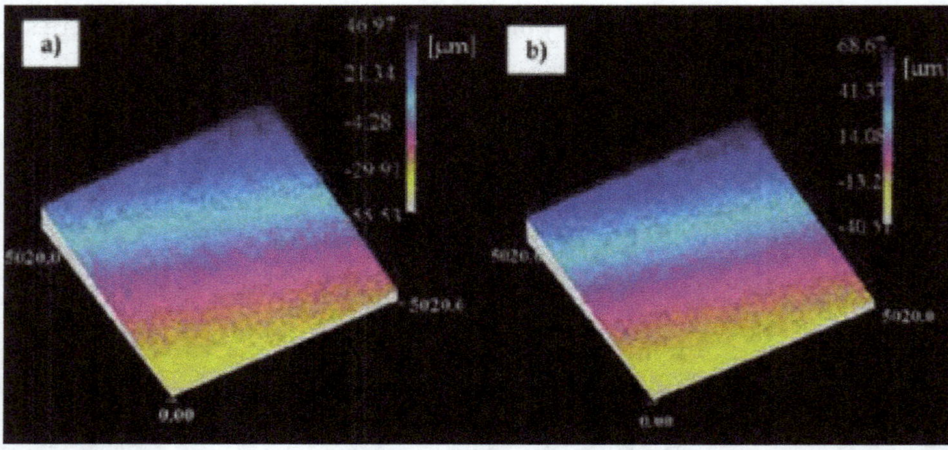

Fig. 20. Thermal grease under crystal glass: (a) at 20 °C and (b) at 95 °C

Images 20a and 20b illustrate the fluctuation in roughness size when the thermal interface material (TIM) is subjected to pressure from an object with a low irregularity coefficient. The comprehensive findings of the Profile Shape size are illustrated in the graphic presented in Figure 21. The differences in the thermal grease layer profile are significantly reduced after the placement of the crystal plate, compared to the free surface condition. To assess the reaction of the same material under a heating process, with significantly reduced beginning roughness, we generated graphs for roughness and waviness using a laser profilometer. The values are depicted in Figure 20.

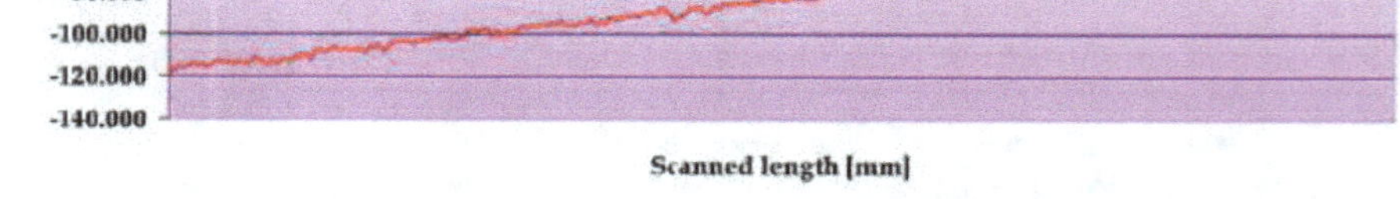

Fig. 21. The change in the Profile Shape size when the crystal glass was applied over the thermal grease

To analyze the varying behaviors of the same material, initially exhibiting significantly lower roughness, after a heating process, we generated graphs for the roughness and waviness values obtained using the laser profilometer. The values can be seen in Figure 22.

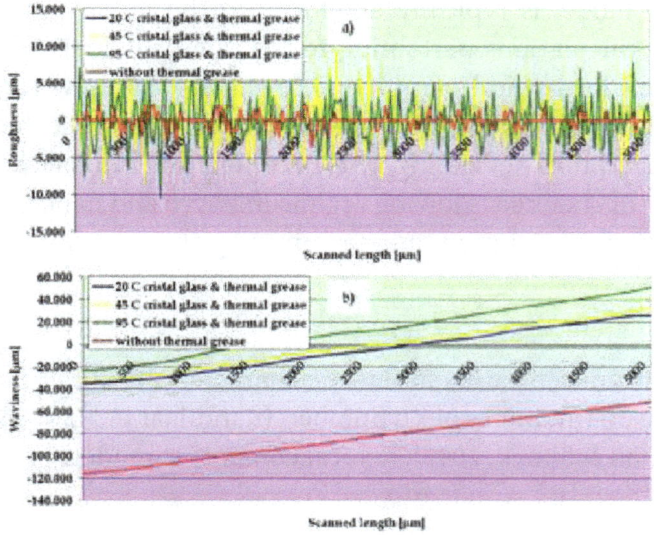

Fig. 22. Results obtained for TIM under a crystal glass. (a) Roughness and (b) Waviness

4. THERMAL TRANSFER IN THE MINI, MICRO, AND NANOCHANNELS OF CPU COOLING SYSTEMS

4.1 Assessment of flowing regime

The current focus has been on the examination of flow dynamics and heat transfer in microdevices. In these systems, the flow and heat transfer processes are at the nano and microscopic scales, differing fundamentally from macroscopic mechanisms due to dimensional properties and molecular phenomena. The initial shift from laminar to turbulent flow, together with the significantly elevated friction factor of a liquid traversing a microchannel compared to conventional ideas. The investigation of heat transfer in a growing compressible flow was conducted by Kavehpour et al. (1997), employing identical first-order slip flow and temperature jump boundary conditions as those utilized in heat transfer for fully developed incompressible flow. The flow was evolving in both hydrodynamic and thermal aspects, with two scenarios examined: uniform wall temperature and uniform wall heat flux. Viscous dissipation was disregarded, whereas the compressibility of the working fluid was considered. The Navier-Stokes and energy equations are unsuitable for micro or nano channels due to the breakdown of the continuity assumption in micro flow. Different computational methods were developed; however, up to date, two main directions in the literature (Kandlikar et al., 2005) are known:

- the working fluid is considered as a collection of molecules;
- the fluid is considered an indefinitely divisible continuum.

The continuum model is typically employed in the modeling of macrosystems. To conduct the calculations pertaining to the microchannels of a heat sink, it is essential to consider several parameters, including the flow rate of the cooling fluid, the temperature of both the fluid and the channel wall, the inlet and outlet pressures required for the cooling fluid, the hydraulic diameter of the channel, and the number of channels. Flowing regime in micro and nano channels is usually estimated by means of Knudsen number, given by the relation:

$$Kn = \frac{k}{H} \tag{1}$$

where *k* is the mean free path and *H* is the characteristic dimension of the flow. In time, depending on the values of Kn number, (Kandlikar et al., 2005, Niu et al., 2007, Hadjiconstantinou & Simek, 2002), several computing models presented in Table 3 were completed.

Table 3. Flow regimes for different Kn Numbers

Knudsen number	Regime	Navier-Stokes equations viability
Kn<0.001	"Continuum"	Accurately modelled by the compressible Navier-Stokes equations; classical no-slip boundary conditions
0.001<Kn<0.1	"Slip-flow"	Navier-Stokes equations applicable; a velocity slip and a temperature jump considered at the walls, as rarefaction effects become sensitive at the wall first.
0.1<Kn<10	"Transition"	The continuum approach is no longer valid; the intermolecular collisions are not yet negligible and should be taken into account.
Kn>10	"Free-molecular"	Intermolecular collisions are negligible compared to collisions between the gas molecules and the walls.

Calculus indicates that the heat transfer coefficient assumes exceedingly high values. The rationale is predicated on the exceedingly small value of the hydraulic flow diameter.

Obviously, a decrease of hydraulic diameter leads to:

- diminished flow and consequently reduced fluid flowing capacity;
- diminished heat transfer phenomenon, leading to increase of temperature values.

As Knudsen number increases, fluid modelling is moving from continuum models to molecular models.

Until now (Kandlikar et al., 2005), a series of collision models were developed for the *inverse power law* (IPL) condition:

- **HS** - The simplest collision model is the *hard sphere* (HS) model, which assumes that the total collision cross-section is constant and for which the viscosity is proportional to the square root of the temperature. Actually, the HS model may be described with the exponent η=∞→ and ω =1/2.
- **VHS** - Bird, who proposed the variable hard sphere (VHS) model for applications to the Monte Carlo method (DSMC), has improved the HS model. The VHS model may be considered as a HS model with a diameter *d* that is a function of the relative velocity between the two colliding molecules.
- **MM** - Another classic model is the Maxwell molecules model with $\eta = 5$ and $\omega = 1$. Actually, the HS and the MM models may be considered as the limits of the more realistic VHS model, since real molecules generally have a behaviour which corresponds to an intermediate value $1/2 \leq \omega < 1$.
- **VSS** - Koura and Matsumoto (1991; 1992) introduced the variable soft sphere (VSS) model, which differs from the VHS model by a different expression of the deflection angle taken by the molecule after a collision. The VSS model leads to a correction of the mean free path and the collision rate values (less than 3%). For α =1, the VSS model reduces to the VHS model.

4.2 Utilization of heat transfer in rectangular micro and nano channels for compressible fluids

As initial statements, on consider that the fluid properties do not modify in time, the fluid is compressible and the flow is of hydrodynamic type. The mathematical model chosen for calculus considers the walls with steady thermal flow. Up to date, two main assumptions are used for computing:

- walls with steady thermal flow;
- constant temperature of the walls.

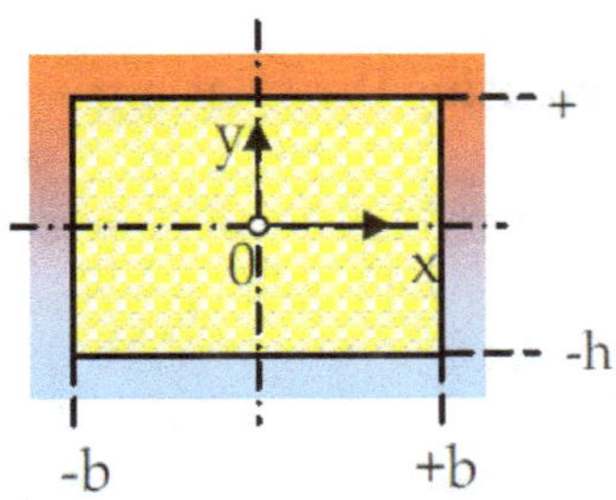

Fig. 23. Rectangular Channel

MathCad application was used in computing. The computing model is VSS and HS as proposed by (Kandlikar et al., 2005). A rectangular channel is considered, as seen in Figure 23, having the dimensions: 2b = 0, 4 ·10^{-6} m, 2h = 0, 2·10^{-6} m, l = 4, 7·10^{-3} m, through which air flows. The inlet and outlet pressures taken into account for computing are presented in Table 4.

Table 4. Inlet and outlet pressure

Pressure [N/m²]								
p outlet	pinlet 1	pinlet 2	pinlet 3	pinlet 4	pinlet 5	pinlet 6	pinlet 7	pinlet 8
1,001 ·10^4	1,006 ·10^4	1,225 ·10^4	1,450 ·10^5	1,675 ·10^5	1,900 ·10^5	2,125 ·10^5	2,350 ·10^5	2,575 ·10^5

The viscosity coefficient for air was consider ω=0.77, the exponent for VSS model α=1,36, the dynamic viscosity μ_0=171,910 x 10^{-7} Pa·s, the molecular mass M=28,9710 x 10^{-3} kg/kmol, and the specific constant R=9,907 x 10^3 J/kg/K. An even temperature of T=365 K was considered for the wall. The viscosity was computed for T, μ(T)= 20,702 x 10^{-6} Pa·s, coefficient k_{2VSS}=1,034, hydraulic diameter D_H=266,667 μm and subsequently, relation (2), the inlet/outlet mean free path is found:

$$\lambda_{i/o,VSS} = \frac{k_2 \cdot \mu(T) \cdot \sqrt{RT}}{p_{i/o}} \quad (2)$$

Consequently, based on *hydraulic diameter, microchannel depth, VSS and HS* model, the Knudsen number is computed with relations:

$$Kn_{i/o} = \frac{\lambda_{i/o\_VSS}}{D_H};\ Kn_{i/o\_P} = \frac{\lambda_{i/o\_VSS}}{2h};\ Kn_{i/oHS} = \frac{\lambda_{i/o\_HS}}{D_H};\ Kn_{i/o\_PHS} = \frac{\lambda_{i/o\_HS}}{2h}; \quad (3)$$

It is regarded that the ratio a*=h/b=55 x 10^{-3} cannot be neglected. From the specialized literature we assumed, for the case of diffusive reflection, the following coefficients: a_1=0,32283, a_2=1,9925 and a_3=4,2853. Using the inlet over outlet pressure ratio Π=p_i/p_o, the mass flow (Kandlikar et al., 2005) we obtained a second-order model with help from the expression:

$$m_{NS2} = \frac{4 \cdot h^4 \cdot p_o^2 \cdot}{a^* \cdot \mu \cdot R \cdot T \cdot l}\left[\frac{a_1}{2} \cdot \left(\Pi^2 - 1\right) + a_2 \cdot Kn_{i/o/^{pr}/HS,} \cdot (\Pi - 1) + a_3 \cdot Kn_{i/o/^{pr}/HS,}^{\ 2} \cdot \ln\Pi\right] \quad (4)$$

The Nusselt number was found using the relation:

$$Nu^{-1} = \frac{\zeta^*}{4} + \frac{a_1 + a_2 \cdot \zeta^* + a_3 \cdot \zeta^{*2}}{a_4 \cdot \left(1 + 3\zeta^*\right)^2} \quad (5)$$

where $a_1 \div a_4$ are constant values.

In the expression above ζ* represents a dimensionless coefficient of slip that results from:

$$\zeta^* = 4 \cdot Kn \frac{2-\sigma}{\sigma} \tag{6}$$

were σ represent the tangential momentum accommodation coefficient, dimensionless.

We will now describe the calculus results that were obtained, for the above referenced data, by using Mathcad. The variation of the parameters Kn and λ_{VSS}, λ_{HS} from relation 3 with respect to inlet pressure, is presented in Figure 24.

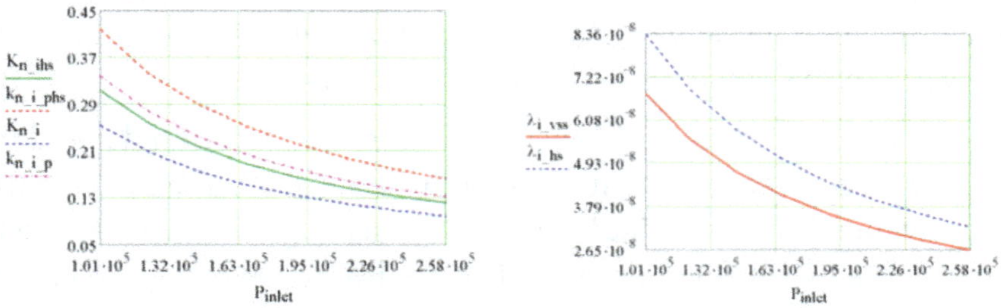

(a) (b)

Fig. 24. Variation of parameters: (a) Kn=f(pinlet) and (b) λi=f(pinlet)

Figure 25a presents the computed mass flow m [kg/s] with respect to Π, and figure 25b the results obtained for the gas: He, 0.05 ≤ Kno ≤ 0.09, *Po* = 1.9·105 Pa.

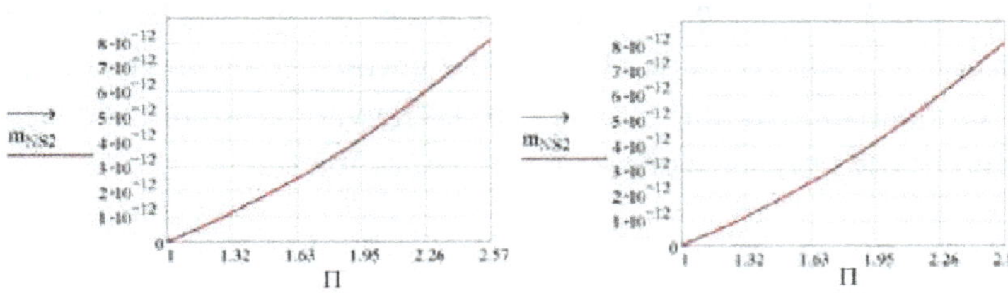

(a) (b)

Fig. 25. Variation of the mass flow depending on Π: (a) values calculated for air and (b) values ascertained for He, by (Colin et al., 2004)

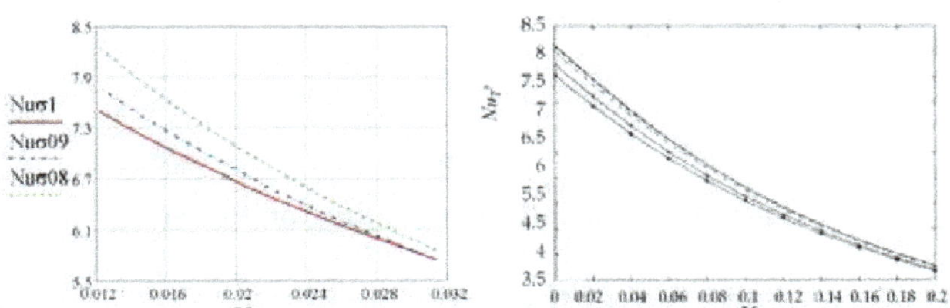

(a) (b)

Fig. 26. Nusselt number variation depending on Knudsen number: (a) for air and (b) for air determinates by (Hadjiconstantinou & Simek, 2002).

By analyzing the diagrams one can notice that there is, in comparison to the inlet over outlet pressure ratio, the same order of magnitude for the mass flow values and almost the same curve shape in both cases, indicating the same variation pattern.

In order to study the Nusselt number dependency with regard to the inlet pressure but, at the same time, modify the tangential momentum accommodation coefficient, we drew, according to the previously resulted calculus, the diagrams described in figure 26a that have σ=1, σ=0,9 and, respectively, σ=0,8. The obtained values can be compared to the ones already existing in figure 26b, with the slip-flow calculation (constant wall), in the range 0<Kn<0.2 and, for air, Pr=0.7.

We can make a representation of the Nusselt number variation depending on the inlet pressure. Figure 27 shows an increase in the Nusselt number together with the pressure value, fact which indicates the intensification of the convective heat exchange together with flow increase. It can also be seen a splay of the curves once the pressure increases. A higher increase in the tangential momentum accommodation leads to a lower coefficient in the Nusselt number values. Currently, several mathematical models are completed, and the VSS and HS models were adopted. The results obtained for rectangular channels with air have the same magnitude order as the ones obtained by (Kandlikar et al., 2005, Niu et al., 2007, Hadjiconstantinou & Simek, 2002) and the shape of the graphs from Figure 23 and Figure 24 is identical with the one obtained by authors (Kandlikar et al., 2005, Niu et al., 2007, Hadjiconstantinou & Simek, 2002). The validation of the mathematical model adopted in the present work is therefore completed.

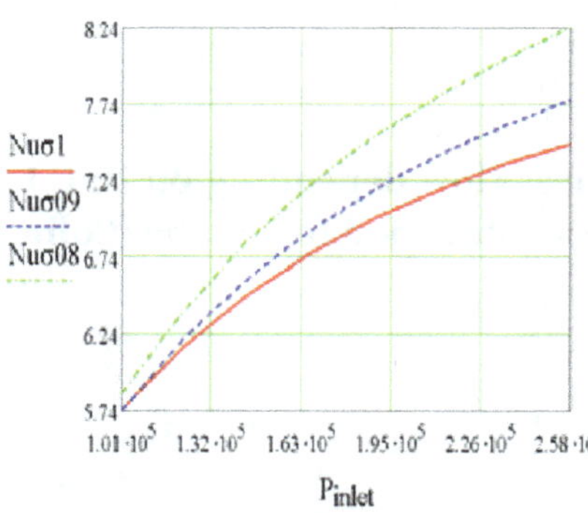

Fig. 27. Nu variation depending on Pinlet

4.3. Thermal conduction over the material interface of CPU cooling systems

Next we will describe the heat exchange calculation at the cooler-CPU interface in the hypothesis of the adequate or inadequate thermal transfer surfaces. Heat exchange at the CPU-cooler interface can be interpreted as conductive, unidirectional with linear heat sources (Lienhard & Lienhard, 2003, Grujicic, 2004).

It is obvious that the heat source is the core of CPU, a case in which the conducted heat flow depends on the processor type (Mihai et al.), the emanation power and the efficiency of the cooling system.

In order to maintain the CPU temperature within a stable working environment, it is desirable that we absorb a major quantity of heat emission, task which is assigned to the thermal compound paste.

The junction-to-coolant thermal resistance is (Yovanovich et al., 1997, Lasance & Simons, 2005, Holman, 1997, Lee, 1998), in fact, composed of an internal, largely conductive resistance Rjc and an external, primarily convective resistance Rex as it can be noticed in figure 28.

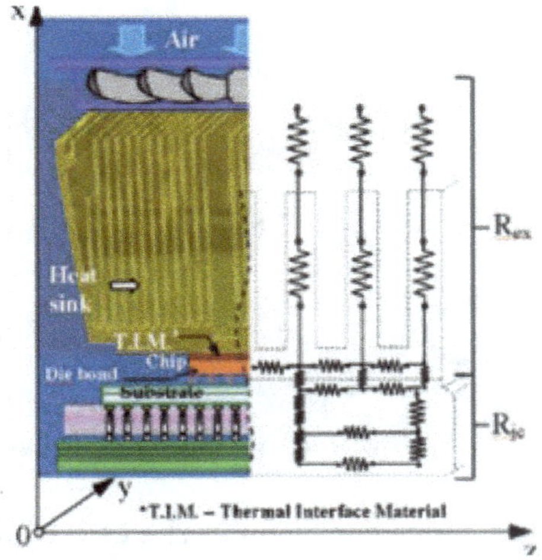

Fig. 28. Primary thermal resistances in a single-chip package

Total thermal spreading resistance for single-chip packages:

$$R_T = R_{jc} + R_{sp} + R_{ex} + R_{fl} \quad \left[\frac{K}{W}\right] \tag{7}$$

were:

- **The internal resistance Rjc** - is encountered in the flow of dissipated heat from the active chip surface through the material used to support and bond the chip and on to the case of the integrated-circuit package;
- **Spreading resistance Rsp** - arising from the three-dimensional nature of heat flow in the heat spreader and heat sink base;
- **External Resistance Rex** - for convective cooling;
- **Flow Resistance Rfl** – the transfer of heat to a flowing gas or liquid that does not undergo phase change is an increase in the coolant temperature from the inlet temperature to an outlet temperature.

In order to calculate the heat exchange within the CPU cooling system we imply using an aluminium heat sink with a number of fins N_f = 28, Height of fins h_f = 0.0355 m, Thickness of fins t_f = 0.0013 m, Shroud spacing h_b = 0.003 m, Base-Plate dimensions LxW=0,06 x 0,06 m, Power dissipation P = 22.5W, Heat sink width w = 0.06 m, Heat sink height h_{hs} = 0.037 m. We will now show the constants that were taken into account for the application that was developed in Mathcad: Ambient temperature (inside the calculator, through measurement) Ta = 316 K, Thermal conductivity of aluminium and air k_{Al} = 170.00 W/mK respectively k_{air}=0.026 W/mK, air density ρ=1.2 kg/m³, heat capacity of air C_p = 1.0 kJ/kgK, kinematics viscosity nu=$1.5\cdot10^{-5}$ m²/sec, Prandtl No. for air Pr = 0.7, Loss coefficient K = 0.9. In Figure 29 we show the graphical representation of the temperature variation depending on the CPU system structure when internal heat sources are present.

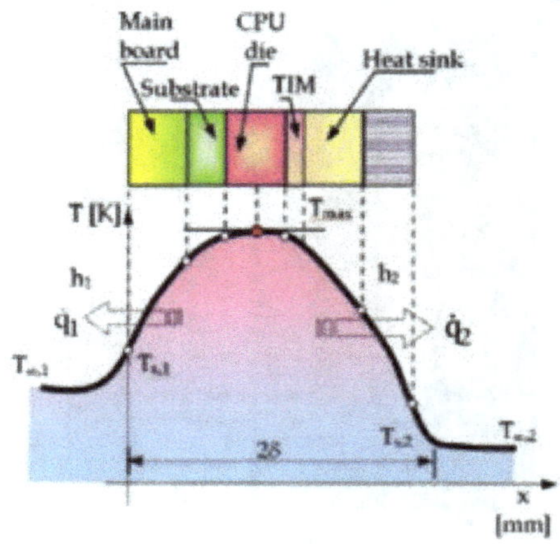

Fig. 29. Temperature variation in the CPU cooling system

In order to conduct the calculations we used the mathematical model proposed by J.P. Holman (Simons, 2004, Guenin, 2003). The area and the perimeter of the radiator are:

$$Area = w\cdot\left(h_b + h_f\right) - N_f \cdot t_f \cdot h_f \qquad \left[m^2\right] \tag{8}$$

$$Perimeter = 2\cdot\left(w + h_b + h_f\right) + N_f\cdot\left(t_f + 2\cdot h_f\right) \quad [m] \tag{9}$$

Calculate hydraulic diameter of heat sink/shroud passage area:

$$D_{hyd} = \frac{4\cdot Area}{Perimeter} \quad [m] \tag{10}$$

Holman indicates the initial use of a certain "guess value", marked as Vel. We gave this Vel an initial value Vel=0.2. The same author indicates the use of the relation below:

$$V = root\left[\left[Vel - \left[2 \cdot \frac{g \cdot P \cdot h_{hs}}{T_a \cdot Area \cdot C_p \cdot \rho\left[\left[\frac{64 \cdot nu}{Vel \cdot D_h}\right] \cdot \frac{h_{hs}}{D_h} + K\right]}\right]^{-\frac{1}{3}}\right] \cdot Vel\right] \tag{11}$$

Calculate Reynolds' number:

$$\mathrm{Re} = V \cdot \frac{D_{hyd}}{nu} \tag{12}$$

The frictional heating factor between two reciproCating parts in contact is:

$$f = \frac{64}{\mathrm{Re}} \tag{13}$$

Reynolds' value on the direction of air flow in the radiator has the following expression:

$$\mathrm{Re}_x = \frac{W \cdot h_{hs}}{nu} \tag{14}$$

Wall heat flux:

$$Q_w = \frac{P}{\left(N_f \cdot 2 \cdot h_f + w\right) \cdot h_{hs}} \quad \left[\frac{W}{m^2}\right] \tag{15}$$

Calculate heat sink temperature rise:

$$\Delta T = \frac{Q_w \cdot \frac{h_{hs}}{k_{air}}}{(0.6795 \cdot \mathrm{Re}_x)^{\frac{1}{3}} \cdot \mathrm{Pr}^{\frac{1}{3}}} \quad [K] \tag{16}$$

Due to the fact that in the heat exchange process the convective effect steps in, Holman suggests for Nusselt number:

$$Nu_{(x)} = 0.453 \cdot \mathrm{Re}_x^{\frac{1}{2}} \cdot \mathrm{Pr}^{\frac{1}{3}} \tag{17}$$

The heat transfer coefficient:

$$h = \frac{\int_0^{h_{hs}} \left[Nu \cdot \frac{k_{air}}{x}\right] \cdot dx}{h_{hs}} \quad \left[\frac{W}{m^2}\right] \tag{18}$$

Calculate fin efficiency:

$$\eta_{fin} = \left(h_f + \frac{t_f}{2} \right)^{\frac{3}{2}} \cdot \left(\frac{h}{k_{al} \cdot t_f \cdot h_f} \right)^{\frac{1}{2}} \tag{19}$$

In order to determine the temperature field we will (Bejan, A. & Kraus A.D., 2003) use the Fourier equation:

$$\frac{\partial T}{\partial \tau} = \alpha \nabla_T^2 + \frac{q_v}{\rho c_p} \tag{20}$$

where $qv = qv\ (x, y, z, \tau\ )$ represents the CPU generated source density, measured in [W/m³]. By integrating the Fourier equation for the unidirectional, stationary regime, we obtain the expression of the temperature distribution in the wall:

$$T_{(x)} = -q_v \frac{x^2}{2k} + \left(\frac{T_{s,2} - T_{s,1}}{2\delta} + \frac{q_v \delta}{k} \right) \cdot x + T_{s,1} \qquad [K] \tag{21}$$

Were $T_{s,1}$ $T_{s,2}$ being the temperatures of the exterior parts of the wall. The maximum temperature Tm in wall is achieved through $x = x_m$, resulting from condition: s$\frac{dT}{dx} = 0$, that is:

$$x_m = \delta + \frac{k}{q_v} \cdot \frac{T_{s,2} - T_{s,1}}{2\delta} \qquad [m] \tag{22}$$

The maximum temperature zone [12] is found within plate ($0 \le x_m \le 2\delta$), providing the following condition is observed:

$$-1 \le \frac{k}{2 \cdot q_v \cdot \delta^2} \cdot \left(T_{s,2} - T_{s,1} \right) \le 1 \tag{23}$$

If we replace $x = x_m$ in equation (6) the maximum wall temperature is obtained:

$$T_{\max} = \frac{q_v \cdot \delta^2}{2k} + \frac{k}{8 \cdot q_v \cdot \delta^2} \left(T_{s,2} - T_{s,1} \right)^2 + \frac{T_{s,1} + T_{s,2}}{2} \ [K] \tag{24}$$

If $T_{\infty,1,2}$ being the coolant temperatures (see figure 7), the limit conditions of third type are:

$$\text{If } x=0,\ ; -k \left. \frac{dT}{dx} \right|_{x=0} = -h_1 \left(T_{s,1} - T_{\infty,1} \right) \tag{25}$$

$$\text{If } x=2\delta,\ -k \left. \frac{dT}{dx} \right|_{x=2\delta} = h_2 \left(T_{s,2} - T_{\infty,2} \right) \tag{26}$$

We can determine the wall surfaces temperature [12]:

$$T_{s,1} = T_{\infty,1} + \frac{T_{\infty,2} - T_{\infty,1} + 2 \cdot \delta \cdot q_v \left(\frac{1}{h_2} + \frac{\delta}{k} \right)}{1 + \frac{h_1}{h_2} + 2\frac{h_1}{k}\delta} \quad [K] \tag{27}$$

$$T_{s,2} = T_{\infty,2} + \frac{T_{\infty,1} - T_{\infty,2} + 2\delta q_v \left(\frac{1}{h_1} + \frac{\delta}{k} \right)}{1 + \frac{h_2}{h_1} + 2\frac{h_2}{k}\delta} \quad [K] \tag{28}$$

We deem that the law of heat spreading throughout the entire volume is observed.

By first using the 1-19 expressions we calculate all the parameters that were previously mentioned. Taking into account the previously calculated measures, we determine, with relations 27 and 28, measures $T_{s,1}$ and $T_{s,2}$. With the help of relation 22, distance x_m which refers to the CPU core, where the temperature is highest, is calculated. The next step allows establishing the maximal temperature value T_{max} with relation 24 for verification of ulterior relations. Using relation 21 the maximum temperature field is determined, in plane z-y of CPU, through insertion of two matrices, which give the distance as well as the square distance in each knot. We thus moved away from a one-dimensional transfer to a bi- dimensional transfer. Knowing the maximum temperature field for each point of the matrix, the same law of heat transfer applies, on direction "x". The mathematic model proposed takes into account the thermal conduction coefficient "k" which is dependent of the type of material, inserting the corresponding values for each knot in the matrix. Sometimes it is common to use the transition from Cartesian coordinates to cylindrical coordinates. In order to validate the suggested model we shall make a comparison between the obtained results and similar cases.

4.4 Results and Calculation Steps

By adhering to the aforementioned calculation procedures facilitated by Mathcad, and considering the internal heat source as directly proportional to the energy produced by each type of processor, we can derive the temperature variation associated with the CPU die area. The computation findings depicted in Figure 30a pertain to scenarios in which TIM remains constant. Concerning TIM flaws manifesting as nano or micro channels, as illustrated in figure 13a, we determine from figure 30b that a temperature increase of around 10 °C occurs, potentially resulting in CPU damage.

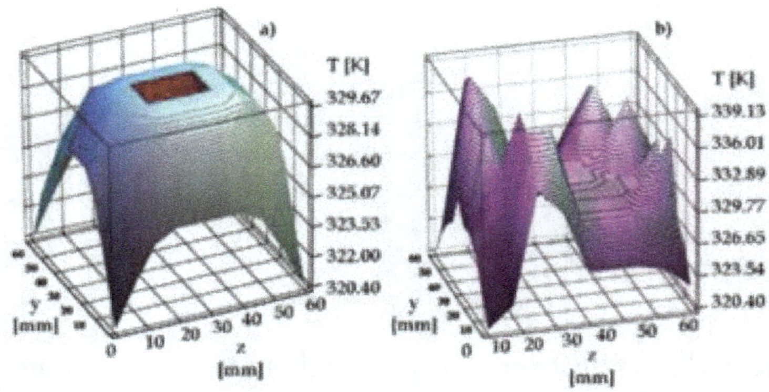

Fig. 30. The field of isotherms that corresponds to interface CPU: (a) for the same thermal conductivity coefficient and (b) in which case the coefficient of thermal conduction is altered.

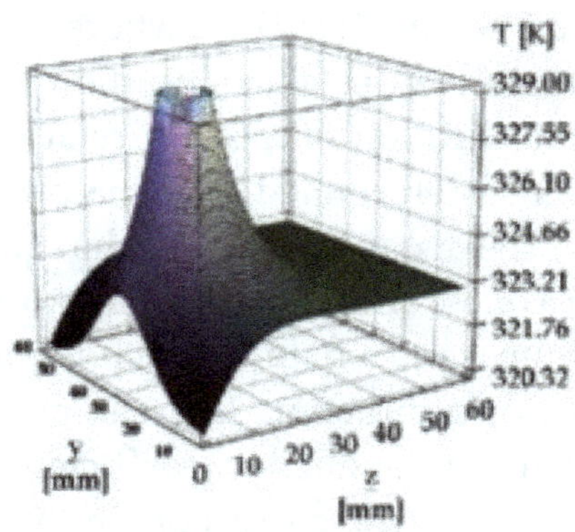

Fig. 31. Calculus in cylindrical coordinates for the field of isotherms that corresponds to for the same thermal conductivity coefficient interface TIM-CPU

Using a different calculation method, when TIM is unchanged, we obtain figure 31, thus noticing the preservation of the parabolic aspect below 323,21 K. However, the CPU area shows a conical shape that is specific to temperature increase. The values that were calculated in Mathcad are significantly close to the Cartesian model, as it can be noticed when comparing the obtained values to those comprised in figure 30a.

In order to study the way in which the temperature changes in the CPU cooling assembly, we conducted simulations using the ANSYS environment. The obtained results (Figure 32) were compared to other similar data. An overview of the CPU die – heat sink that was obtained by (Meijer, 2009) is referred to in figure 33. We can see that there is a uniform temperature field distribution and that the maximal value obviously relates to the CPU die area.

Fig. 32. The temperature field in the cooling assembly – view towards the heat sink obtained by Mihai

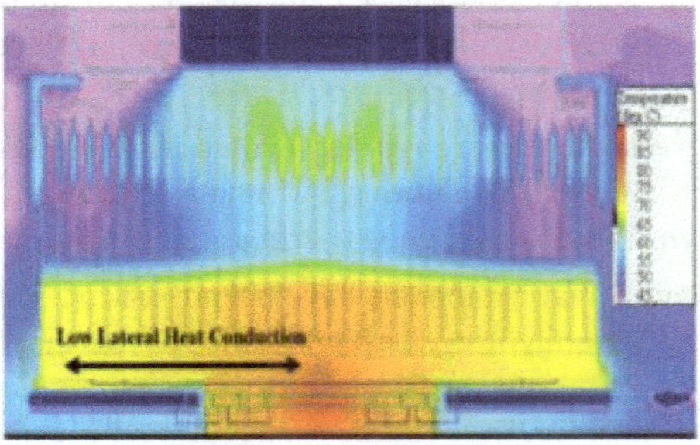

Fig. 33. Thermal modelling of the heat exchange for the CPU die – Heat Sink assembly (Meijer, 2009)

5. CONCISE SUMMARY

Based on the facts presented, we may ascertain that there exists a substantial diversity of small, macro, and even nano channels within CPU cooling systems. In most instances, they serve a practical purpose to facilitate the evacuation of the maximum quantity of heat, employing various criteria and effects like as Joule-Thomson or Peltier. We demonstrated that the thermal interface material (TIM) is crucial for facilitating heat exchange. The AFM images of the CPU-cooler interface indicate the presence of channels with intricate geometries or stagnant patches, which can disrupt thermal transfer. Experimental observations demonstrated (figure 13) that even in the first phase, microchannels with widths ranging from 0.05 to 0.01 μm emerge in the TIM, at depths not exceeding 1000 Å, a phenomenon attributed to the plastic properties during deposition on the CPU surface. Despite the channels that form inadvertently being of nanometric dimensions, they can induce anomalies in CPU performance, namely those resulting from overheating. The objective of the tests performed using laser profilometry was to ascertain if the parameters of profile, waviness, and roughness exhibit varying changes under load, as well as to assess dilatation in response to increasing temperature.

These kind of experimental determinations allow us to make the following assessments:

- Unwanted dilatation phenomena were experimentally outlined. This leads to a "pump up" effect for the material trapped at CPU – cooler interface, phenomenon also illustrated in (Viswanath et al., 2000);
- No surface discontinuities (localized lack of material) were observed during or after heating;
- It was clearly showed that shape deviations can appear when the material is freely applied on CPU surface, before cooler positioning (figure 17), but most of these variations are flattened after cooler placement as shown in figures 21.
- Thermal grease surface roughness evolution was monitored and it was illustrated that its mean values show no major changes after temperature increase, which indicates a good thermal stability of the used material .

At now, multiple mathematical models have been finalized, with the VSS and HS models implemented, highlighting the significance of thermal contact resistance. The performed computations are crucial for examining the effects of TIM deterioration. The mathematical findings unequivocally demonstrate that any deformation in the interface material results in a variation in thermal contact resistance, consequently affecting CPU overheating. The results for rectangular channels with air have a similar magnitude to those reported by Colin (2006), and the graphs are identical to those produced by Niu et al. (2007). The validation of the proposed mathematical model is thus complete. Future study is necessary about TIM stability to mitigate the formation of nano or micro channels.

6. REFERENCES

Banton, R. & Blanchet D. (2004). Utilizing Advanced Thermal Management for the Optimization of System Compute and Bandwidth Density, *ProCeeding of CoolCon* MEECC *Conference, pp. 1-62,* PRINT ISSN #1098-7622 online ISSN #1550-0381, Scottsdale, Arizona, (May 2004), Publisher ACM New York, NY, USA

Bejan, A. & Kraus A.D. (2003). *Heat transfer handbook,* Publisher John Wiley & Sons Inc. Hoboken, ISBN 0-471-39015-1, New Jersey, USA

Colin, S.; Lalonde, P. & Caen, R. (2004). Validation of a Second-Order Slip Flow Model in Rectangular Microchannels, *Heat Transfer Engineering,* Volume 25, No. 3., (mars 2004) 23 – 30, ISSN 0145-7632 print / 1521-0537 online

Colin S. (2006). Single-phase gas flow in microchannels, In: *Heat transfer and fluid flow in minichannels and microchannels,* Elsevier Ltd, 9-86, ISBN: 0-0804-4527-6, Great Britain

Escher, W.; Brunschwiler, T., Michel, B. & Poulikakos, D. (2009). Experimental Investigation of an Ultra-thin Manifold Micro-channel Heat Sink for Liquid-Cooled Chips, *ASME Journal of Heat Transfer*, Volume 132, Issue 8, (August 2010) 10 pages, ISSN 0022-1481

Escher, W.; Michel, B. & Poulikakos, D. (2009). A novel high performance, ultra thin heat sink for electronics, *International Journal of Heat and Fluid Flow*, Volume 31, Issue 4, (August 2010), 586-598, ISSN 0142-727X

Grujicic, M.; Zhao, C.L. & Dusel, E.C. (2004). The effect of thermal contact resistance on heat management in the electronic packaging, *Applied Surface Science*, Vol. 246 (December 2004), 290–302, ISSN 0169-4332

Guenin, B. (2003). Calculations for Thermal Interface Materials, *Electronics Cooling*, Vol. 9, No. 3, (August 2003), 8-9, Electronic Journal

Hadjiconstantinou, N. & Simek, O. (2002). Constant-Wall-Temperature Nusselt Number in Micro and Nano-Channels, *Journal of Heat Transfer,* Vol. 124, No. 2, (April 2002) 356- 364, ISSN 0022-1481

Holman, J.P. (1997). *Heat transfer*, 8th ed., published by McGraw Hill, pp. 42-44, New York:, 1997. ISBN 0-07-029666-9

Kandlikar, S. & Grande, W. (2003). Evolution of Microchannel Flow Passages—Thermohydraulic Performance and Fabrication Technology, *Heat Transfer Engineering*, Vol. 24, No. 1, (Mars 2003), 3-17, ISSN 1521-0537

Kandlikar, S.; Garimella, S., Li D., Colin, S., King, M. (2005). *Heat transfer and fluid flow in minichannels and microchannels,* Elsevier Publications, ISBN: 0-08-044527-6, Great Britain

Kavehpour, H. P.; Faghri, M., & Asako, Y. (1997). Effects of compressibility and rarefaction on gaseous flows in microchannels, *Numerical Heat Transfer* part A, Volume 32, Issue 7, November 1997, 677–696, ISSN 1040-7782, Online ISSN: 1521-0634

Kim, D-K. & Kim, S. J. (2007). Closed-form correlations for thermal optimization of microchannels, *International Journal of Heat and Mass Transfer*, Vol. 50, No. 25-26. (December 2007) 5318–5322, ISSN 0017-9310

Lasance, C., & Simons, R. (2005). Advances in High-Performance Cooling For Electronics, *Electronics Cooling*, Vol.11, No. 4, (November 2005), 22-39, Electronic Journal

Lee, S. (1998). Calculating spreading resistance in heat sinks, *Electronics Cooling*, Vol. 4, No. 1., (January 1998), 30-33, Electronic Journal

Lienhard, J.H.IV. & Lienhard, J.H.V. (2003). *A heat transfer textbook*, 3 rd ed., published by Phlogiston Press, ISBN/ASIN: 0971383529, Cambridge-Massachusetts, USA

Meijer, I.; Brunschwiler T., Paredes S. & Michel B. (2009). Advanced Thermal Packaging, *IBM Research GmbH Presentation*, (nov.2009), pp.1-52, *Zurich Research Laboratory*

Mihai, I.; Pirghie, C. & Zegrean, V. (2010). Research Regarding Heat Exchange Through Nanometric Polysynthetic Thermal Compound to Cooler–CPU Interface, *Heat Transfer Engineering*, Volume 31, No. 1. (January 2010) 90 – 97, ISSN 1521-0537

Niu X.D.; Shu C. & Chew Y.T. (2007). A thermal lattice Boltzmann model with diffuse scattering boundary condition for micro thermal flows, *Computers & Fluids*, No. 36, (March 2006) 273-281, ISSN 0045-7930

Pautsch G. (2005). Thermal Challenges in the Next Generation of Supercomputers, *ProCeeding of CoolCon* MEECC *Conference, pp. 1-83,* PRINT ISSN #1098-7622 online ISSN #1550- 0381, Scottsdale, Arizona, (May 2005), Publisher ACM New York, NY, USA

Simons, R.E. (2004). Simple Formulas for Estimating Thermal Spreading Resistance, *Electronics Cooling,* Vol. 10, No. 2, (May 2004), 8-10, Electronic Journal

Viswanath, R.; Wakharkar, V., Watwe, A., & Lebonheur, V. (2000). Thermal Performance Challenges from Silicon to Systems, *Intel Technology Journal, Vol. Q3*, (Mars 2000), pp. 1-16, ISSN 1535-864X

Yovanovich, M.M.; Culham, J.R., & Teertstra, P. (1997). Calculating Interface Resistance, *Electronics Cooling,* Vol. 3, No. 2, (May 1997), 24-29, Electronic Journal

5 Chapter HEAT TRANSFER IN MICROCHANNELS

1. INTRODUCTION: AN OVERVIEW

Microchannel heat transfer possesses significant promise for extensive applications in cooling high power density microchips within CPU systems, micropower systems, and various large-scale thermal systems necessitating efficient cooling capabilities. The micro-size of the cooling system not only substantially decreases the weight load but also improves its capacity to dissipate a significantly bigger volume of heat compared to large-scale cooling systems. It has been recognized that for flow in a large scale channel, the heat transfer Nusselt number, which is defined as hD/k, is a constant in the thermally developed region where h is the convective heat transfer coefficient, k is thermal conductivity of the fluid and D is the diameter of the channel. As the channel size diminishes, the convective heat transfer coefficient, h, must increase to sustain a constant Nusselt number. As the channel dimensions decrease to the micron or nanoscale, the heat transfer coefficient may increase by a factor of thousands or millions compared to its original value. This can significantly enhance heat transfer and has sparked considerable interest in the study of microchannel heat transfer, both experimentally and conceptually. Conversely, the lab-on-chip system has experienced swift advancements in fabrication techniques and the components—microchannels functioning as conduits, together with structures that include valves, mixers, and pumps—that are vital components of microChemical 'factories' on a chip. Consequently, several microchannels are used for the conveyance of fluids in chemical or biological processes. A specially built conduit is utilized for the mixing of various fluids or the separation of distinct species. The mass or momentum conveyance process within the channel is crucial. The mass transfer process closely resembles the heat transfer process due to the analogous governing equations for both mass and heat (Incropera et al., 2007). The Nusselt number divided by the Prandtl number raised to the nth power is equivalent to the Sherwood number (defined as the product of the convective mass transfer coefficient and the characteristic length, divided by the mass diffusivity) divided by the Schmidt number (defined as the ratio of kinematic viscosity to mass diffusivity) raised to the nth power. Comprehension of heat transfer can facilitate the understanding of mass transfer and momentum transfer within the microchannel (Incropera et al., 2007). Conventional theories, including the constitutive equations that delineate stress and deformation rates in flow, as well as the Fourier conduction law, are predicated on macroscopic observations of flow and heat transfer processes. However, they neglect numerous micro-scale phenomena, such as gas flow rarefaction and compressibility, as well as the electric double layer effect in liquid flow, which can substantially influence both flow dynamics and heat transfer within microchannels. Consequently, the flow and heat transmission processes in a microchannel markedly differ from those in a large-scale channel. A comprehensive examination and analysis of both the fluid dynamics and thermal transfer processes in the microchannels is necessary. Furthermore, an experimental research to corroborate and validate the analysis is important. Accurate measurements of flow and heat transfer in a microchannel are heavily dependent on the precise construction of both the microchannel and the microsensors using MEMS techniques. The successful manufacturing of these intricate microchannel systems necessitates a comprehensive understanding of MEMS technology. The precise measurement of heat transfer within a microchannel is contingent upon the effective fabrication of the microchannel, which must incorporate arrays of miniature temperature and pressure sensors, as well as micro heaters to elevate the flow temperature.It appears that microfluidics has become an emerging science and technology of systems that process or manipulate small (10-9 to 10-18 liters) amounts of fluids, using channels with dimensions of tens to hundreds of micrometres (George, 2006; Vilkner et al., 2004; Craighead, 2006). Various long or short micro or nanoChannels have used in the system to transport fluids for chemical or biological processing. The basic flow behavior in the microchannel has been studied in certain depth (Bayraktar & Pidugu, 2006; Arkilic & Schmidt., 1997; Takuto et al., 2000; Wu & Cheng, 2003). The major problem in the past is the difficulty to install micro pressure sensors inside the channel to obtain accurate pressure information along the channel. Therefore, almost all of the pressure information is based on the pressures measured at the inlet and the outlet outside of the channel, which is used to reduce to the shear stress on the wall. The measurements have either neglected or subtracted an estimated entrance or exit pressure loss. These lead to serious measurement error and conflicting results between different groups (Koo & Kleinstreuer, 2003). The friction factor or skin friction coefficient measured in microchannel may be either much greater, less than or equal to the one in large scale channel. Different conclusions have been drawn from their measurement results and discrepancies are attributed to such factors as, an early onset of laminar-to turbulent flow transition, surface roughness (Kleinstreuer & Koo 2004; Guo & Li 2003), electrokinetic forces, temperature effects and microCirculation near the wall, and overlooking the entrance effect. In addition, when the size or the height of the microchannel is much smaller than the mean free path of the molecules or the ratio of the mean free path of the molecules versus the height of the microchannel, i.e. Kn number, is greater than 0.01, one has to consider the slip flow condition on the wall (Zohar et al. 2002; Li et al. 2000; Lee et al., 2002. More precise measurements of pressure distribution within the microchannel and enhanced control over wall surface conditions are essential to resolve differences across various studies. The absence of technology for integrating sensors into the microchannel also pertains to the monitoring of heat transfer data. All provided heat transfer data is derived from an average of the heat transfer across the whole microchannel. By monitoring the bulk flow temperature at both the inlet and outlet of the channel, the average heat transfer for the channel can be determined. Temperature sensors cannot be put into the channel to provide local heat transfer data. Consequently, comprehensive data regarding the local heat transfer distribution within the channel is not provided. Furthermore, there is an absence of information on entrance length and the heat transfer process in the thermally completely formed region. Furthermore, the wall roughness within the channel could not be directly adjusted or assessed within the tube. Consequently, its impact on heat transmission remains ambiguous. This was ascribed to significant variations in heat transport across different studies (Morini 2004; Rostami et al., 2002; Guo & Li, 2003; Obot, 2002). Accurate measurements of local heat transport are necessary to elucidate the inconsistencies among various studies. This chapter will present a thorough discussion of microchannel flow and heat transfer, encompassing both historical and contemporary findings. A comprehensive analysis of the impact of the aforementioned surface forces on microchannel flow and heat transfer will be provided. An overview of MEMS fabrication techniques will be provided. We have devised MEMS techniques to construct a microchannel system capable of integrating arrays of miniature pressure and temperature sensors. The created tiny sensors will undergo testing to verify dependability and will be calibrated for precise measurements. The manufacturing of this microchannel system necessitates intricate procedures, as noted by Chen et al. in 2003a and 2003b. The successful manufacture of this channel, appropriate for measuring both local pressure drop and heat transfer data, is a challenging endeavor. The production of this complex system can be significantly streamlined through the utilization of polymer materials (Ko et al., 2007). This necessitates the creation of a pressure sensor utilizing polymer materials (Ko et al., 2008). Polymer materials with low thermal conductivity can be manufactured as channel walls to offer excellent thermal insulation and substantially diminish streamwise heat transmission along the wall. This facilitates precise measurements of local heat transport within the channel. The channel height can be regulated to the desired thickness by spin coating the polymer accordingly. The channel's configuration can be easily fabricated using photolithography. This chapter will address the design and fabrication processes for the channel and sensor arrays. Measurements of the local pressure drop and heat transfer within the channel will be presented and analyzed. Consequently, the chapter's contents are succinctly outlined as follows:

- Gas flow and the assoCiated heat transfer characteristics in microchannels.
- Liquid flow and heat transfer characteristics in microchannels including (a) the single phase and (b) the two phase flows.
- MEMS fabrication techniques
- Discussion on recent developments and challenges faced for MEMS fabrication of the microchannel system.
- Working principle and fabrication of the miniaturized pressure and temperature sensors.
- Fabrication of the complicated microchannel system integrated with arrays of either or both the miniaturized pressure and temperature sensors.
- local heat transfer and pressure drop inside the microchannels.

2. CHARACTERISTICS OF GAS FLOW IN MICROCHANNELS

The recent advancement of micromachining processes utilized for the miniaturization of fluidic devices has garnered significant industrial attention, including applications like as micro cooling devices, micro heat exchangers, micro valves and pumps, and lab-on-chips, leading to increased research in this domain. The fluid flow in microscale capillary tubes can be attributed to Knudsen in 1909. Nonetheless, conducting experiments on microscale flow and making precise observations within a microchannel has proven challenging due to the inadequacy of techniques for fabricating microchannels and creating arrays of miniature sensors on the channel surface. To yet, the majority of critical information regarding micro-scale thermal and flow parameters within microchannels remains unmeasured and inaccessible. The previous investigations on micro-scale flow regarding flow and heat transfer predominantly relied on measurements of pressure or temperature at the channel's entrance and exit, as well as mass flow rate, or on measurements taken from the surface of a comparatively larger channel. Consequently, certain unique transport processes that are insignificant in large-scale channels may significantly influence the flow and heat transfer mechanisms in micro-scale channels, such as the rarefaction impact of gas flow. Consequently, the rarefaction of gas flow within the microchannel must be considered in the analysis.

2.1 Theoretical analysis

In order to describe the rarefaction of gaseous flow, a ratio of the mean free path to the characteristic length of the flow called Knudsen number (Kn) has been used as a dimensionless parameter. The Knudsen number is defined as λ/Dc, where "λ" denotes the mean free path of gas molecules and "Dc" denotes the characteristic dimension of the channel. For convenience, it has been suggested (Tsien, 1948) that the rarefaction in gases can be typically classified into three flow regions by the magnitude of the Knudsen number, which are "the continuum flow regime", "the free-molecular flow regime" and "the near- continuum flow regime", as described as follows.

1. Continuum flow regime: This regime is defined for flow with $Kn < 0.001$. In this regime, the theories of the gas flow and fluid properties completely conform to the continuum assumption, and the Knudsen numbers approach to zero. In addition, the modified classical theories of the liquid flow are also suitable in this regime.
2. Near-continuum flow regime: this flow regime is defined in the range with $0.001 \le Kn < 10$. The Knudsen number in this flow regime is still large enough that the flow is subject to a slight effect of rarefaction. The flow can be considered as a continuum in the core region except in the region adjacent to the wall where a small departure from the continuum such as velocity-slip or temperature jump is assumed. For convenience, one can further subdivide the flow into two regimes, i.e. the slip-flow regime and the transition-flow regime. In the slip-flow regime, the macroscopic continuum theory, therefore, is still valid due to small departures from the continuum. However, in order to conform to the real-gas behavior, it is necessary to adopt some appropriate corrections for the slip of fluid at the boundary. The slip-flow regime is defined in the range of $0.001 \le Kn < 0.1$ while the transition-flow regime is defined in the range of $0.1 \le Kn < 10$. In the transition-flow regime, the intermolecular collisions and the collisions between the gaseous molecules and the wall are of more or less equal importance. The flow configuration can be regarded as neither a continuum, nor a free- molecular flow. There is no simplified approach to attack this problem. Some conventional methods, such as, directly solving the complete sets of Boltzmann equations or using the empirical correlations from the experimental data, have been adopted.
3. Free-molecular flow regime: This flow regime is defined in the region with $10 \le Kn$. The rarefaction effect dominates the entire flow field. The gas is so rarefied that intermolecular collisions can be negligible. Hence, the flow characteristic is described by the kinetic theory of gas. Only interaction between gas molecules and boundary surface is considered.

Meanwhile, it has also been suggested (Tsien, 1946) that one can employ the kinetics theory of gases or the conventional heat transfer theory to study the gas flow in the continuum flow regime. When the gaseous rarefaction is within the range of the free-molecular flow regime, the kinetics theory of gases is suitable for use. However, in the range of the near-continuum flow regime, there has been no well-established method. In the slip-flow regime the gas flow can be considered as continuum. Hence, we can employ the macroscopic continuum theory to study the heat transfer in gases by taking account the velocity-slip and temperature-jump conditions at the wall. In the transition-flow regime the transport mechanisms in the rarefied gas are between the continuum and the free molecule flow regime, it is incorrect to consider the gas as a continuum or free molecule medium. Therefore, the theoretical study in the transition regime is very difficult. Many of the works (Ko et al., 2008, 2009, 2010; Bird et al., 1976a; Eckert and Drake, 1972; Yen, 1971; Ziering, 1961; Takao, 1961; Kennard, 1938) intend to develop some convenient methods to solve this problem, such as enlarging the validation of macroscopic continuum theory by using some corrections in boundary conditions or developing mathematical schemes to directly solve the highly nonlinear Boltzmann equation. However, these approaches are still not successful.

For theoretical study of the rarefied-gas flow, Kundt and Warburg (1875) have been the first to propose an important inference by experimental observation. They found an interesting phenomenon that the gaseous flow exhibits a velocity-slip on solid wall when the pressure in the system is sufficiently low. This phenomenon later has been confirmed by the analytical results from kinetics theory of gas by Maxwell (1890). In addition, Maxwell also defined a parameter "*fS*" called tangential momentum accommodation coefficient to modify the departures from the theoretical assumptions and real-gas behavior in molecular collision processes. The value of *fS* will presumably depend upon the character of the interaction between the gaseous molecules and the wall, such as the surface roughness or the temperature etc. In the observations of wall slip, Timiriazeff (1913) made the first direct measurements of wall slip. However, the most accurate measurements of velocity slip are undoubtedly made by Stacy and Van Dyke, respectively. Hence, a sound theory used to describe the rarefied gas behaviors has been established successfully. In the heat transfer studies, Smoluchowski (1910) has performed the first experiments for a heated rarefied gas flow and found the temperature-jump occurring on the solid wall.

Kennard (1938) has suggested that it could be analogous to the phenomenon of velocity slip and thus developed an approximate expression to describe this temperature discontinuity. In a flow field with a temperature of the gas flow different from the neighboring solid wall, there exists a temperature difference in a small distance "g", which is called temperature jump distance, between the gas and the solid wall. The jump distance "g" is inversely proportional to the pressure but directly proportional to the mean-free-path of the gas. Due to the very small jump distance, it looks as having a discontinuity in the temperature distribution between the gas flow and the neighboring solid wall. By using the thermal accommodation coefficient proposed by Knudsen (1934) and the concepts of heat transfer mechanism between gas molecules defined by Maxwell, a theory for the microscopic heat transfer occurred in the rarefied gas flows has been successfully established.

In addition, the gas flow in a micro-channel also involves other problems, such as compressibility and surface roughness effects. Therefore, other dimensionless parameters, such as the Mach number, Ma, and the Reynolds number, Re, should also be adopted. The relationship among these parameters has been derived and can be expressed as follows.

$$\mathrm{Re} = \sqrt{\frac{k\pi}{2}}\frac{Ma}{Kn} \tag{2-1}$$

where k is the specific heat ratio (cp/cv) of the gas. Since both Ma and Kn vary with compressibility of gas in the channel, the value of Re should vary according to the above equation. The full set of governing equations for two dimensional, steady and compressible gas flows can be written as follows (Khantuleva et al., 1982):

$$\frac{\partial(\rho u)}{\partial x} + \frac{\partial(\rho v)}{\partial y} = 0 \tag{2-2}$$

$$\rho u\frac{\partial u}{\partial x} + \rho v\frac{\partial u}{\partial y} = -\frac{\partial p}{\partial x} + \mu\left[\frac{\partial^2 u}{\partial x^2} + \frac{\partial^2 u}{\partial y^2} + \frac{1}{3}\left(\frac{\partial^2 u}{\partial x^2} + \frac{\partial^2 v}{\partial x \partial y}\right)\right] \tag{2-3}$$

$$\rho u\frac{\partial v}{\partial x} + \rho v\frac{\partial v}{\partial y} = -\frac{\partial p}{\partial y} + \mu\left[\frac{\partial^2 v}{\partial x^2} + \frac{\partial^2 v}{\partial y^2} + \frac{1}{3}\left(\frac{\partial^2 v}{\partial x^2} + \frac{\partial^2 u}{\partial x \partial y}\right)\right] \tag{2-4}$$

$$\rho u C_p\frac{\partial T}{\partial x} + \rho v C_p\frac{\partial T}{\partial y} = u\frac{\partial p}{\partial x} + v\frac{\partial p}{\partial y} + k\left(\frac{\partial^2 T}{\partial x^2} + \frac{\partial^2 T}{\partial y^2}\right) + \mu\left[2\left(\frac{\partial u}{\partial x}\right)^2 + 2\left(\frac{\partial v}{\partial y}\right)^2 + \left(\frac{\partial v}{\partial x} + \frac{\partial u}{\partial y}\right)^2 - \frac{2}{3}\left(\frac{\partial u}{\partial x} + \frac{\partial v}{\partial y}\right)^2\right] \tag{2-5}$$

$$p = \rho RT = nkT \tag{2-6}$$

The boundary conditions for the velocity slip and temperature jump on the top and bottom walls are shown as follows (Wadsworth et al., 1993):

$$u - u_w = \frac{2-\sigma_u}{\sigma_u}\lambda\left(\frac{\partial u}{\partial y}\right)_w + \frac{3}{4}\frac{\mu}{\rho T}\left(\frac{\partial T}{\partial x}\right)_w; \quad y = \pm h/2 \tag{2-7}$$

$$T - T_w = \frac{2-\sigma_T}{\sigma_T} \frac{2\gamma}{(\gamma+1)} \frac{\lambda}{\text{Pr}} (\frac{\partial T}{\partial y})_w; \quad y = \pm h/2 \tag{2-8}$$

where σ_u and σ_T are the momentum and the energy accommodation coefficient, respectively. λ, γ and h are the mean free path, the specific heat ratio and the height of the microchannel, respectively. Review of the recent literature indicates that compressible gas flow problems have been studied from the slip to the continuum flow regimes, however, different results are obtained in the micro-channels as described in the following paragraphs.

To analyze the rarefied gas characteristics in the near-continuum flow regime, the methods used (Takao, 1961; Kennard, 1938) in the classical kinetics theory of gas include (1) the small-perturbation approach, (2) the moment methods and (3) the model equation. The mathematical procedures of the small-perturbation approach are to use the perturbation technique to linearize the Boltzmann equation. Since this method can be used in both the near-continuum regime and the near free-molecules regime, therefore, it is suitable for practical applications. The moment methods are first to make adequate assumptions in the velocity distribution *f* such as to express *f* in terms of a power series, i.e. $f = fo(1 + a1(Kn) + a2(Kn)2 + ...)$ as proposed by Chapman and Enskog. Then, substitute the assumed velocity distribution into the Boltzmann equation. The methods of the model equation are to construct a physics model, such as the B-G-K model proposed by Bhatnagar, Gross and Krook (1954), to simplify the expression of Boltzmann equation. Since the governing equation of the system is greatly simplified by the appropriate assumptions in the previous two methods, these approaches can be used for limited ranges of flows. In the numerical simulation (Bird, 1976a; Yen, 1971; Ziering, 1961), a very efficient computational scheme, i.e. DSMC (Direct Simulation Monte Carlo) method, has been developed. However, this method still suffers from the highly nonlinear behavior in the Boltzmann equation. Meanwhile, the use of different approach to solve even the same physical problem will encounter different difficulties due to the different advantages and limitations faced by each method. In addition, the predictions from the analysis should be confirmed by the experiments.

In the studies of numerical calculation, Beskok and Karniadkis (1994) have developed a scheme called "spectral element technique" to simulate the momentum and heat transfer processes of a rarefied gas subjected to either a channel-flow or an external-flow condition. The results have indicated that when the gas passes through a micro-channel at velocity-slip condition, it can cause a significant reduction in drag coefficient CD on the walls. This is mainly caused by the thermal-creep effect when the Knudsen number increases significantly. Meanwhile, they have also addressed that the thermal-creep effect of the gas flow in a uniformly heated micro-channel can increase the mass flow rate, and the increase can be greatly enhanced by raising the inlet velocity. In addition, other effects, i.e. the compressibility and the viscous heating effects that may be occurred in the rarefied gas flow should also be considered. Chu et al. (1994) has used numerical analysis to evaluate the efficiency of heat removal when gas flows through an array of micro-channel under continuum or the velocity-slip condition. This numerical simulation is intended to study the cooling performance inside a micro-channel array that fabricated in a silicon chip. The numerical approaches have adopted the finite-difference methods incorporated with SOR (Successive over-relaxation) techniques to solve the problem with Neumann boundary conditions. The assumptions used include fully developed hydrodynamic condition, fully developed thermal condition and uniform heating on the bottom wall with the top wall well insulated. From the numerical results they have found that even though the temperature- jump causes decrease in Nusselt number that is contrary to continuum flow, the entire heat transfer performance were still higher than the case of continuum flow; this peculiar phenomenon is mainly due to the velocity-slip effects that induce greater mass flow per unit time into the channel. Therefore, the design of gas flow through a micro-channel array at the slip-flow regime as cooling is suggested. Fan and Xue (1998) have used the numerical method of the "DSMC" to simulate the gas flow in micro-channels at the slip-flow regime. They have assumed that the gas flow is simultaneously subjected to the effects of the velocity-slip and the compressibility. In addition, the effects of pressure ratio "Po" between two ends of the micro-channel on the flow are also studiedSimulation research was conducted at various Po ratios, revealing that the velocity profiles of the flow near both ends of the channel depart from the parabolic profile. The average flow velocity near the channel outlet significantly increases with an elevated Po ratio. The departure from the parabolic profile is mostly attributed to the entrance and exit effects of the microchannel; only the flow field distant from the microchannel's terminus may satisfy fully developed flow conditions. The secondary account of flow acceleration is substantially influenced by velocity-slip and is also induced by gas compressibility. The compressibility effect results in a reduction of both density and pressure at the channel exit, with a more significant decrease in exit pressure potentially accelerating the flow to compensate for the density reduction. Consequently, the acceleration of flow in a microchannel can be enhanced by augmenting the pressure ratio Po. The simulation results reveal the slip flow features in the channel, specifically regarding shear stress and velocity distributions in the wall region. The results demonstrate that the compressibility resulting from the elevation of Po significantly influences gas flow behavior in the slip-flow regime within the microchannel.

2.2 Experimental measurements

For experiments of gas flow in micro-channels, Wu and Little (1983) have measured the friction factors for both laminar and turbulent gas flows in trapezoidal channels. The widths of the channels are from 130 to 200 μm and the depths are from 30 to 60 μm, respectively. The working fluids used include nitrogen, helium and argon gases. The friction factors, f, obtained in his experiment are larger than the theoretical prediction for the critical Reynolds number less than 400. The deviations of the data form the prediction are attributed to the very high degree of surface roughness and measurement uncertainty. For a nitrogen gas flow in micro-tubes, the effects of wall surface roughness on the pressure drop or the friction factors are studied by Choi et al. (1991) for both laminar and turbulent flow. The micro-tube diameters are from 3 to 81 μm and the wall roughness is from 0.00017 to 0.0116. It is found that the Poiseuille number, Po, which is defined as f × Re, is 53 in the laminar region when the diameter of the tube is less than 10 μm. The Po of 53 in his experiment is lower than the theoretical value of 64 for fully developed laminar flow in the large scale tube, where the Po is kept as a constant. In the experiments of turbulent flow region, the results indicate that the Colburn analogy is not valid when the diameter of micro-tubes is less than 80 μm.

Some of pressure drop measurements have a good agreement with the predictions of the conventional theory. Acosta et al. (1985) has measured the friction factors in rectangular micro-channels, and the results are very close to the friction factor predicted by the conventional theory in small aspect ratios channels. Lalonde et al. (2001) has studied the friction factor of air flow in a micro-tube with a diameter of 52.8 μm. The experimental data has a good agreement with the predictions from the conventional theory. Turner et al. (2001) has performed an experiment to measure the friction factor with different working fluids, such as nitrogen, helium and air in microchannels with hydraulic diameters varying from 4 to 100 μm. The walls of the rectangular channels consider both the rough and the smooth wall conditions. The results indicate that the friction factors in laminar region for both the rough and the smooth wall conditions have good agreement with the conventional theory. In contrast to the results that agree with the conventional theory, Pfahler et al. (1990a, 1990b) and Pfahler et al. (1991) have performed experiments to obtain the friction factor for working fluids of helium and nitrogen in micro-channels with the heights varying from 0.5 to 40 μm. The results indicate a significant reduction of Cf (Poexp/Potheo) which is a function of channel depth. The Cf decreases with decreasing Re in the smallest channel. Yu et al. (1995) has performed the experiments of gas flow in a micro-channel with either a trapezoidal or a rectangular cross section. The hydraulic diameter varies between 1.01 and 35.91 μm. They have observed a friction factor smaller than the prediction of the conventional theory, and conclude that the deviation may be caused by both effects of compressibility and rarefaction of the gas. Harley et al. (1995) has performed the experiments for subsonic, compressible flow in a long micro-channel. The working fluids used are nitrogen, helium and argon gases. The channels are fabricated by silicon wafer, and the dimensions of the channels are 100 μm wide, 10 mm long with depths varied from 0.5 to 20 μm. The experimental data have been presented in terms of the Po with hydraulic diameter from 1 to 36 μm. The measured friction factors agree with the theoretical prediction, but become smaller when the depth of channel decreases to 0.5 μm. The reduction in the friction factor is attributed to the occurrence of slip flow. The compressibility effects are also found by Li et al. (2000) who have performed an experiment of nitrogen gas flow in five different micro-tubes with diameters from 80 to 166 μm. The pressure drop along the tube became nonlinear when the Much number is higher than 0.3. In order to understand more detailed pressure information inside a micro-channel, arrays of the pressure sensors should be integrated in the micro-channel for measurement of pressure distribution. Pong et al. (1994) are the first to present that a rectangular micro-channel can be fabricated with integrated arrays of pressure sensors for pressure distribution measurements. Both the helium and the nitrogen gas are used as the working fluid in his study. The channels are from 5 to 40 μm wide, 1.2 μm deep and 3000 μm long. The experimental results indicate that the pressure distribution is not linear and is lower than the prediction based on the continuum flow analysis in the micro-channel. The non-linear effects are caused by both effects of rarefaction and compressibility of the gas due to the high pressure loss. Liu et al. (1995) have used the similar channel as in Pong et al. (1994) but having different shapes to perform the experiments. The channel has a uniform cross section and has the dimensions of 40 μm wide, 1.2 μm deep and 4.5 mm long. The pressure drop distribution found is also nonlinear. For the channel with non-uniform cross section, sudden pressure changes are found at locations where variations of the cross section occur. In the meantime, analysis of the channel flow has also been performed with the assumptions of a steady, isothermal, and continuum flow with wall slip condition. However, the analysis cannot explain the small pressure gradients measured near the inlet and the outlet of the channel.

Shih et al. (1996) has repeated the experiments of Pong by using a similar micro-channel with dimensions of 40 μm wide, 1.2 μm deep and 4000 μm long to measure the pressure distribution and mass flow rate for helium or nitrogen gas flow. The results of helium have a good agreement with the analysis based on the Navier-Stokes equations with slip boundary condition. The boundary condition of a slip flow on the wall is given by

$$u_w = \psi Kn(\partial u / \partial y) \tag{2-9}$$

where ψ is momentum accommodation coefficient. In general, $\psi = 1$ has been used for engineering calculation. All the experimental data indicate a non-linear dependence of the pressure drop with the mass flow rate. Li et al. (2000) and Lee et al. (2002) have performed experiments for channels with orifice and venture elements. The dimensions of channels are 40 μm wide, 1 μm deep and 4000 μm long. The working fluid used is nitrogen which has an inlet pressure up to 50 Psig. The mass flow rates are measured as a function of the pressure drop. The results indicate that the pressure distribution is non-linear and the pressure drop is a function of mass flow rate. The experimental data are used to compare with the prediction from the Navier-Sotkes equation with a slip boundary condition. The friction factors for both channels with either the orifice or the venture are all lower than theoretical prediction.

It appears that contradictory results have been found in the previous studies. More accurate measurements of the pressure drop and heat transfer inside a microchannel are required. This requires fabrication of a micro-channel system, integrated with arrays of micro pressure sensors or temperature sensors, fabricated by surface micromachining process. However, the microchannel fabricated previously with arrays of pressure sensor is limited to a channel height of 1.2 μm due to the use of oxide sacrificial layer which is deposited by chemical vapor deposition (CVD) process. Much thicker deposition of the oxide layer is not possible with the current technology. Furthermore, the channel structure is significantly compromised due to the construction of the channel wall with an exceedingly thin coating, permitting only gas movement for the experiment. Consequently, to establish a channel with significantly greater height that accommodates liquid flow under robust wall conditions, a whole new manufacturing procedure for the channel must be contemplated and engineered.

3. CHARACTERISTICS OF LIQUID FLOW IN MICROCHANNELS

The liquid flow can be considered a continuous even within a little channel. Liquid flow may transition to boiling when the wall temperature exceeds the liquid's vaporization temperature. Consequently, liquid flow regimes can be categorized into single-phase flow and two-phase flow regimes. The actual heat transfer patterns in laminar or transitional flow regimes diverge markedly from predictions made using continuum theory, owing to the nonlinear surface force factors in the Navier-Stokes equations. Surface forces greatly influence micro-scale liquid flow, which can be markedly impacted by geometry, electrokinetic transport processes, and the hydrophilic or hydrophobic nature of the surface conditions within the microchannel.

3.1 Experimental results

In a micro-channel, single-phase liquid flow is regarded as incompressible. Nonetheless, geometric configurations, including the aspect ratio, channel cross-section, and surface roughness, can profoundly influence the flow properties and heat transfer processes within a microchannel. Harms et al. (1997, 1999) have observed a friction factor well predicted by the conventional theory in the laminar region. Webb et al. (1998) have observed that the conventional theory is able to predict the single phase heat transfer and the friction factor for a rectangular channel. Pfund et al. (1998) have studied the water flow in rectangular micro-channels at Reynolds numbers between 40 and 4000. The friction factor has a good agreement with the conventional theory in the laminar flow region, but increase by the surface roughness in the turbulence region. Xu et al. (1999, 2000) have fabricated the rectangular micro-channels by bonding an aluminum plate or a silicon wafer with a Plexi glass. The channels were etched on a silicon or aluminum substrate. The hydraulic diameters of the aluminum channels are from 46.8 to 344.3 μm and for silicon channels are from 29.59 to 79.08 μm, respectively. The experimental results for liquid flow in micro-channels have very good agreement with the prediction from the Navier-Stokes equation for a Newtonian flow in laminar region. Qu et al. (2000, 2002) has performed experiments for water in silicon micro-channels with trapezoidal cross section having hydraulic diameter from 51 to 169 μm. The pressure drop measured has a good agreement with the prediction based on conventional theory. More experiments have indicated that the deviation from the prediction is attributed to the roughness of the channel wall and viscosity of the fluid. The friction factors obtained from these experiments are higher than the predictions from the conventional theory. Li et al. (2000, 2003) have fabricated different micro-tubes made by glass, silicon or stainless steel with diameters ranging from 79.9 to 166.3 μm, 100.25 to 205.3 μm and from 128.6 to179.8 μm, respectively. The results of the friction factor measured for DI water, in glass and silicon micro-tubes where tube wall can be considered smooth, has good agreement with the conventional theory. The deviation of the data in the stainless steel tube is attributed to the surface roughness. They have concluded that the relative roughness of the wall can not be neglected for micro-tube in the laminar flow region. Sharp et al. (2000) have considered laminar flow of water in micro-tubes with hydraulic diameters ranging from 75 to 242 μm. Their data agree with the conventional theory. Wu et al. (2003) have provided the experimental data of friction factor for DI water in smooth silicon micro-channels with trapezoidal cross section having hydraulic diameter from 25.9 μm to 291 μm. The results of their data have a good agreement with the prediction from the conventional theory. They conclude that the Navier-Stokes equations are still valid for laminar flow of DI water in microchannel with smooth wall and hydraulic diameters as small as 26 μm.

Some work reported the friction factors that are very different from the theoretical prediction. Yu et al. (1995) has performed experiments of water flow in silica micro-tubes with diameters ranging from 19 to 102 μm and the Reynolds numbers between 250 and 20000. The friction factors are lower than the theoretical predictions. Jiang et al. (1995, 1997) have studied water flow through rectangular or trapezoidal channels. The dimensions of the channels are 35 to 120 μm wide and 13.4 to 46 μm deep. The friction factor data are greater than the theoretical prediction, but become lower when the Reynolds numbers are between 1 and 30The discrepancies in the measured friction factor from the predicted values may be ascribed to the surface characteristics of the liquid flow, particularly the channel wall's surface roughness, surface potential, and the electrokinetic effects induced by the electrical double layer (EDL), as elaborated in the subsequent section.

3.2 Analysis of electric double layer effect

If the liquid contains a minimal quantity of ions (e.g., contaminants), the electrostatic charges on the non-conductive solid surface will attract the counter-ions in the liquid flow. The reconfiguration of charges on the solid surface and the compensatory charges in the liquid is referred to as the electrical double layer. The thickness of the electric double layer is substantially influenced by ion concentration, liquid flow polarity, surface roughness, and surface potential. A thicker electrical double layer, potentially caused by reduced ion concentration, a polar solvent, inadequate surface roughness, or elevated surface potential, may result in an increased friction factor and pressure gradient. This can significantly reduce the flow velocity, and the heat transfer of a liquid flow in the microchannel. This is true for infinitely diluted solution such as the millipore water, the thickness of the EDL is considerably large (about 1 μm). However, for solution with high ionic concentration, the thickness of the EDL becomes very small, normally a few nanometer. In this case, therefore, the EDL effects on the flow in microchannels can be negligible.

To account for the EDL effect for polar liquid flow in the microchannel, most of the work performed in the past is the theoretical simulation where the physical models can be formulated based on (1) the Poisson-Boltzmann equations for the EDL potential, (2) the Laplace equations with the applied electrostatic field, and (3) the Navier-Stokes equations modified to include effects of the body force due to the interaction between electrical and zeta potential. However, the numerical results are always lower than the empirical data due to the unusual and complex surface behaviors described above. In addition, the aspect ratio and the geometric cross-section of the channels can also affect the thickness of the EDL. In general, the friction factor increases with decreasing the aspect ratio of the channels. A microchannel with a cross section of circular shape usually has the lowest friction factor. The friction factor in a silicon channel is larger than in a glass channel due to the different surface potential of the channel walls with millipore water.

The Poisson-Boltzmann equations for the EDL potential in a rectangular microchannel are described as follows (Beskok & Karniadakis, 1994):

$$\frac{\partial^2 \psi}{\partial y^2} + \frac{\partial^2 \psi}{\partial z^2} = -\frac{\rho_e}{\varepsilon_0 \varepsilon} = \frac{2 n_\infty z e}{\varepsilon_0 \varepsilon} \sinh\left(\frac{z e \psi}{k_b T}\right) \tag{3-1}$$

where ψ and ρ_e are the electrical potential and the net charge density per unit volume. ε is the dielectric constant of the solution. εo is the permittivity in vacuum.

ni∞ and *zi* are the bulk ionic concentration and the valence of type *i* ions, respectively; *e* is the charge of the proton; *kb* is the Boltzmann constant; *T* is the absolute temperature.

To account for the electric field effect, the Navier-Stokes equation describing the flow motion can be rewritten as following:

$$\frac{\partial^2 u}{\partial y^2} + \frac{\partial^2 u}{\partial z^2} = \frac{1}{\mu}\frac{dp}{dx} - \frac{1}{\mu} E_x \rho_e \tag{3-4}$$

where E_x is an induced electric field (or called electrokinetic potential) and *p* is the hydraulic pressure in the rectangular microchannel.

At a steady state, the net electrical current is zero, which means:

$$I = I_s + I_c = 0 \tag{3-5}$$

$$I_s = 4 \int_{h-1/k}^{h} \int_{w-1/k}^{w} u(y,z)\, \rho_e(y,z)\, dy\, dz \tag{3-6}$$

where I_s and I_c are the streaming and the conduction currents, respectively. In addition, the net charge density is non-zero essentially only in the EDL region whose characteristic thickness is given by 1/*k* (*k* is the Debye-Huckel parameter).

The conduction current, that is the transport of the excess charge in the EDL region of a rectangular microchannel, driven by the electrokinetic potential is given by:

$$I_c = 4 \lambda_o E_x (h + w) \frac{1}{k} \tag{3-7}$$

$$k = [2 z^2 e^2 n_\infty / (\varepsilon \varepsilon_o k_b T)]^{1/2} \tag{3-8}$$

where λ_o is the bulk electrical conductivity (1/Ω m). *h* and *w* are the height and the width of the microchannel, respectively. Substituting Eq.(3-6) and Eq.(3-7) into Eq.(3-5), the electrokinetic potential (E_x) can be written as follows:

$$E_x = -\frac{\int_{h-1/k}^{h}\int_{w-1/k}^{w} u(y,z)\,\rho_e(y,z)\,dy\,dz}{\lambda_o\,(w+h)(1/k)} \tag{3-9}$$

Both the Poisson-Boltzmann equation, Eq.(3-1) and Navier-Stokes equation, Eq.(3-4), can be solved numerically in order that both the EDL and the velocity fields in the rectangular microchannel can be determined.

3.3 Comparison with the data

Despite the theoretical prediction, some work presents occurrence of the electrical double layer of water flow in a micro-channel. Ren et al. (2001) have performed experiments to measure the interfacial electrokinetic effects of a liquid flow through rectangular silicon micro-channels with diameters of 28.1, 56.1 and 80.3 μm. Both the DI water and the KCl solutions with two different concentrations of 10-4 and 10-2 M are used as working fluid. The measured pressure drops for the pure DI water and the lower KCl concentration solution are significantly higher than that for higher concentration solution and the theoretical prediction. The authors have concluded that a significant increase in the friction factor is attributed to occurrence of the electrical double layer (EDL) which increases the pressure drop in the small micro-channels. Similar results have also been obtained by Li et al. (2001).

To compare with the experimental results, the analytical predictions for both the flow and the heat transfer developed from continuum assumption indicate large discrepancy when the characteristic length of the micro-channel becomes small enough. In the studies of liquid flow, many investigators (Ren et al., 2001; Fan et al., 1998; Chen, 1996; Chu et al., 1994; Choi et al., 1991; White et al., 1991; Pfahler et al., 1990, 1991) have concluded that even though the liquids can be regarded as a continuum in a very small system, the real behaviors of heat transfer at the laminar or the transition (before turbulent) condition are deviated from the predictions based on the conventional theory. Usually, for the data published, the uncertainties of flow rate measured and friction factor estimated are 2-5 % and 10-15 %, respectively. For most heat transfer studies, the uncertainties are under ± 20 %. In summary, the geometric effects, such as the aspect ratio, the cross-section shape or the surface roughness etc., can significantly affect the characteristics of both the flow and the heat transport in a microchannel. The onset of transition to turbulent flow in smooth microchannels does not occur if the Reynolds number is less than 1000. For a laminar flow, the Nusselt number varies as the square root of the Reynolds number. In turbulent flow, however, the numerical studies are not applicable and thus many empirical correlations have been proposed, but were not verified. However, satisfactory estimates of the heat transfer coefficients can be obtained with sufficient accuracy by using either experimental results in smooth channels with large hydraulic diameter or conventional correlations.

Tso and Mahulikar (1998) have obtained the heat transfer for laminar liquid flow through a microchannel in both the thermal-developing region and the thermal-developed region. The Nusselt number diminishes as the Reynolds number increases, both in the thermally developed zone and in the thermal entering region. The findings further demonstrate that the pressure distribution throughout the microchannel displays a non-linear characteristic. Although numerous investigations have indicated that liquid flow exhibits a highly complex relationship between the Nusselt number and the Reynolds number, all findings are fundamentally predicated on the assumption of continuum flow. Consequently, a more thorough analysis coupled with experiments is necessary to elucidate the function of the EDL and the varying outcomes across different studies.

3.4 Two-phase flow phenomenon in the microchannel

The two-phase flow or flow-boiling phenomenon in the microchannel exhibits some unusual characteristics. It is found that the bubbles are not rapidly generated even at a very high heat flux from the heated microchannel (Qu et al., 2000). Therefore, further experimental investigations on the flow boiling in microchannels were made by others (Ren et al., 2001; Peng & Wang, 1993; Lin and Pisano, 1991, 1994). In addition, the effect of microchannel scale, geometric configuration, liquid velocity, liquid sub-cooling and liquid concentration on the flow boiling were investigated. It is found that the heat transfer enhanced by a large more volatile component concentration is greater than the pure more volatile liquid. The heat transfer coefficient at the onset of flow boiling and in the partial nucleate boiling region was greatly influenced by the liquid concentration, the geometric configuration, the size of microchannel, and the flow velocity and sub-cooling, but not in the fully nucleate boiling region. Peng and Wang (2001), and Hu (1998) found the so-called "bubble extinction" behavior due to an induced vigorous nucleate boiling mode on a normal-sized heater or abnormal-sized channels. The normal bubbles could not successfully grow and form, if the channel height is less than a critical liquid space required. In order to interpret the unusual behavior observed in microchannel boiling, Peng and Wang (1994) proposed the concepts of "evaporating space" and "fictitious boiling". In fact, the small bubbles that can form initially in microchannel will eventually collapse since the size of the bubble could not grow up exceed the critical radius of bubble (*rc*) formulated by conventional nucleation theory. The fictitious boiling occurred was attributed to the crowded tiny bubbles that grow and then collapse rapidly in a cyclic manner, and thereby mimicking a boiling state that can transfer large amount of heat. The observations suggest that close to bubble nucleation temperature the liquid will vigorously oscillate in the microchannel due to the emergence of tiny bubble embryos. More detailed explanations are given in (Jiang et al., 2001; Peng et al., 1998).

The experiments by Peng and Wang (1993) for flow boiling of water have been carried out in a stainless steel microchannel with rectangular cross-section of 600 μm×700 μm. In a much smaller channel array, with hydrodynamic diameter of 40 and 80 μm, made on a silicon substrate by wet etch, three stable phase-change modes, i.e. local nucleation boiling, large bubble formation and annular flow, were observed depending on the input power level (Qu & Mudawar, 2003). However, bubbly flow, commonly observed in macroChannels, could not be developed in the microchannels. A stable annual flow was also observed in a micro-channel heat sink contained 21 parallel channels having a 231 μm × 713 μm cross-section (Lee et al., 2003).

Lee et al. (2003) proposed that a nearly rectangular microchannel heat sink with 14 μm in depth integrated with a local heater and array of temperature sensors on silicon substrate was made to investigate the size and shape effects on the two-phase patterns in microchannel forced convection boiling. It is found that when the heat input power increases, the downstream movement of the transition region increases the void fraction and causes a lower devices temperature. However, at the high flow rate, the transition region almost oCcupies the entire channel, the increase in the heat input power results in a higher devices temperature. An annular pattern induced by flow boiling appears stably in triangular microchannels, but not in rectangular microchannels. Two-phase boiling or superheated flow has numerous promising applications such as in cooling of electronic components. The principle advantage of two-phase flow lies in the utilization of latent heat absorbed by the working fluid due to phase change from liquid to vapor without increasing the flow fluid temperature. In fact, two-phase flow heat transfer in microCahnnel is a very important and interesting problem indeed.

However, much of the attention at later time has been given to the study of dynamic flow boiling instability in microchannels (Cheng et al., 2009; Wang et al., 2008; Wang et al., 2007; Kandlikar, 2006; Wu & Cheng, 2003, 2004; Brutin et al., 2003; Hetsroni et al., 2002; Hetsroni et al., 2001). A periodic annular flow and the periodic dry steam flow were observed for boiling of water in 21 silicon triangular microchannels having a diameter of 129 μm in (Hetsroni et al., 2001, 2002). However, two types of two-phase hydrodynamic instabilities, i.e. severe pressure drop oscillation and mild parallel channel instability were identified (Qu & Mudarwar, 2003) in the similar microchannels as in other work (Hetsroni et al., 2001). A simultaneous flow visualization and measurement was made on flow boiling of water in two parallel silicon microchannels of trapezoidal cross-section having hydraulic diameters of 158.8 μm and 82.8 μm, respectively (Wu & Cheng, 2003). The results shows that two-phase flow and single-phase liquid flow appear alternatively in microchannels, which leads to large amplitude/long-period fluctuations with time in temperatures, pressures and mass flux. The flow pattern map in terms of heat flux versus mass flux showing stable and unstable flow boiling regimes in a single microchannel has been identified (Wu & Cheng, 2004). It is found that stable and unstable flow-boiling modes existed in microchannels, depending on four parameters, namely, heat/mass flux ratio, inlet water subcooling, channel geometry, and physical properties of the working medium (Wang et al., 2007). In addition, the magnitudes of temperature and pressure fluctuations in the unstable flow- boiling mode depend greatly on the configurations of the inlet/outlet connections with the microchannels (Wang et al., 2008). By fabricating an inlet restriction on each microchannel or the installation of a throttling valve upstream of the test section, reversed flow of vapor bubbles can be suppressed resulting in a stable flow-boiling mode. Based on the exit quality of the flow from a microchannel, more detailed flow regimes are identified (Cheng et al., 2009).

In the past, however, a very important issue, i.e. the surface wettability effect, has been overlooked in the study of boiling flow heat transfer in a microchannel. The boiling flow phenomenon found in the microchannel is only for certain surface wettability. By changing the material of the microchannel or surface wetting property, the boiling flow phenomenon may be completely different. This may cause discrepancy of flow patterns observed in different channels made by different materials. Phan et al. (2009) have found that the wettability of a surface has a profound effect on the nucleation, growth and detachment of bubbles from the bottom wall in a tank. For hydrophilic (wetted) surfaces, it has been found that a greater surface wettability increases the vapor bubble departure radius and reduces the bubble emission frequency. Moreover, lower superheat is required for the initial growth of bubbles on hydrophobic (un-wetted) surfaces. However, the bubble in contact with the hydrophobic surface cannot detach from the wall and have a curvature radius increasing with time. At higher heat flux, the bubble spreads over the surface and coalesces with bubbles formed at other sites, causing a large area of the surface to become vapour blanketed.

The wettability of channel surface has been studied by Liu et al. (2011) who have fabricated three different microchannels with identical sizes at 105 x 1000 x 30000 μm but at different wettability. The microchannels were made by plasma etching a trench on a silicon wafer. The surface made by the plasma etch process is hydrophilic and has a contact angle of 36o when measured by dipping a water droplet on the surface. The surface can be made hydrophobic by coating a thin layer of low surface energy material and has a contact angle of 103o after the coating. In addition, a vapor-liquid-solid growth process was adopted to grow nanowire arrays on the wafer so that the surface becomes super-hydrophilic with a contact angle close to 0o. Different boiling flow patterns on a surface with different wettability were found, which leads to large difference in temperature oscillations. Periodic oscillation in temperatures was not found in both the hydrophobic and the super- hydrophilic surface. During the experiments, the heat flux imposed on the wall varies from 230 to 354.9 **kW/m**^{2} and the flow of mass flux into the channel from 50 to 583 kg/m^{2}s. Detailed flow regimes in terms of heat flux versus mass flux are also obtained.

4. MEMS FABRICATION TECHNIQUES

4.1 Chemical vapor deposition

Chemical vapor deposition (CVD) is a conventional method for producing a thin coating on a substrate. In a chemical vapor deposition (CVD) procedure, gaseous reactants are injected into a heated reaction chamber. The chemically reactive gases diffuse onto and are absorbed by the substrate. The thermal dissolution process of the reactive gases occurs, resulting in the deposition of a thin solid coating on the heated substrate surfaces. Depending upon the relative pressure and the temperature used the CVD processes are categorized as: (1) the atmospheric pressure chemical vapor deposition (APCVD), (2) the low pressure chemical vapor deposition (LPCVD), and (3) the plasma-enhanced chemical vapor deposition (PECVD). The process temperatures of APCVD and LPCVD are ranged from 500°C to 850°C. In PECVD processes, a part of thermal energy is shared from the plasma source. Therefore, the process temperatures of the PECVD are lower on the order of 100°C to 350°C. The silicon based thin films such as poly-silicon, amorphous silicon, silicon dioxide, tetraethoxysilane (TEOS, Si(C2H5O)4) or silicon nitride film can be fabricated by using the CVD process. The chemicals used and the reaction occurred in the CVD process for different kinds of films are listed in Table 1. The poly-silicon film is applicable in the production of pressure or temperature sensors and micro-heaters. The TEOS oxide layer is constructed as an insulator between each sensor layer. Furthermore, the deposition of the silicon nitride film can serve to inhibit moisture infiltration into the sensors during liquid flow experiments, which could potentially damage the micro-sensors or microelectronics incorporated within the micro-channel.

Table 1. Chemical reactions used in the CVD process for different kinds of films.

Films	Chemical reactions
Poly-silicon	$SiH_4 \rightarrow Si + 2\ H_2$
Silicon dioxide	$SiH_4 + O_2 \rightarrow SiO_2 + 2\ H_2$ $SiCl_2H_2 + 2\ N^2o \rightarrow SiO_2 + 2\ N_2 + 2\ HCl$

Films	Chemical reactions
TEOS (tetraethoxysilane)	$Si(oC_2H_5)_4 \rightarrow SiO_2$ + by-products
Silicon nitride	$3\ SiH_4 + 4\ NH_3 \rightarrow Si_3N_4 + 12\ H_2$ $3\ SiCl_2H_2 + 4\ NH_3 \rightarrow Si_3N_4 + 6\ HCl + 6\ H_2$

4.2 Evaporation and sputtering deposition

Both evaporation and sputtering deposition are classified as physical vapor deposition (PVD) process which can form different kinds of films on a substrate directly from a source material. PVD is typically used for deposition of electrically conducting layers such a metal or silicide. Evaporation deposition of a thin film on a substrate is done by sublimation of a heated source material in a vacuum chamber. The vapor flux from the source can be condensed and coated on the substrate surface. The evaporation methods can be further categorized as the vacuum thermal evaporation (VTA), the electron beam evaporation (EBE), and the molecular beam epitaxy (MBE).

The simplest evaporator consists of a vacuum chamber with a crucible which can be heated to a high temperature, as shown in Figure 1(a) and 1(b) by a filament. The filament is used as a heater, which is made of Tungsten (W), a refractory (high temperature) metal. Evaporation is accomplished by gradually increasing the temperature of the filament until the source material melts. Filament temperature is then further raised to evaporate the source material from the crucible. The substrates are mounted on top of the crucible and are deposited with a thin film of evaporated material.

In the electron beam (E-beam) evaporation system, the high-temperature filament is replaced with an electron beam, as shown in Figure 1(c). A high-intensity beam of electrons, with energy up to 15 keV, is foCused on the source material to be evaporated in a crucible. The energy from the electron beam only melts a portion of the source material, which eventually evaporates and condenses on the substrate to form a thin layer.

Sputtering deposition requires generation of plasma gas between high voltage electrodes, as shown in Figure 2, where positively ions can be accelerated and bombards on a target material (a cathode) so that flux of atoms can be sputtered and collected on the substrate. Usually, a physically inert gas, such as argon gas, is made into plasma by knoCking out electrons of the molecules with high speed electrons emitted from the cathode. The sputtering deposition has the advantages of depositing various materials include not only for pure materials or metals, but also for compounds, alloys, refractory materials, or piezoelectric ceramics. In addition, puttering deposition has no shadowing effect as that occurred in evaporation deposition, which causes non-uniform deposition of a film. Therefore, sputtering deposition has been widely used for deposition of different kinds of films.

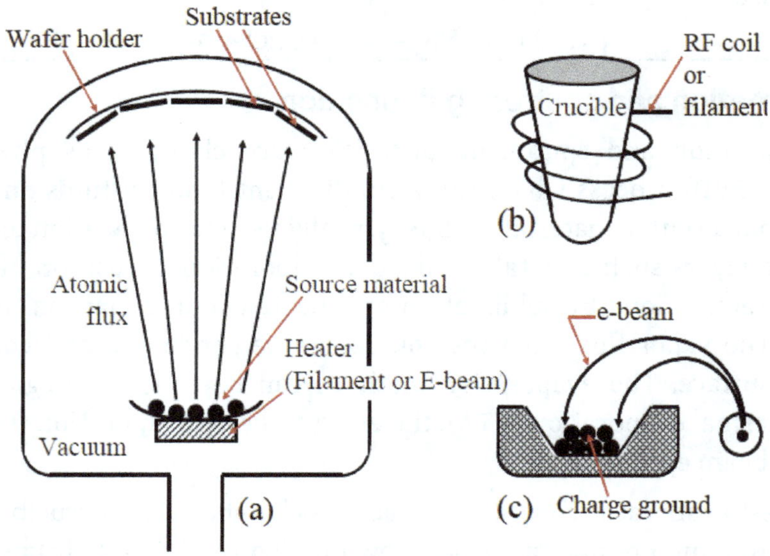

Fig. 1. (a) Schematic of the thermal evaporation system, (b) the use of filament or RF coil as heating source and (c) the use of electron beam as the heating source.

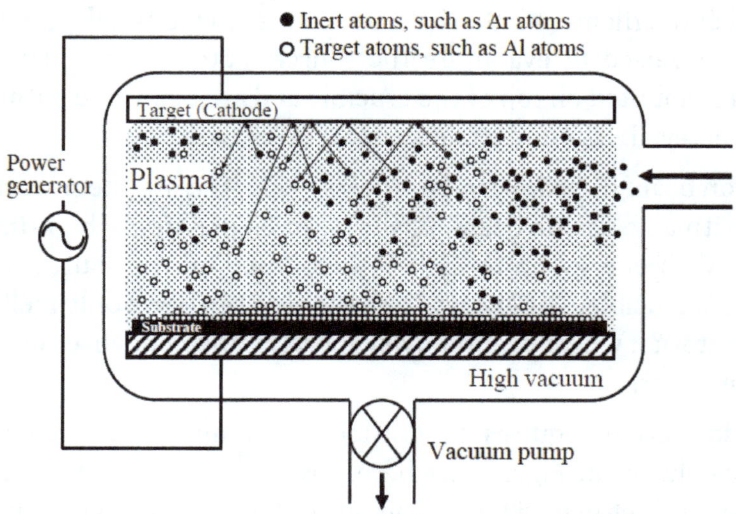

Fig. 2. Schematic of a sputtering system.

4.3 Photolithography Technique

Lithography is the most important technique for transferring micro-patterns onto substrate. Depending upon resolution required, the light sources used for lithography process can be a mercury lamp, a laser light, an electron beam, X-ray or an ion beam. However, photolithography using mercury lamp or laser light is the most popular method for low-cost and fast prototyping of micro-fluidic fabrications. This technique uses a photosensitive polymer layer, the so called photo-resist (PR), to transfer a desired pattern from a photomask to the substrate. The mask is a transparent glass plate or a plastic sheet with metal (chromium, Cr) or ink patterns. The photolithography process is shown in Figure 3. First, the photo-resist is spin coated onto the substrate. After light exposure and a developing process of the PR a desired pattern can be transferred from the photomask to the PR, and then transferred to the underneath layer by a wet or a dry etch process. The resolution of a proximity photolithography (R) process which depends on the wavelength (λ) of the light source can be written as follows:

$$R = K_1 \frac{\lambda}{NA}$$

where K_1 is the optical system constant and NA is the numerical aperture of image lens system. The reduction of the resolution can be made by reducing the wavelength or increasing the NA. However, increasing the NA can lead to large reduction in the depth of the foCus, which is very detrimental to the images. It appears that the resolution of photolithography is primarily determined by the wavelength of the light source.

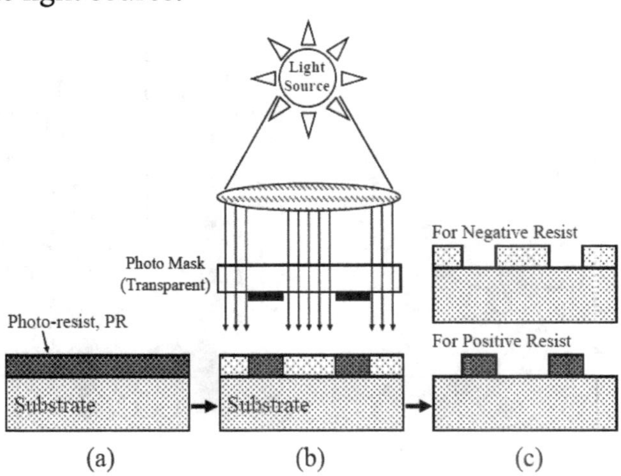

Fig. 3. process of photolithography: (a) spin coat the PR, (b) light exposure and (c) developing the PR.

4.4 Anisotropic wet etching

Single crystal silicon can be anisotropically etched. In general, the etching rate is highly dependent on the crystal's orientation in the single-crystal silicon as shown in Figure 4. Most of the chemical solutions have a distinct slow etching rate on the crystal face of (111), which practically causes a etch stop on this surface. This allows formation of microstructures with sharp edges and corners that can be defined in a single crystal substrate. For the different orientation silicon substrates, the angle between other crystal faces with respect to the (111) crystal face are shown in Figure 5. Different etchants, such as KOH, NaOH, TMAH (Tetramethyl ammonium hydroxide) and EDP (Ethylenediamine pyroChatechol), may be used for anisotropic wet etch for the silicon. The angles between different facets are 54.7o and 90o for {100} to {111} and {110} to {111}, respectively. However, using KOH solution the etching selectivities for different facets are around 400 and 600 for {100}/{111} and {110}/{111}, respectively. Therefore, V-shaped or rectangular cavities can be readily fabricated and obtained by this method.

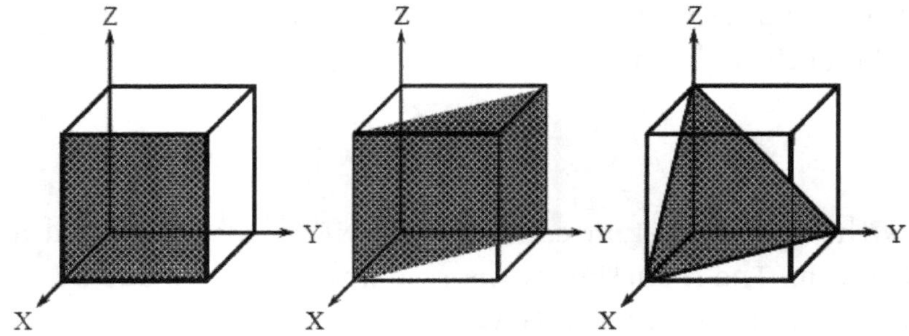

Fig. 4. The crystal planes in silicon lattice structure.

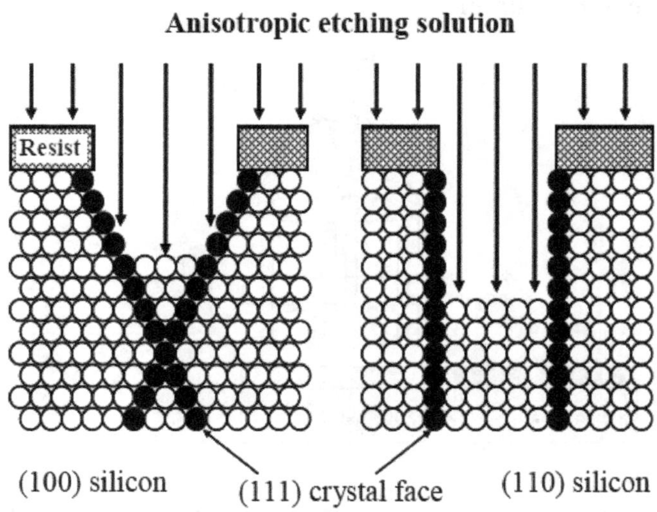

Fig. 5. Anisotropic etching in the single crystal silicon. [Madou, 2002]

4.5 Deep reactive ion etching

The procedure that eliminates materials from the surface of a substrate or a bulk substrate is termed "ETCHING."The anisotropic dry etching method is independent of the crystal orientation of the silicon wafer. This micromachining procedure entails exposing the substrate to an ionized gas. Etching occurs through chemical or physical interaction between the ions in the gas and the atoms of substrate. The most often applied techniques can be divided into three groups: (1) physical sputter etching or ion beam etching, (2) chemical plasma etching and (3) combined physical/chemical etching. The physical method as so called "sputtering etching or ion beam etching" can achieve an anisotropic profile structure. As opposed to sputtering etching, chemical plasma etching is completely isotropic etching profile and has excellent selective properties. The etching rate of plasma etching is much faster than sputtering etching and is uncontrollable. However, the reactive ion etching (RIE) method combines the physical and chemical etchings to achieve excellent high selectivity and anisotropy for micromachining of silicon substrate. The etching rate ranges from 20 to 200 nm/min. However, this low etching rate cannot satisfy the micromachining of microchannel with a high aspect ratio structure.

The problems of reactive ion etching (RIE) used in the micromachining process are that the sidewalls of the trenches are not vertical and the etching speeds are too slow. Therefore, two modified RIE techniques, i.e. the Bosch process and the cryogenic process, are developed to solve this problem. In the first approach, the Bosch process consists of cyclic etching and deposition process, as shown in Figure 6. In the etching step, silicon is etched by SF6 plasma. In the deposition step, the supply gas is switched to C4F8 so that a fluoroCarbon polymer thin film with a thickness of 10 nm can be deposited on the side wall of the trench. In the next cycle, the polymer film at the bottom surface of the trench is removed by Ar ion bombardment, while the film at the sidewalls is intact, which can protect the sidewalls from the attack of the SF6. Usually, the etching rate of Bosch process ranges from 1 to 12 μm/min. In the second approach, i.e. the cryogenic process, the substrate is cooled down to -110 to - 195°C with liquid Helium. Cryogenic temperatures facilitate the condensation of reactant gases, such as SF6 or O2, on the trench surface. The condensed layer can safeguard the sidewall, but the condensed film on the bottom can be eliminated through ion bombardment. The exposed bottom wall may be further etched into the substrate, as illustrated in Figure 7.

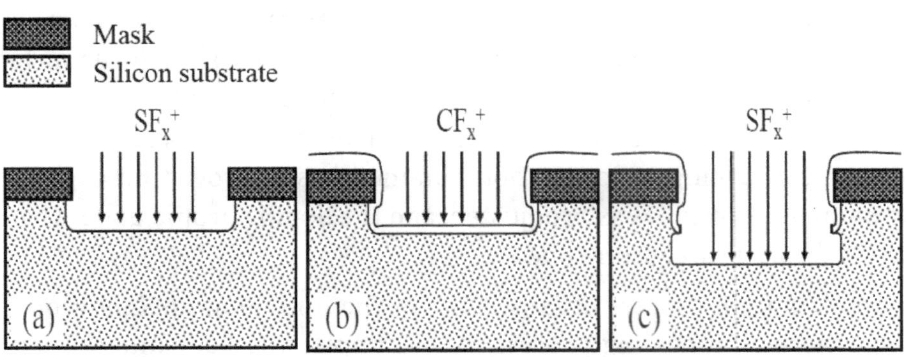

Fig. 6. Anisotropic dry etching: Bosch process. (a) SFx+ etching, (b) CFx+ deposition and (c) SFx+ etching.

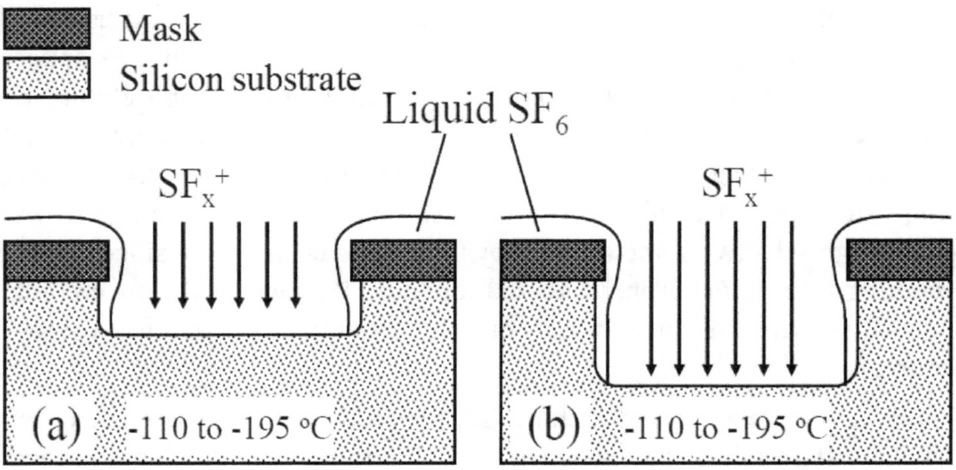

Fig. 7. Anisotropic dry etching: cryogenic process.

4.6 Doping: Semiconductor Technique

Doping is probably the best-known semiconductor technique. Here, doping atoms are introduced into a silicon substrate in a defined way so that either n-type or p-type semi- conductor layer can be formed. The techniques play an important role in fabrication of semiconductor devices. In addition to varying the electrical properties, doping can also improve wear and corrosion of a semiconductor material. The doping atoms, like boron or phosphorous, can create an etching stop layer in a silicon substrate, allowing fabrication of a thin film or microstructure at desired locations by wet etch when doping atoms can be placed in the desired location.

4.6.1 Ion implantation

Ion implantation is one of the most expensive technology processes, after photolithography, due to the complex systems involved. Ion implantation involves shooting charged ions, which are externally accelerated in a vacuum, into the silicon wafer. The ions can penetrate up to a few micrometers below the surface. The silicon disc is irradiated uniformly with scanning foCused ion beam. By measuring the beam current, the amount of dopant implanted can be precisely controlled. In addition, the doping concentration obtained has improved homogeneity, the doping parameters can be easily adjusted, and the doping profile under the wafer's surface can be accurately controlled.

An ion implantation process usually includes the initial ion implantation and the following thermal annealing process. For example, after dopants are implanted into the central portion of a LPCVD polysilicon layer, the implanted layer requires annealing at 950°C for 30 minutes in a standard anneal furnace to re-crystallize the polysilicon layer and to diffuse the dopants uniformly across the layer (Ko et al., 2007; Liu, 2004; Qu et al., 2003). After annealing process, the doped polysilicon layer can be patterned. The variation of the resistivity of the doped LPCVD polysilicon with the boron concentration is shown in Figure 8. This empirical data can be used to design arrays of polysilicon sensors or heaters integrated with the microchannel system.

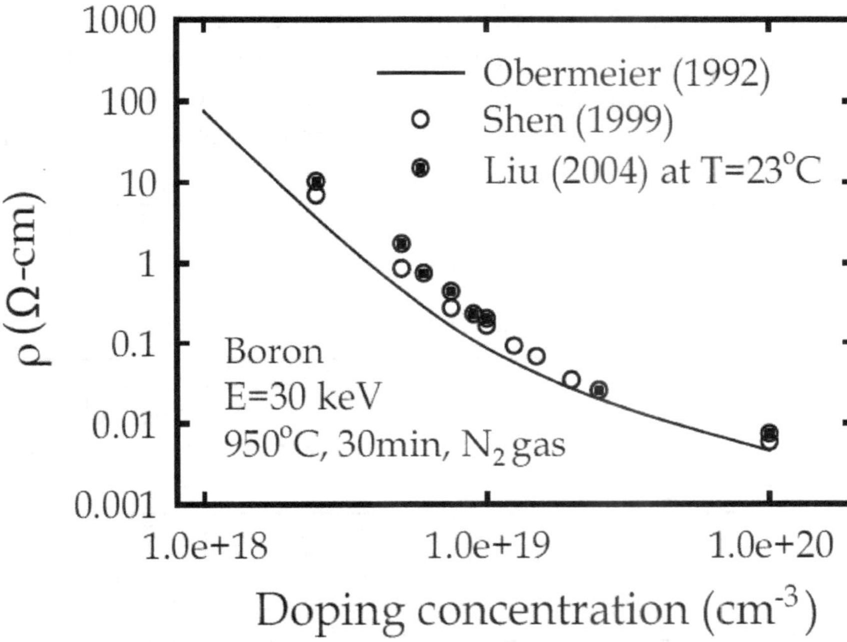

Fig. 8. The resistivity of boron-doped LPCVD polysilicon at different boron concentrations.

4.6.2 Dopant diffusion

In the diffusion method, silicon wafers are put into a furnace under a high temperature range from 850 to 1050°C, where dopant can be thermally diffused into the substrate. The sources of dopants can be either gases, liquids or solids. The solid dopant source can be BN, AS^{2}o3, P^{2}o5, the liquid dopant source can BBr3, AsCl3, PoCl3 while the gas dopant source can be B2H6, AsH3, PH3. However, the liquid dopant sources are the most frequently used. More detailed description can be found in the ref. (Sze, 2002). However, either of them has to be vaporized first and then the vapor can diffuse into the silicon substrate and form covalent bonding with the silicon lattice atoms. The main difficulty is the determination of the amount of dopant diffused into the substrate and the concentration of the doping atoms bonded in the silicon. Despite of the low cost, the method can only make a doping profile on the surface of a wafer, which restrains the wide applicability of this technique.

5. MICROCHANNEL FABRICATION WITH MEMS TECHNOLOGY

5.1 Silicon based microchannel fabrication

5.1.1 Bulk micromachining process

Either isotropic or anisotropic etch process can be used to fabricate microchannels in a bulk material, such as a silicon or a glass substrate. A channel with a variety of cross-sectional shapes can be made by different etch processes, as shown in Figure 9. Different microchannels can be formed by either anisotropic etch of the {100} or {110} silicon wafers or by isotropic etch of silicon or glass substrate, respectively. Finally, either anodic or fusion bonding can be employed to bond and enclose the channels with a glass plate or a silicon wafer.

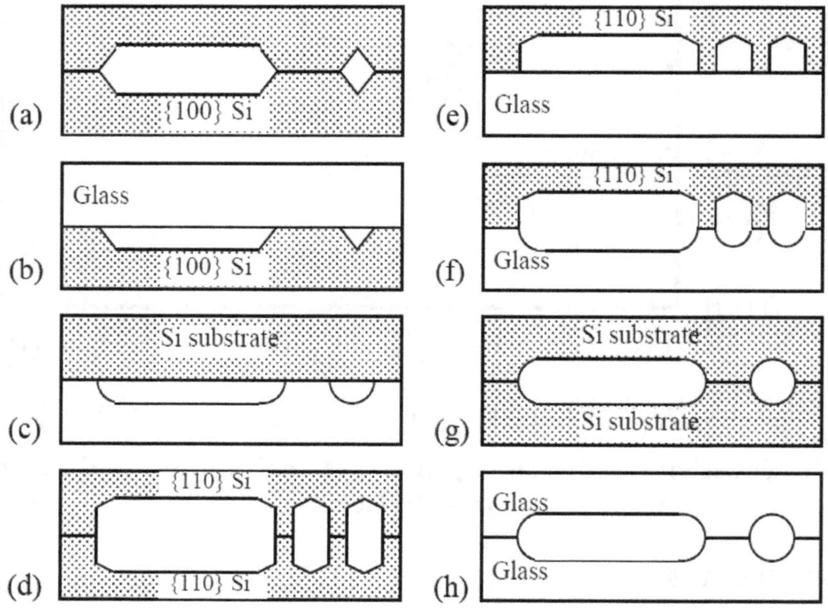

Fig. 9. Different microchannels made by bonding of bulk micromachined substrates in: (a) silicon-silicon, (b) glass-silicon, (c) silicon-glass, (d) silicon-silicon, (e) silicon-glass, (f) silicon-glass, (g) silicon-silicon, and (h) glass-glass.

5.1.2 Surface micromachining process

The surface micromachining is originated for deposition and patterning of thin layers into different structures, which may possibly use sacrificial layers to form microstructures. In this section, surface micromachining processes combined with deposition and etch of a polysilicon layer and the use of a sacrificial layer can be used to fabricate a microchannel. The fabrication process is described as follows:

1. Deposit and pattern the sacrificial layer, as shown in Figure 10(a).
2. Deposit and pattern the polysilicon layer as a microstructure, as shown in Figure 10(b).

3. Remove the sacrificial layer, followed by rinsing and drying the microstructure, as shown in Figure 10(c).
4. Seal the microstructure, as shown in Figure 10(d).

However, by using the surface micromachining the maximum height of the channel allowed for fabrication is only within a few micrometers (1 to 2 μm) due to the limitation of CVD deposition for the sacrificial layer. In comparison with the bulk micromachining process, the channel fabricated by the bulk micromachining process can be as large as a few hundred micrometers in height.

5.2 Polymer based microchannel fabrication

5.2.1 SU-8 resist

Photolithography of a thick resist is very beneficial for fabrication of a complicated microchannel system. Microchannel system can employ thick resists directly as a channel structure or a template for channel molding or electroplating of metal to form a channel with high aspect ratios. The most popular thick photoresist used is EPON SU-8 epoxy resin which is a negative and transparent PR and can form a layer with a thickness from less than 1μm to few mm. The SU-8 appears to be the most suitable one since it can be readily spin coated on the substrate at desired thickness, and patterned into required shape of channel by photolithography. Therefore, different sizes and shapes of channel can be readily made. However, the use of SU-8 layer as part of the substrate has precluded any high temperature fabrication process since the material cannot withstand processing temperature higher than 200°C. The formation of a SU-8 channel structure should be at a low temperature process.

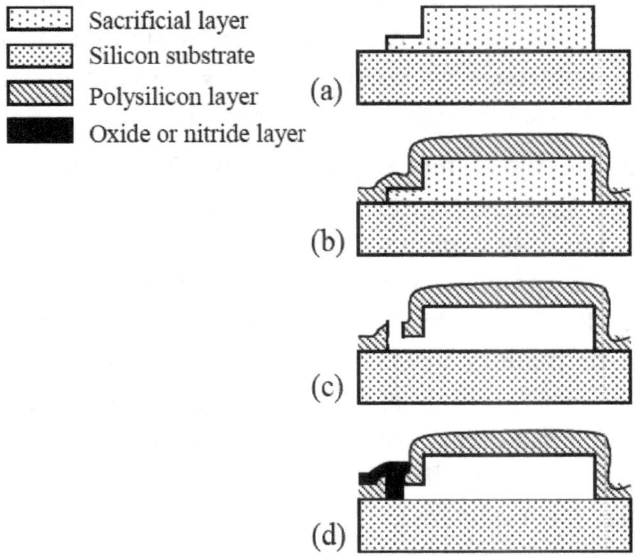

Fig. 10. The use of surface micromachining process to fabricate a channel with the use of a sacrificial layer.

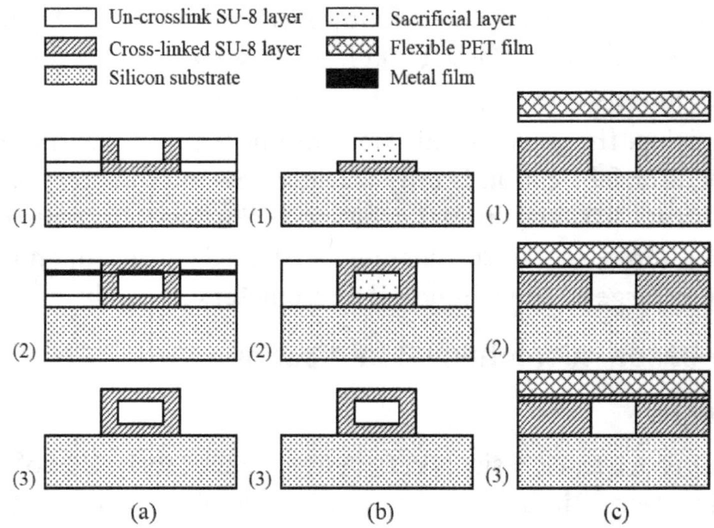

Fig. 11. Fabrication of SU-8 based microchannels with: (a) embedded metal mask, (b) removal of sacrificial layer and (c) direct use of the fusion bonding by SU-8.

In numerous microfluidics applications, a singular SU-8 layer can be utilized to create the requisite number of microchannels, which may be integrated with arrays of microfluidic components or sensors. The channels' upper and lower sections can be sealed with an SU-8 coated glass or plastic plate, employing subsequent blanket exposure as illustrated in Figure 11.

5.2.2 Poly-dimethylsiloxane (PDMS) molding

Poly-dimethylsiloxane (PDMS) is an excellent bio-compatible and an elastomeric polymer material, which is frequently used for applications in μTAS (micro total analysis system) or lab-on-a-chip systems. This is attributed to a number of useful properties of this material, such as low cost, low toxic, transparent from visible into near ultraviolet in wavelengths and chemical inertness.

PDMS is made by mixing different prepolymers. The weight ratio of the base and the curing agent can be 10:1 or 5:1. Then, the PDMS mixture is poured into a master and stays for a few minutes to self-level. The whole set is then cured at a low temperature in the range from 60 to 80°C for several hours. After peeling off, the structured PDMS membrane is bonded with the other substrates such as silicon, glass, plastic plate or PDMS plate etc. to complete the channels. Due to its simplicity in fabrication and rapid prototyping, PDMS molding is also called microCasting, which is a direct transfer of patterns. In many applications, the elastomeric PDMS can be used directly as a microfluidic device with microchannel connecting different components. Fabrication of PDMS into a microCannel is described as follows:

1. Etch the silicon or glass master, as shown in Figure 12(a-1).
2. Molding the PDMS microstructures, as shown in Figure 12(a-2).
3. Seal and laminate the PDMS microstructures, as shown in Figure 12(a-3).

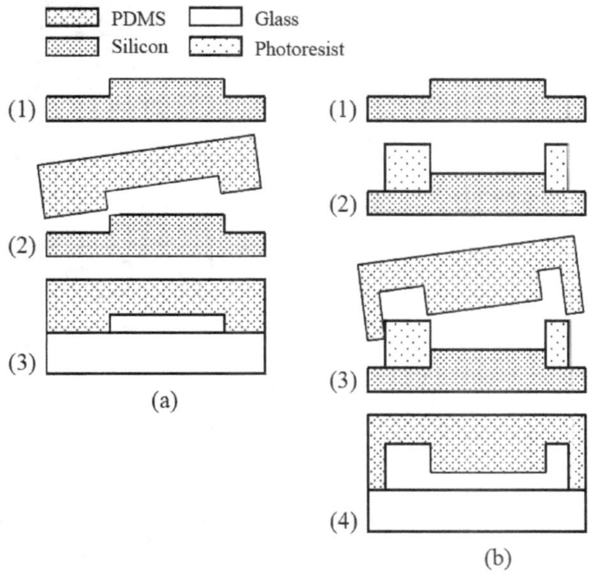

Fig. 12. Fabrication process for a (a) one-staged or (b) two-staged PDMS microchannels.

5.2.3 Parylene

Parylene is a polymer that can be deposited by a CVD process at a low or room temperature. The CVD process allows coating a conformal film with a thickness ranging from several micros to several millimeters. The basic types of Parylenes are Parylene N, C and D, all of which have good dielectric, electrical, physical and hydrophobic properties. However, Parylene D is modified from the same monomer by the substitution of the chlorine atom with two of the aromatic hydrogens. Parylene D is similar in properties to Parylens C, but with an additional property to withstand higher temperature. Deposition rates are fast, especially for Parylene C, which is normally deposited at a rate of about 10 μm/min. The deposition rates of Parylene N and Parylene D are slower. Parylene can be used in microfluidic devices as a structural material, which has a very low Young's modulus. Fabrication of microchannel using Parylene is described as follows:

1. Deposit a Parylene base on a provisional substrate, as shown in Figure 13(a-1).
2. Deposit and pattern a sacrificial layer on the Parylene base, as shown in Figure 13(a-2).
3. After Secondary deposition of a Parylene film on the top, then remove the sacrificial layer, as shown in Figure 13(a-3).

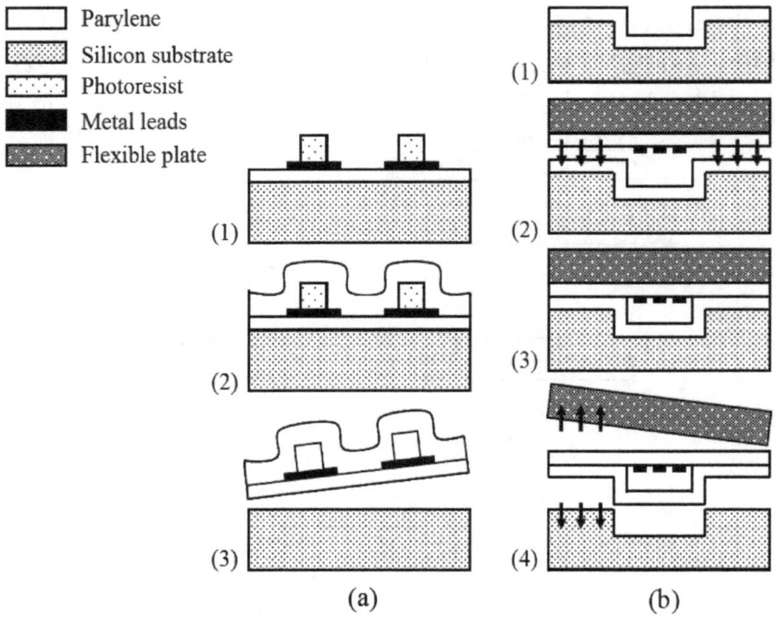

Fig. 13. Fabrication of Parylene based microchannels with: (a) removal of the sacrificial layer and (b) direct fusion bonding using Parylene.

6. MICROSENSORS FOR THE MEASUREMENT OF PRESSURE AND TEMPERATURE

6.2 Miniaturized temperature sensors and heaters

6.1.1 Working principles

When the polysilicon is doped heavily with boron, a relatively low resistivity that is independent of temperature can be obtained. This heavily doped polysilicon layer can be used as a heater. However, when the polysilicon is doped with a less amount of boron, a linear relationship between the resistivity and the temperature of the layer can be obtained and this doped polysilicon layer can be used as a temperature sensor. The temperature variation of the resistivity for the doped polysilicon layer at different concentrations has been demonstrated (Ko et al., 2007; Liu, 2004; Qu et al., 2003). Shen proposed that the resistance change with temperature is slightly nonlinear at low boron concentration. The temperature coefficient of the resistance (TCR) of the doped polysilicon layer is negative over the range covered and the temperature dependence increases with decreasing doping concentration, as shown in Figure 14 made by Shen (2003). It appears that for the boron concentration of 1020 atoms/cm^3, the resistivity of the doped polysilicon layer is almost independent of temperature with a near zero TCR. The polysilicon doped at this concentration can be used as a heater which can provide a constant heating power for a long period of operation. Otherwise, for the boron concentration at 1019 atoms/cm^3, the resistivity of the doped polysilicon layer has a linear relationship with temperatures. Therefore, the polysilicon doped at this concentration can be used as a temperature sensor. All the temperature sensors and heaters required and made in a microchannel system should be based on different doping conditions of the concentrations described above.

6.1.2 Design consideration

The total resistance of a device is proportional not only to the resistivity (ρ) and thickness (t) of the doped polysilicon layer, but also to the ratio of lR/wR, i.e. shape effect. The definition of resistance is written as follows:

$$R = (\rho/t)\cdot(l_R/w_R) = R_S\cdot(l_R/w_R) \quad (6\text{-}1)$$

where R_S = (ρ/t) is called the sheet resistance of the material. l_R and w_R are the length and the width of the resistor, respectively.

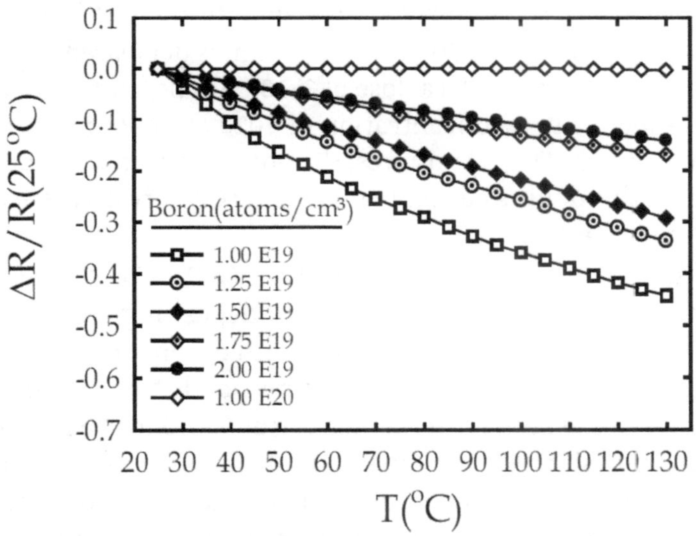

Fig. 14. The relative resistance variation with temperature at different concentrations of boron doped in polysilicon (Shen, 2003).

The TCR, α, is defined as follows:

$$\alpha = (\Delta R / R_0) / \Delta T = \Delta R / (R_0 \cdot \Delta T) \quad (6\text{-}2)$$

$$\Delta R = R - R_0$$

$$\Delta T = T - T_0$$

where Ro is the resistance of the material at a reference temperature. R is the resistance of the material measured at certain temperature. The zero TCR means that the Δ*R* is near zero and the resistance of the material is constant within the range of the temperatures measured. The resolution (R_{esoul}.) of temperature sensor is defined as follows:

$$R_{esoul.} = \Delta R / (R_o \cdot \Delta T) \times R \tag{6-3}$$

For example, the size of temperature sensor designed is 80 μm×20 μm×0.4 μm in length, width and thickness, respectively. The material used as the sensor is polysilicon. The concentration of boron used for doping is 10^{19} atoms/cm^3. The ρ is found to be 0.2034 Ω-cm, as shown in Figure 8. After a calculation, the sheet resistance (RS), total resistance (R) and TCR of temperature sensor are 5.085×10^3 Ω/squ., 20.34 kΩ at 23°C and -4.286×10^{-3}/°C. Therefore, the resolution of the polysilicon sensor can be calculated and is found to be -87.18 Ω/°C. It appears that the resolution of the temperature sensor made by the doped polysilicon is much better than any types of the thermocouples.

6.1.3 Fabrication processes (Ko et al., 2007, 2009; Liu,2004)

1. A 0.3 μm thick LPCVD tetraethoxysilane (TEOS) oxide is deposited on the (100) wafer as a insulation layer between sensors to silicon substrate.
2. Then, a 0.3 μm thick LPCVD polysilicon film is deposited and then implanted heavily with boron with a dose of 3×10^{15} atoms/cm^2. This amount of dosage corresponds to a concentration of 1020 atoms/cm^3 in the layer. After annealing at 950°C for 30 minutes, the doped polysilicon is patterned as the heaters. In this doping concentration, the TCR of resistors are near zero, as shown in Figure 15(a).
3. Next, after a deposition of a 0.3 μm thick LPCVD TEOS layer as insulation, the temperature sensors were made by depositing a 0.3 μm thick LPCVD polysilicon layer and then implanting with boron ions with a dose of 3×10^{14} atoms/cm^2. This amount of dosage corresponds to a concentration of 1019 atoms/cm^3 in the layer. After annealing at 950°C for 30 minutes to re-crystallize the polysilicon layer and to uniformly diffuse the dopant ions across the layer, the doped polysilicon was patterned. The concentration of 1019 atoms/cm^2 will give a linear relationship between the resistance and the temperature of the doped polysicon layer. Thus, after proper calibration, the doped polysilicon layer can be used as temperature sensors, as shown in Figure 15(b).
4. Before metallization, a 0.3 μm thick LPCVD TEOS oxide is deposited as insulation. Then, contact holes were opened in this layer for metallization. The metallization was made by sputtering standard IC four layers of metals, i.e. Ti/TiN/Al-Si-Cu/TiN with a thickness of 0.04 μm/0.1 μm/0.9 μm/0.04 μm, respectively, onto the substrate surface. The metal layers were then patterned into circuits, as shown in Figure 15(c).

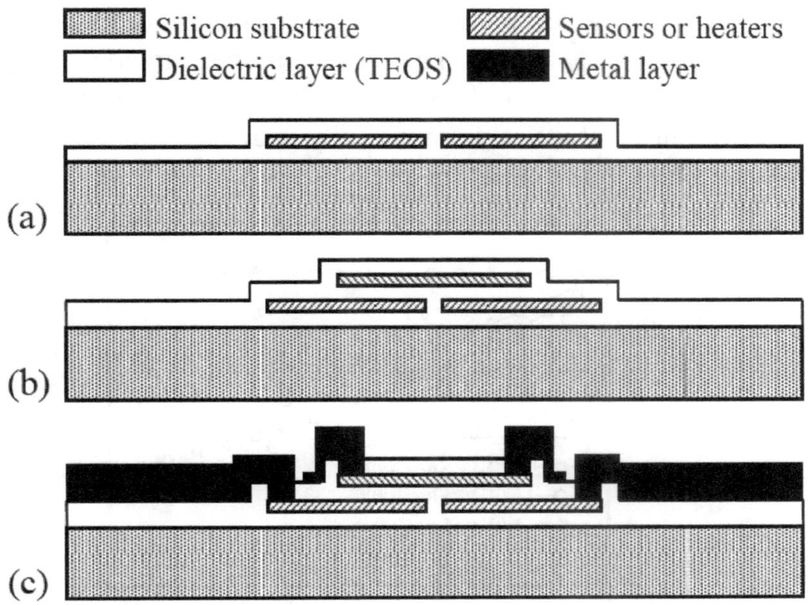

Fig. 15. Fabrication process of temperature sensor and heater (Ko et al., 2007; Liu, 2004).

6.1.4 Characteristics features of temperature sensors and heaters

All the temperature sensors and the heater fabricated on the silicon substrate have to be calibrated in a constant temperature oven to ensure a measurement accuracy of ± 0.1°C. In fact, all the temperature sensors and the heaters can be moved onto a glass substrate for other purpose. The procedure to move all the sensors and heaters onto a glass substrate will be described in section 6.2 or 7.2. In that case, the substrate now is a very low thermal conductivity material, it is difficult for the heat in the oven to conduct into the substrate, especially in the region for the channel formation, and to provide a uniform and constant temperature condition required for the calibration. Therefore, the calibration procedure will take a much longer time than the case with silicon substrate. The electric resistance measured in the doped polysilicon layer at various boron concentrations and temperatures is compared with the published data (Ko et al., 2007; Liu, 2004), as shown in Figure 16. It is found that the resistance in the current work for the concentration at 1020 atoms/cm^3 agrees well with the published results. However, the data for the current case with the concentration at 1019 atoms/cm^3 is slightly different from other work, as indicated by the red circle as shown in Figure 16. This difference is attributed to the use of different implantation equipments from different companies, and the batch deposition of LPCVD polysilicon layer with a 5-10 % uniformity can occur in a single wafer or from wafer to wafer. The characteristic curves of the polysilicon temperature sensor and the heater used in the present channel system are also shown in Figure 17 (Liu, 2004). It is found that the output signal of the temperature sensor has a linear relation with the temperatures and a high resolution of 1.26 KΩ/°C. Therefore, this can provide a very accurate measurement in the temperature. In addition, the resistance of the heater does not vary with the temperature. This can readily provide a uniform and constant heating power required in the experiments.

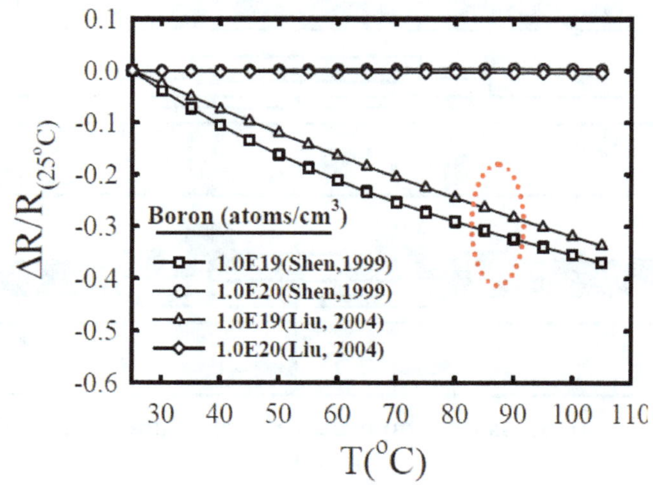

Fig. 16. Comparison of the resistance variation with temperatures between works for the polysilicon doped at different concentrations (Ko et al., 2007; Liu, 2004).

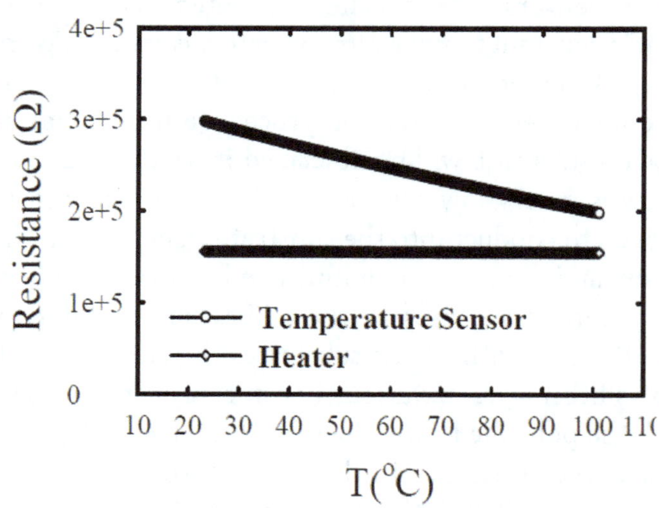

Fig. 17. Characteristic curves of the polysilicon temperature sensor and the heater that can be used in the present channel system (Liu, 2004).

6.2 Miniaturized pressure sensors

6.2.1 Working principles

Fabrication process of the micro pressure sensor has attracted much attention due to its wide industrial applications in various areas (Sze, 1994). In general, fabrication of pressure sensor adopts inorganic materials, e.g. piezoresistive sensor which is fabricated on a silicon substrate. Fabrication of these kinds of piezoresistive pressure sensors can be classified into bulk (Sze, 1994; Peake et al., 1969; Tufte et al., 1962) and surface (Guckel, 1991; Sugiyama et al., 1991; Clark et al., 1979) micromachining process. In the conventional designs of pressure sensor, both the bulk and the surface micromachining can be used. For the bulk micromachining technique, KOH solution is usually used to etch into the silicon substrate to form cavities, allowing for diaphragm deformation, in trapezoid shape with a 54.7o inclined wall. Since the silicon substrate is relatively thick and the bottom region consumed is much larger than the top region where pressure diaphragm defines. The number of sensors per wafer fabricated with bulk micromachining is much less than that fabricated with surface micromachining. Fortunately, this sensor is readily made and the yield of sensor is very high.

For the surface micromachining process, however, the number of sensors per wafer fabricated is much higher due to the fact that much smaller size of cavities for the pressure sensors can be made. However, the height of the cavities which allows deformation of diaphragm is made much smaller due to the use of sacrificial layer for later formation of the diaphragm. The sacrificial layer is deposited by chemical vapor deposition (CVD) process, which is relatively thin and is usually less than 2-3 μm. This has limited the height of the cavity defined by the sacrificial layer. Therefore, the pressure measurement range for the sensor made by the surface micromachining is much narrow. In addition, stiction of diaphragm can readily occur in the drying of DI water rinse process after removal of the sacrificial tetraethyl orthosilicate (TEOS) or phosphorus silicate glass (PSG) oxide. Special attention should be made to avoid the problem of stiction (Kim et al., 1998; Komvopoulos, 1996; Core et al., 1993; Legtenberg et al., 1993). In addition, the height of the cavity cannot be made large enough which will not give enough space for diaphragm deformation and allow the pressure measurement in a wider range. This is attributed to the use of sacrificial material, such as silicon oxide, that cannot be deposited thick enough by the CVD process.

In view of the disadvantages of the inorganic pressure sensor fabricated by bulk or surface micromachining of silicon material, the current work proposes fabrication of pressure sensor with polymer material, such as SU-8. In a review of literature, there are few other polymer materials that were found for fabrication of sensor system (Madou, 2002; Martin et al., 1998; Shirinov et al., 1996). A schematic diagram for a cross-section view of the sensor is shown in Figure 18 where the cavity can be filled with pressurized gas that causes deflection of the diaphragm on the top. All the wall material used, including the diaphragm for the sensor, is SU-8 except for the sensing material and circuit system on the top. The fabrication process presented in this paper can make arrays of pressure sensors and has the advantages of both the bulk and the surface micromachining process. The process can fabricate a greater number of sensors per wafer, and the sensors fabricated allow wider measurement range of pressure. The substrate can be any material. Here we select Pyrex glass. The fabrication process is very simple and the sensor can be readily made by spin coat SU-8 layer and patterning with lithography. In addition, the height of the cavity which allows deformation of diaphragm can be varied readily by spin coat different thickness of the SU-8 layer. Especially, the SU-8 diaphragm is formed without use of sacrificial layer and subsequent DI water rinse processes after removal. Therefore, fabrication process has completely absence of diaphragm stiction and has a much higher yield.

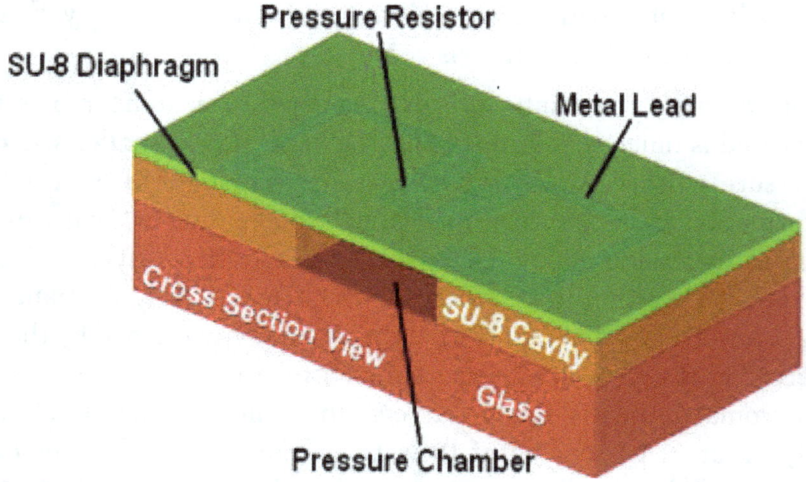

Fig. 18. A schematic diagram for a cross-section view of the pressure sensor designed.

The SU-8 material has been widely used for formation of microfluidic system (Arshak et al., 2006; Ribeiro et al., 2005; Lee et al., 2003), and relatively few in sensor system (Giordani et al., 2007). Therefore, current pressure sensor can also be readily applied and integrated into a microfluidic system or lab-on-chip (Pelletier et al., 2007; Trung et al., 2005; Vilkner et al., 2004) where pressure information is required. Arrays of current pressure sensors can be readily fabricated along the channel or in a micromixer for more detailed flow information. They can also be fabricated at inlet, outlet or inside of a micropump for evaluation of its performance.

Theoretical modeling for the stress and deformation of the diaphragm is derived. Numerical calculation is provided for diaphragm design consideration. Finally, the current pressure sensors made can provide a much better thermal insulation than the ones made by the previous surface micromachining process of silicon because of the use of the extremely low thermal conductivity materials, such as the SU-8 and the Pyrex glass. The SU-8 material has a thermal conductivity of 0.2 W/mK (Monat et al., 2007) where the Pyrex glass has a value of 1.4 W/mK. The fabrication of a micro thermal system necessitates careful consideration of thermal insulation. Comprehensive fabrication processes and performance assessment of this sensor are also presented and analyzed.

6.2.2 Design consideration

Before fabrication process, the size of the SU-8 diaphragm used in the pressure sensor should be determined from a proper analysis and design. The size of the SU-8 diaphragm is actually determined by the pressure range to be measured, the maximum strain that the diaphragm can sustain and the sensitivity required. The deformation of SU-8 diaphragm can be modeled as a square shell or plate with four edge clamped under a uniform normal pressure force as shown in Figure 19.

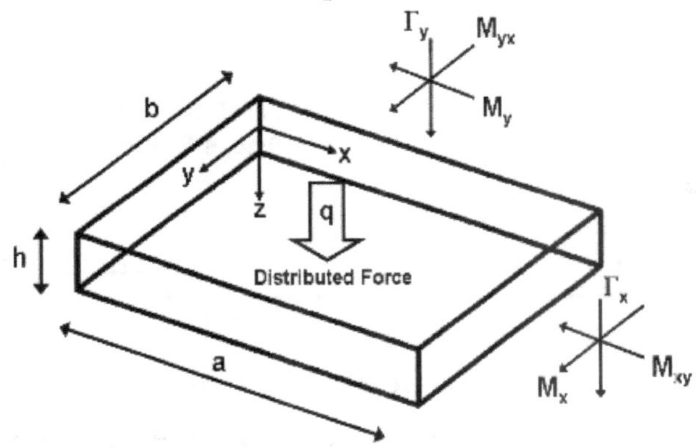

Fig. 19. Schematic diagram for the square plate deformed under pressure force.

From the solving of the governing equation for the deflection of the plate, the deflection of plate can obtain:

$$u = \frac{q_n}{D}\left(\frac{b}{\pi n}\right)^4 \sum_{n=1,3,5,\dots}^{\infty} X_n(x)\sin\frac{n\pi y}{b} \tag{6-4}$$

By letting n= 1, 3 and 5 only, the maximum deflection can be found as follows:

$$u_{\max} = C\,(1-\nu^2)\frac{pb^4}{Eh^3} \tag{6-5}$$

where C is $\frac{0.032}{1+\alpha^4}$, with the assumption that all edges of the plate are clamped, and α is ratio of the width to the length of the plate. However, C obtained by Guckel (1990) is 0.0152. The comparison is shown in Figure 20, which shows that for the diaphragm size of 200μm by 200μm and the thickness of the diaphragm varying from three to thirteen micron, the numerical result from ANSYS calculation agrees very well with the approximate solutions of Westergaard and Guckel, respectively. Therefore, the ANSYS is adopted for calculation in the design for proper size of diaphragm.

For a pressure range from 1 to 4 atms, the prediction for the maximum strain occurred in the plate versus the applied pressure force is presented in Figure 21 for different thickness of diaphragm. The maximum strain the SU-8 layer can sustain is found 0.77% (Guckel, 1991). Therefore, the rectangle presented in Figure 21 represents the safety region where the maximum strain is less than 0.77% for the diaphragm size designed. For a better sensitivity, i.e. more deflection, the size of 150μm×150μm wide at a thickness of approximately 9 μm is selected.

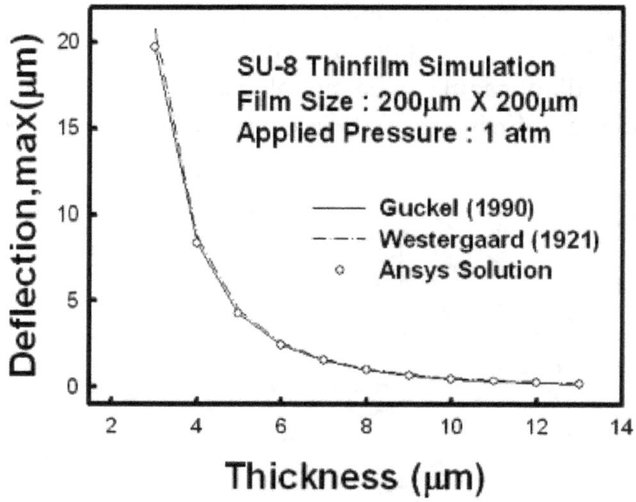

Fig. 20. Comparison of the numerical result from the ANSYS with the analytical results and other published data.

Polymer pressure sensor does not have the problems of narrow depth of cavity and stiction of diaphragm made by surface micromachining method. The most promising candidate for pressure sensor fabrication is the use of the photoresists, i.e. SU-8. The SU-8 can be readily spin coated on the substrate from a few microns to a hundred microns thick. This can give us a much deeper cavity depth if diaphragm layer can be properly deposited on the top to enclose the cavity, as shown in Figure 18. The diaphragm designed uses the same material as the cavity side wall in order to reduce difference of the thermal expansion coefficient for the cavity wall and the diaphragm which may cause deflection of diaphragm and significantly affect measurement accuracy. In this way, one would face another problem of selecting suitable sacrificial material to fill into the cavity to create a flat surface in order to spin coat the SU-8 diaphragm on the above. In addition, the sensor material currently available, which is to be deposited on the top of diaphragm, is piezoresistive film that is a doped polysilicon layer deposited by LPCVD at a high temperature process. The use of high temperature process of doped polysilicon for sensor formation has precluded the possible use of SU-8 as both diaphragm and side wall of cavity.

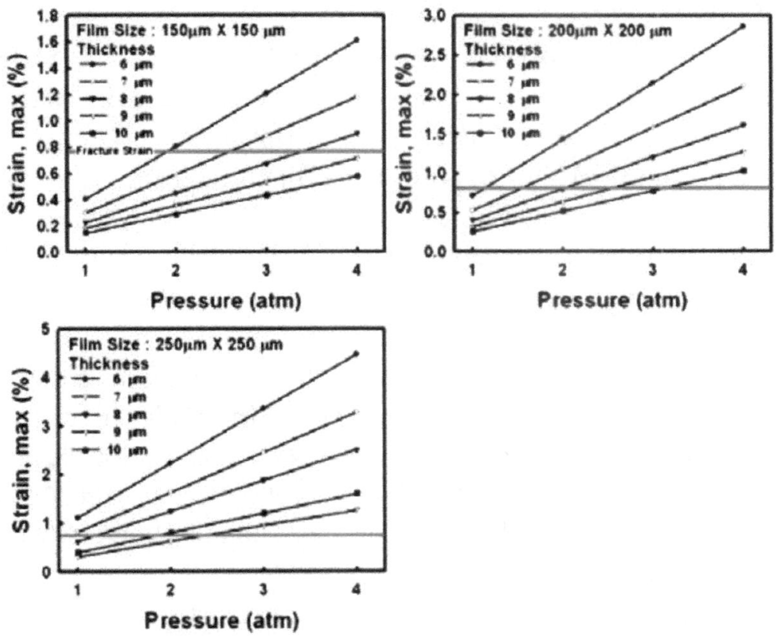

Fig. 21. The maximum strain variation with pressure for different sizes and thicknesses of SU-8 diaphragm.

However, this difficulty can be readily overcome by reversing the fabrication process of the pressure sensor. That is, one can first deposit the the piezoresistive layer and the metal lines on the silicon substrate which is a high temperature process, and then the diaphragm and the cavity wall which is a low temperature process. The piezoresistive layer can be made with the polysilicon layer implanted with very high concentration of boron or phosphorous.

In the current process, one selects very high concentration of boron such that the resistivity variation with the temperature in the polysilicon layer can be minimized. From the experimental data plotted in (Shen, 2004; Kanda, 1982; Mason, 1969), for boron concentration greater than 1020 atms/cm^3 the resistivity variation with temperature can be negligible small. Thus, this concentration of boron is adopted in the implantation process for the polysilicon layer. After implantation, the polysilicon is annealed for 30 mins at 950°C. The next step is to spin coat SU-8 diaphragm, which has the flexibility to readily control the thickness. It is then followed by spin coat another layer of SU-8 for cavity wall at desired thickness. Thus, the formation of SU-8 layer will not go through a high temperature process. Finally, a Pyrex glass can be bonded with the patterned SU-8 layer on the top to enclose the cavities. Once the silicon substrate is completely removed by wet etch, a successful pressure sensors can be readily achieved.

6.2.3 Fabrication processes (Ko, 2009)

1. A 0.3 μm thick LPCVD TEOS oxide is deposited on the (100) wafer and used as protection mask for the upper layer devices during the later long period of TMAH wet etch to completely remove the Si wafer.
2. Next, a 0.3 μm thick LPCVD polysilicon film is deposited and then implanted heavily with boron with a dose of 3×10^{15} atoms/cm^2. This amount of dosage corresponds to a concentration of 10^{20} atoms/cm^3 in the layer. After annealing at 950°C for 30 minutes, the doped polysilicon is patterned as the pressure sensors.
3. Before metallization, a 0.3 μm thick LPCVD TEOS oxide is deposited as insulation. Then, contact holes were opened in this layer for metallization. The metallization was made by sputtering standard IC four layers of metals, i.e. Ti/TiN/Al-Si-Cu/TiN with a thickness of 0.04μm/0.1μm/0.9μm/0.04μm, respectively, onto the substrate surface. The metal layers were then patterned into circuits, as shown in Figure 22(a).
4. A 9 μm thick SU-8 layer is spin coated on the substrate as the pressure sensing diaphragm. Then, a 50 μm thick SU-8 layer is spin coated again on the substrate and patterned into the active cavity to allow for movement of the pressure diaphragm. It is noted that there is a soft bake before layer exposure to evaporate the solvent and a post exposure bake to make the edge between the exposed and unexposed region more sharp and clear. Finally, instead of using the hard bake, a high intensity of light is used to illuminate to complete the cross-linking of the resin since the hard bake will need a temperature at 200°C that will damage the underneath thick epoxy layer. For the soft bake, the SU-8 layer is first maintained at 65°C for 7 minutes with a 5°C/minute ramping rate starting from room temperature, and then baked at 95°C for 15 minutes with a 5°C/minute ramping rate starting from 65°C to release the internal residual stress of SU-8 thick layer. In fact, the success of the SU-8 channel strongly depends on the baking process after light exposure.
5. It is now ready to move the devices made on the Si wafer onto a low thermal conductivity Pyrex glass. This is done first by bonding the silicon wafer with the Pyrex glass, as shown in Figure 22(b).
6. After the bonding process, the silicon substrate is ready for removal and cleared off. This is done by wet etch the silicon with TMAH solution at 90°C for 5-6 hours. Instead of using KOH, the selection of TMAH is attributed to its relatively high selectivity for silicon versus oxide. This can avoid the sensors attacked by the etchant during the long period of wet etch process since the protection layer of the sensors are made of TEOS oxide. A suc-

cessful movement of the pressure sensors onto the Pyrex glass substrate is shown in Figure 23(a). The cavities of the pressure sensors are successfully made, as shown in Figure 23(b). No distortion was found.

7. Next, a 80 μm thick SU-8 layer is spin coated on the substrate and patterned by lithography to form a test section for the pressurized gas.
8. A PMMA plate is then bonded, using epoxy resin, to enclose the test section, as shown in Figure 22(c). The pressure sensor is completed.

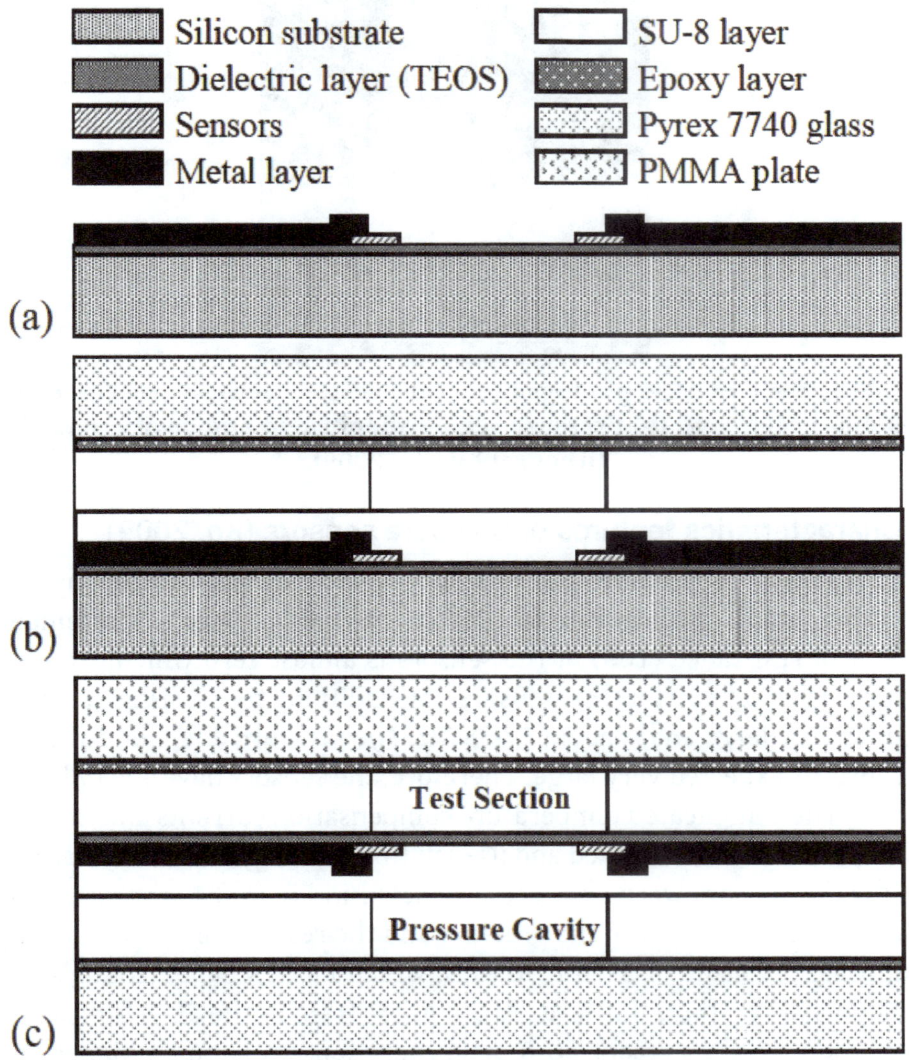

Fig. 22. Fabrication process of pressure sensor (Ko, 2009).

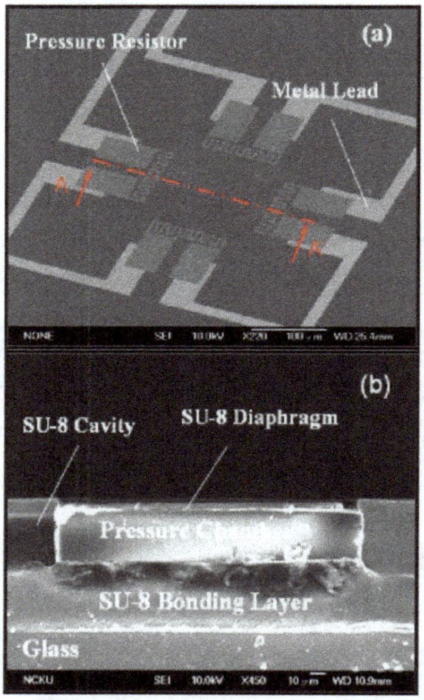

Fig. 23. Side view of SEM photographs. (a) Pressure sensor resistor embedded in SU-8 diaphragm and (b) pressure sensor diaphragm and cavity made by SU-8 lithography (Ko, 2009).

6.2.4 Characteristics features of pressure sensors (Ko, 2009)

Since the arrays of pressure sensors were made of polysilicon layer doped with a designated concentration of 1020 atoms/cm^3 of boron, the temperature coefficient of resistance (TCR) of the sensors is almost zero which can eliminate the temperature effect for the sensors. In addition, the resistance of sensors designed is very large and is about 57-58 KΩ. The pressure effect on the resistivity of the sensor is expected very large. Therefore, the sensors have a very high signal resolution. The complicated temperature compensation currents and the electrical signal amplifier are not required and used in this sensor system. The characteristic curves for the sensors were obtained, as shown in Figure 24. The results show a very linear variation between the pressure and the resistance of the sensors. Since the diaphragm is made of SU-8 materials, which has a smaller Young's modulus than the polysilicon film, this leads to a higher strain in the SU-8 film. Therefore, the resolution for the present SU-8 diaphragm sensors is much higher than the polysilicon diaphragm sensors and is found to be 20.88 ohm/psig.

In order to test the reliability, i.e. the response, recovery and life time or fatigue of the current pressure sensors, a pressurized air is supplied to the sensor cavity for a period of 30 seconds and then released. The repeated sharp rise and descend of the pressure, as shown in Figure 25, suggests that the present SU-8 pressure sensor has very good reliability. Higher frequency of pressure oscillation still indicates that response and recovery test in the current sensor is still reliable except a slight drop of the oscillation signal is observed, as shown in Figure 26. This slight drop in oscillation signal is attributed to the frictional heating of the polymer diaphragm due to high frequency of vibrations. The frication heating increases the temperature and leads to reduction of the resistance of the polysilicon sensor. After cooling the sensor back to the ambient temperature, oscillation signal recovers back to the original value. The frictional heating may not be readily seen in polysilicon sensor due to the very high thermal conductivity of the material that can rapidly dissipate the frictional heating. The test for membrane under high frequency oscillation of pressure has proCeeded for more than three thousand cycles which is more than 24 hours and does not indicates creeping or fatigue of the SU-8 membrane.

In addition, the sensor is put into oven at controlled temperatures for calibration at different temperatures. The results indicate that resistance variation at different temperatures is small and is less than 1% of the measured value. As one plots pressure variation versus temperature, as shown in Figure 27, the pressure variation is very small and less than 1 psig. The small pressure variation versus temperature may be attributed to the difference in the thermal expansion coefficient between the SU-8 and the Pyrex glass which causes small stress in the SU-8 side wall and the diaphragm. It is expected that replacing the Pyrex glass with other high temperature polymer plate, such as Poly(ether-ether-ketone) (PEEK), that can endure the high temperature annealing process of the SU-8 and have a closer thermal expansion coefficient to the SU-8, can further reduce the temperature effect on the pressure measurements.

Finally, in order to ensure that the SU-8 side wall or the diaphragm of the sensor is not permeable to the air, permeability test for the SU-8 to the air is performed. This is done by sending pressurized air into the test section (chamber), as shown in Figure 22(c), of the sensor at desired pressure level and closing both the inlet and outlet valves. The pressure variation in the test section is monitored by the sensor for more than 7 days, and the signals are sent into computer for plotting. It appears that the pressure inside the test section is almost kept constant except for a slight variation due to noise from the instrument. The variation of the pressure due to noise is less than 1.76% of the total pressure during a long period of 7 day observation. This result indicates that the pressurized air inside the test section will not leak slowly through the SU-8 diaphragm into the pressure chamber, nor the pressurized air (in either the test section or the pressure chamber) will leak slowly through the SU-8 side wall into the ambient. This permeability test also indicates that the SU-8 diaphragm will not creep under a constant pressure load of 350 kPa for 7 days. Therefore, one may expect that the SU-8 diaphragm will not creep as long as the pressure load is within the elastic deformation of the material unless when the material becomes aged.

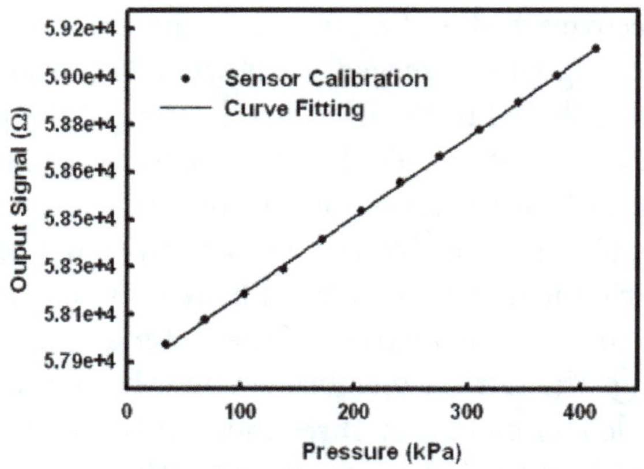

Fig. 24. A very linear variation between the pressure and the resistance of the sensors (Ko, 2009).

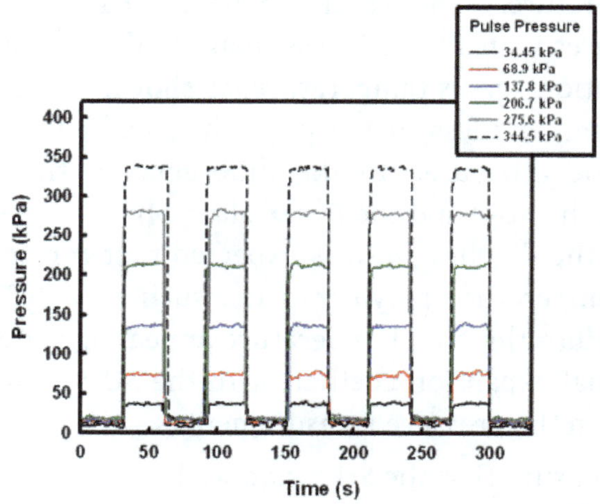

Fig. 25. Response and recovery test for pressure sensor at different pulsed pressures (Ko, 2009).

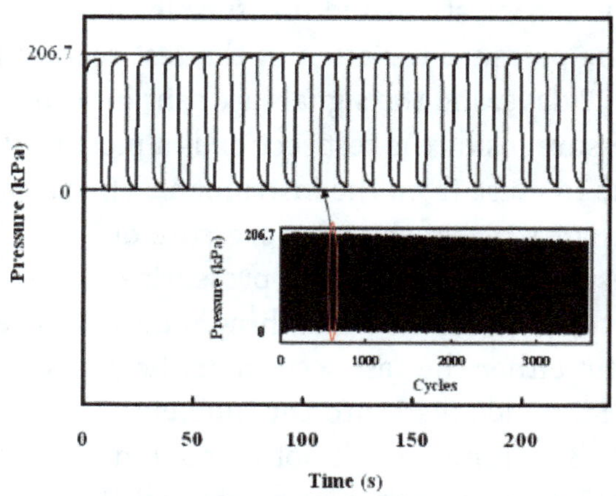

Fig. 26. Response and recovery test for pulsed pressure at high frequency (Ko, 2009).

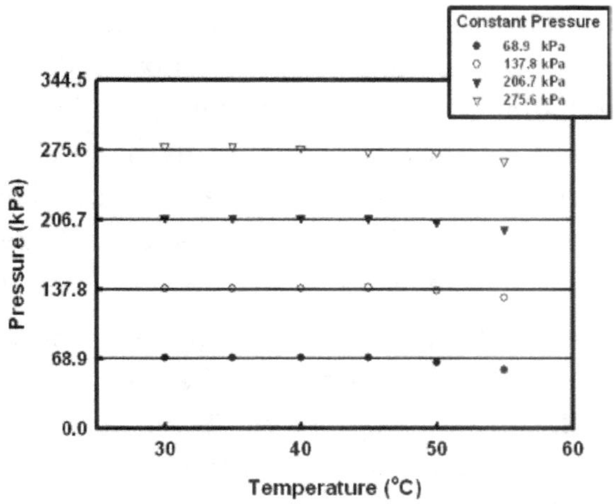

Fig. 27. Thermal stability test shows that the pressure signal is almost independent of temperature variation in the system (Ko, 2009).

7.MANUFACTUREOFMICROCHANNELSINTEGRATEDWITHARRAYS OF MINIATURE PRESSURE AND TEMPERATURE SENSORS

The development and fabrication of a micro-flow heated channel for studies of local flow and heat transfer process is very important not only for the basic research but also for practical application in MEMS or NEMS thermal system. Unfortunately, most of the current micro-channel used (Choondal and Suresk, 2000; Yu et al., 1995) or fabricated (Jiang et al., 1999; Tao and Mahulikar, 1998; Chu et al., 1994) could not provide a local or detailed flow and heat transfer information, but a global information (only for flow information at inlet and outlet). How to prevent or control heat loss are the most important considerations for heat transfer study. Most of the MEMS or NEMS techniques developed up to the present are to use silicon wafer, a very good thermal conductor that has a very high thermal conductivity (148 W/mK), as substrate to fabricate a microchannel (Pfahler et al., 1990; Pfahler et al., 1991; Arkilic and Breuer, 1994; Arkilic and Schmidt, 1997; Shih et al., 1996; Qu et al., 2000a and 2000b; Ren and Li, 2001; Ren et al., 2001a and 2001b). This will lead a large amount of heat loss from the inside channel to the outside ambient, a uniform heat flux boundary condition on the heated wall could not be maintained. Therefore, local heat transfer information in the channel cannot be obtained and understood.

The idea to fabricate a suspended channel over an air layer (Wu et al., 1998) to reduce the heat loss may be adopted to fabricate a heat transfer channel. However, this kind of channel that has a very thin wall can be readily deformed by the high-pressure flow. This deformation is expected to significantly affect the detailed flow and heat transfer process inside the channel. In addition, the techniques involved in the fabrication are suitable for fabrication with a very small height (in a few micrometers) channel.

The current work is to develop a novel fabrication procedure for a micro-channel system that can be used to study and obtain the flow and the local heat transfer coefficient inside the channel. One side of the channel wall can be heated uniformly and all the other sidewalls can be well insulated. In the meantime, the heated wall is integrated with an array of micro-temperature sensors and pressure sensors that can provide measurements of local temperature and local pressure of the heated wall and study the local flow and heat transfer process along the channel.

Initially, both the heater and the array of temperature sensors and pressure sensors that are made of polysilicon layer doped with different concentration of boron are deposited on a (100) silicon wafer by LPCVD. Then, an ultra thick SU-8 layer is spin coated on the substrate and patterned into the active cavity of the pressure diaphragm. After SU-8 lithography process, both the heater and the array of the sensor layers are moved to the surface of a low thermal conductivity Pyrex glass. This was made by bonding the deposited silicon wafer with the Pyrex glass to use a thick layer of epoxy resin. Then, the TMAH (25 wt%) solution at 90°C was used to etch and completely remove the silicon wafer. The next step is to spin a thick layer of SU-8 on the epoxy-glass substrate and form a micro-channel structure by lithography. The last step is to bond the epoxy-glass substrate with a thick PMMA plate to form a well-insulated and compact micro-channel. It is noted that both the SU-8 layer and PMMA plate are the very low thermal conductivity materials, as shown in Table 2. Design consideration and fabrication techniques involved in this processes will be discussed in the following sections.

7.1 Design consideration

In the design of a micro-channel for current flow and heat transfer analysis, the primary issue is to minimize or regulate heat loss to the environment and to ensure a uniform heat flux input on the wall that can be easily quantified. Ensuring a consistent heat flux on the wall is not challenging. This can be accomplished by choosing a resistor material that is temperature-independent and uniformly dispersing the resistor material throughout the channel wall. To achieve a consistent current via the resistor, uniform heat generation along the wall must be attained. Nevertheless, the heater may experience variable heat loss at various points, resulting in some of the generated heat dissipating into the substrate, leading to an uneven heat input into the channel flow. Consequently, a uniform heat flux boundary condition on the wall cannot be established. Furthermore, the thermal energy within the channel flow may dissipate to the wall and subsequently to the substrate if the wall exhibits high conductivity. Regrettably, the majority of MEMS techniques established to date utilize silicon wafers, which possess excellent thermal conductivity, as substrates for micro-channel fabrication. These challenges render the measurement and study of heat transfer a challenging undertaking.

Chen et al. (2001) created an air layer beneath the channel system to minimize heat loss. This air layer is sufficiently dense to significantly diminish heat loss. However, the thinness of the heated wall may cause the channel to deform under the flow of high-pressure gas or liquid. Furthermore, the constructed channel is exceedingly complex. While the concept of utilizing the air layer beneath the channel for insulation is commendable, the fabrication process is exceedingly complex and challenging, resulting in a significantly poor yield.

Table 2. Thermal physical properties of the fabrication materials (Touloukian, 1970-1979).

Material	Thermal Conductivity (W/mK) at 300K	Thermal Expansion Coefficient (ppm/K) at 300K
Silicon	148	2.6
Al-Cu-Si alloy	161	19.2
Polysilicon	34	2.4
Silicon nitride	16	0.8
Silicon oxide	1.38	10.3
Pyrex glass	1.4	2.8
Epoxy resin	0.15	62.5
PMMA plate	0.16	62
SU-8	0.2*	52*
Air	0.026	

Data can be found in data sheet for NATOTM SU-8 negative tone photoresists, formulations 50 and 100, released by Micro Chem Corp.

In general, the heat input into the channel flow system can be moved away from the bottom heated wall of the channel by the convective flow, i.e. air or water flow through the channel. However, some portion of the heat input, i.e. the heat loss, can be moved away to the ambient by the conduction of the substrate or the channel structure. In fact, most of heat loss occurs by the substrate conduction, especially for a substrate with a higher thermal conductivity. Thus, in order to effectively reduce the heat loss and maintain a uniform heat flux boundary condition on the heated wall, several alternatives can be selected. One can use the thermal insulation materials on the back of the silicon substrate to reduce the heat loss. In addition, to reduce the spanwise heat conduction along the substrate, the thickness of the substrate can be reduced by mechanical polishing or particular wet etching techniques. In addition, instead of using the air layer underneath the channel, one was thinking of using other low thermal conductivity materials, such as Pyrex glass, as substrate. However, by a careful calculation even the Pyrex glass could not provide enough thermal insulation for both the channel and the heater. Other materials such as PMMA, polyimide or epoxy resin that have a much lower thermal conductivity, as shown in Table 2, can be used to avoid the heat loss from the heater to the ambient outside. An alternative insulation is to move the devices above the silicon wafer onto the Pyrex glass. This requires bonding the device with the Pyrex glass using epoxy resin, which is a low thermal conductivity material. Therefore, the epoxy resin can be served as both the bonding agent and a part of the substrate. This can significantly reduce both the normal and the spanwise conduction along the substrate. A schematic diagram designed for the microchannel integrated with arrays of micro pressure and temperature sensors is shown in Figure 29.

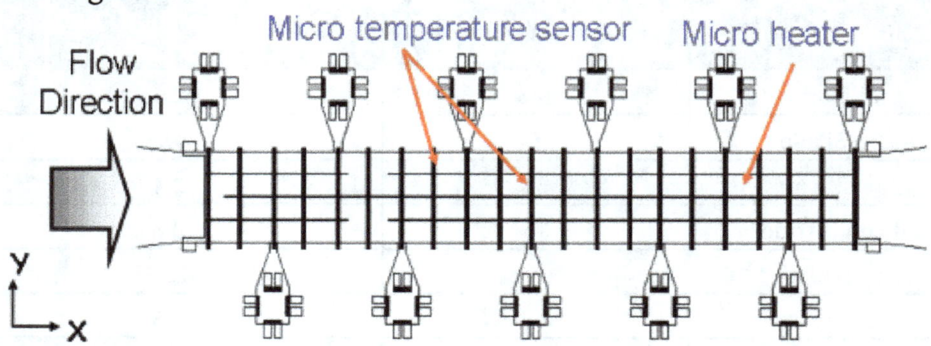

Fig. 28. Micro-channel system with arrangements of arrays of micro pressure sensors, micro temperature sensors and two sets of heaters used for study of micro scale heat transfer and pressure drop of the flow.

7.2 Fabrication process flow

The entire fabrication process, as shown in Figure 29, is described in the following:

1. A 0.3 μm thick LPCVD tetraethoxysilane (TEOS) oxide is deposited on the (100) wafer and used as protection mask for the upper layer devices during the later long period of TMAH wet etch to completely remove the Si wafer.
2. The temperature sensors were made by depositing a 0.3 μm thick LPCVD polysilicon layer and then implanting with boron with a dose of 3×10^{14} atoms/cm^2. This amount of dosage corresponds to a concentration of 10^{19} atoms/cm^3 in the layer. After annealing at 950°C for 30 minutes to re-crystallize the polysilicon layer and to uniformly diffuse the dopant ions across the layer, the doped polysilicon was patterned. The concentration of 10^{19} atoms/cm^2 will give a linear relationship between the resistivity and the temperature of the doped polysicon layer. Thus, after proper calibration, the doped polysilicon layer can be used as temperature sensors.
3. Next, after a deposition of a 0.3 μm thick LPCVD TEOS layer as insulation, a 0.3 μm thick LPCVD polysilicon film is deposited and then implanted heavily with boron with a dose of 3×10^{15} atoms/cm^2. This amount of dosage corresponds to a concentration of 10^{20} atoms/cm^3 in the layer. After annealing at 950°C for 30 minutes, the doped polysilicon is patterned as the pressure

sensors.

4. Before metallization, a 0.3 μm thick LPCVD TEOS oxide is deposited as insulation. Then, contact holes were opened in this layer for metallization. The metallization was made by sputtering standard IC four layers of metals, i.e. Ti/TiN/Al-Si-Cu/TiN with a thickness of 0.04 μm/0.1 μm/0.9 μm/0.04 μm, respectively, onto the substrate surface. The metal layers were then patterned into circuits, as shown in Figure 29 (a).
5. A 9 μm thick SU-8 layer is spin coated on the substrate as the pressure sensing diaphragm, as shown in Figure 29 (b). Then, a 50 μm thick SU-8 layer is coated again on the substrate and patterned into the active cavity to allow for movement of the pressure diaphragm. It is noted that there is a soft bake before layer exposure to evaporate the solvent and a post exposure bake to make the edge between the exposed and unexposed region more sharp and clear. Finally, instead of using the hard bake, a high intensity of light is used to illuminate to complete the cross-linking of the resin since the hard bake will need a temperature at 200°C that will damage the underneath thick epoxy layer (refers to the epoxy layer mentioned in item 5 below). For the soft bake, the SU-8 layer is first maintained at 65°C for 7 minutes with a 5°C/minute ramping rate starting from room temperature, and then baked at 95°C for 15 minutes with a 5°C/minute ramping rate starting from 65°C to release the internal residual stress of SU- 8 thick layer. In fact, the success of the SU-8 channel strongly depends on the baking process after light exposure.
6. It is now ready to move the devices made on the Si wafer onto a low thermal conductivity Pyrex glass. This is done first by bonding the silicon wafer with the Pyrex glass, as shown in Figure 29 (c), using epoxy resin as the bonding agent. Since the epoxy resin has a significantly lower thermal conductivity than the Pyrex glass, this can better reduce the heat loss from the channel to the ambient.
7. After the bonding process, the silicon substrate is ready for removal and cleared off. This is done by a wet etch process. The silicon is etched with TMAH solution at 90°C for 5-6 hours. Instead of using KOH, the selection of TMAH is attributed to its relatively high selectivity for silicon versus oxide. This can avoid the sensors attacked by the etchant during the long period of wet etch process since the protection layer of the sensors are made of TEOS oxide. After successful movement of the pressure sensors onto the Pyrex glass substrate, no distortion was found, as shown in Figure 29(d).
8. Next, another SU-8 layer is spin coated on the epoxy-glass substrate at desired thickness and patterned by lithography to form a rectangular microchannel and cavities of the pressure sensors that connect with channel through narrow size tunnels. The side walls of the channel, the narrow tunnels, or the cavities of the pressure sensors are successfully made, as shown

in Figures 30 (a), (b) and (c). It is noted that a similar post exposure bake should be made in order to obtain a nice shape of channel and cavities of the pressure sensor.

9. A PMMA plate is then bonded, using epoxy resin, with the epoxy-glass substrate to enclose the channel and the cavities, as shown in Figure 29 (e).

A picture for the completed channel is shown in Figure 30 (d).

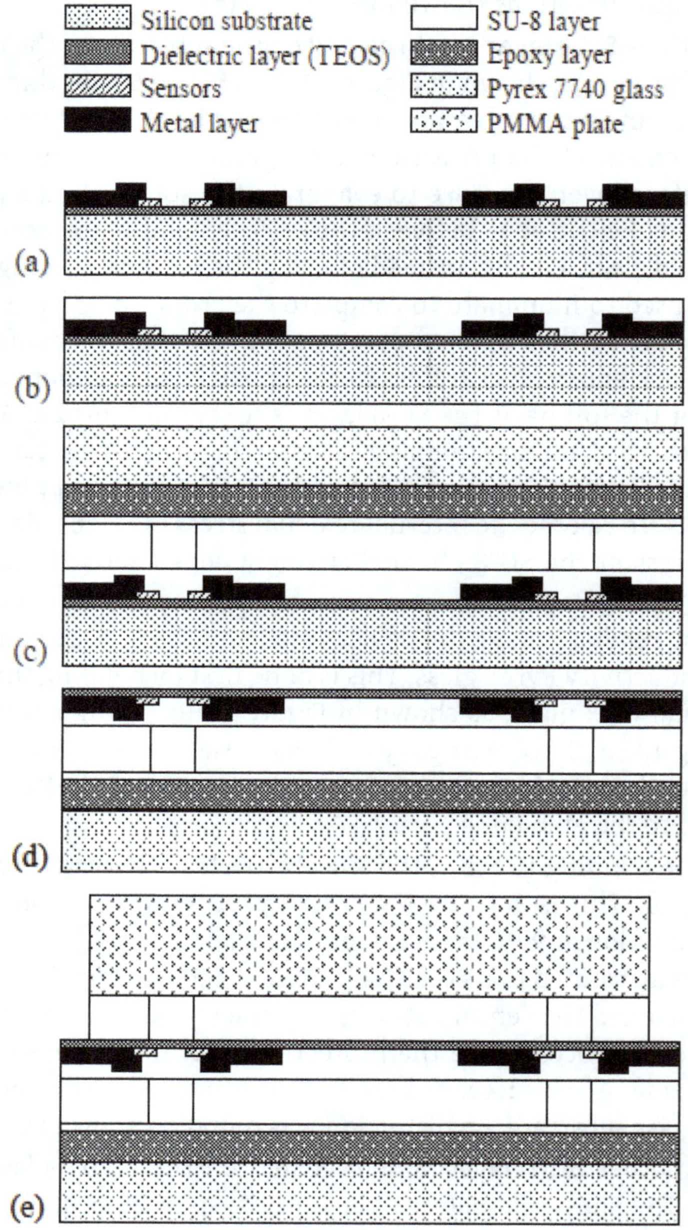

Fig. 29. Fabrication process for the entire microchannel system.

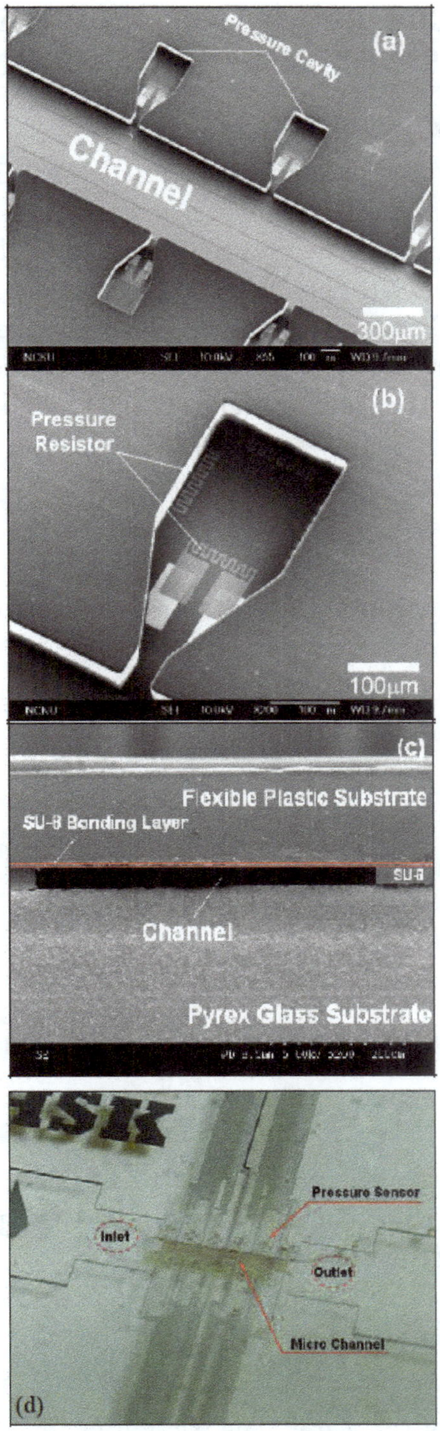

Fig. 30. SEM photographs of SU-8 channel integrated with arrays of micro pressure sensors: (a) global view, (b) close view and (c) cross section view and (d) the completed microchannel (Ko, 2009).

8. LOCAL PRESSURE REDUCTION AND THERMAL TRANSFER CHARACTERISTICS WITHIN A MICROCHANNEL

8.1 Characteristics of liquid flow

A theoretical analysis of the flow characteristics within the microchannel is conducted to compare with the experimental findings. Given that the microchannel's width significantly exceeds its height, the flow within the microchannel can be approximated as parallel plate channel flow. As the microchannel flow can quickly reach a fully developed state, as evidenced by subsequent data measurements, the governing equation for fully developed flow in the microchannel can be significantly simplified. The solution can be readily found as follows (Incropera et al., 2007):

$$u(y) = \frac{y^2}{2\mu}\frac{dp}{dx} - \frac{H}{2\mu}\frac{dp}{dx}y = \frac{1}{2\mu}\frac{dp}{dx}(y^2 - Hy) \quad (8\text{-}1)$$

To integrate the equation transversely, the mean velocity of the flow in the channel can be obtained as follows:

$$U = -\frac{H^2}{12\mu}\frac{dp}{dx} \quad (8\text{-}2)$$

where U is the mean velocity of the flow. From Eq. (6-2), the longitudinal pressure distribution can be found as follows:

$$p_x = p_i - \frac{12\mu U}{H^2}x \quad (8\text{-}3)$$

or

$$\Delta p = \frac{12\mu U l}{H^2}$$

The Reynolds number here is defined as the liquid density times the mean velocity of the flow and the hydraulic diameter (Dh) of the rectangular channel divided by the dynamic viscosity of the liquid as follows:

$$\mathrm{Re} = \frac{\rho U D_h}{\mu} \quad (8\text{-}4)$$

where is ρ the density of the water. The friction factor of flow can be calculated with a given pressure drop.

For the liquid flow, Qu et al. (2000, 2002) has found that the pressure drop in microchannel is significantly higher than the theoretical analysis due to the roughness effect. To eliminate the wall roughness effect, therefore, the roughness on the wall surface of the current micro- channel is minimized, and is less than ± 5 nm to 15 nm. This roughness is negligibly small in comparison with the height of the channel, which is either 68.2 μm or 23.7 μm. During the experiment, the pressure distributions measured at different Reynolds numbers are compared with the analysis, as shown in Figure 31(a) for the channel height at 34.7 μm. The pressure drop of the water is found significantly higher than the prediction by Eq. (8-3). A similar result was also found by Mala and Li (1998) who showed that the pressure drop of water in micro-tubes is significantly higher than the analysis. Under the same flow rate and tube diameter, the pressure drop in metal tubes is lower than in silica glass tubes. This can be attributed to the electric surface potential in the channel which attracts counterions in the liquid and causes formation of electric double layer on the wall. The thickness of the EDL in the micro channel for these three different solutions can be estimated by the inverse of the square root of the Debye-Huckel parameter (Mala et al. 1997) as follows:

$$\tau = \left(\frac{\varepsilon\varepsilon_o\kappa_b T}{2z_i^2 e^2 n_\infty}\right)^{1/2} \tag{8-5}$$

where ε is the dielectric constant of the solution and ε_o is the permittivity of vacuum, κ_b is the Boltzmann constant, T is the absolute temperature, n_∞ and z_i are the bulk concentration and the valence of type *i* ions. Therefore, the EDL thickness is approximately 1 μm for water. In order to minimize the EDL effect, the working fluid used is changed to KCl solution at concentration of 10^{-4} M or 10^{-2} M, which can effectively remove the surface charges by the metal ions in the solution and significantly reduce the thickness of EDL. The EDL thickness is approximately 100 nm for the KCl solution with concentration of 10^{-4}M and 1 nm for the KCl solution with concentration of 10^{-2}M. The result indicates, as shown in Figure 31(a), that the pressure distribution for a higher KCl concentration (10^{-2} M) is significantly lower than for the pure DI water and is close to the analysis. It appears that the deviations for the experimental data of water from the prediction are attributed to surface charges and the EDL on the channel wall. The thicker the EDL, the greater is the deviation of the pressure drop from the analysis. The current finding for the reduction of EDL thickness which makes the pressure drop in the microchannel approach to the analysis agrees with the results in Ren et al. (2001). When the channel height increases to 68.2 μm, the pressure drops for the case of the pure water and the cases of different concentrations of KCl solutions collapse into a single line, as shown in Figure 31(b), and are closer but higher than the analysis. It appears that the EDL of different fluids does not have a significant effect on the pressure drop in a greater height of channel even though the EDL of different fluids on the wall has different thickness.

For a large scale channel, the entrance region can be clearly identified by the non-linear distribution of pressure distribution. However, in the entry region, no clear indication of non-linear distribution of pressure can be found, as shown in Figure 31(a) or 31(b). For large scale channel, the entry length can be estimated by the equation as follows Ren et al. (2001):

$$L_{in} = 0.02\ D_h\ Re \tag{8-6}$$

For the channel height at 68.2 μm and Reynolds number at 128, the hydrodynamic entry length calculated from Eq. (8-6) is approximately 300 μm. However, pressure drop within 300 μm does not indicated greater or nonlinear variation, as shown in Figure 31(b). It shows that the hydrodynamic entry region in microchannel is very short and could not be identified from the current pressure distribution measurements. It appears that the boundary layer in the entrance region of a microchannel develops very fast and becomes fully developed very close to the entrance.

The same kinds of fluids are also performed in channels at the height of 23.7 μm In the channel at the height of 23.7 μm, more deviations of the pressure drop data for different fluids occur, as shown in Figure 32(a) for pressure drop variation with the Reynolds numbers in the fully developed region. The pressure drop for water flow has the largest slop, and decreases when the concentration of KCl solution increases. These results clearly indicate the effect on the thickness of EDL relative to the height of channel. For water flow with a thicker EDL on the channel wall, the appearance of the EDL can increase the pressure drops by 13.4 % in the smaller height of channel (23.7 μm), but increase only 8.6 % in the greater height of channel (34.7 μm). For the channel height at 68.2 μm, all the pressure drop variations with the Reynolds number for different fluids collapse into a single line, as shown in Figure 32(b), and are closer but higher than the analysis. The higher pressure drop data for different fluids than the prediction by Eq. (8-3) can be attributed to the aspect ratio effect of the channel. The prediction by Eq. (8-3) is based on an analysis in a parallel plate channel, while the current channel with channel height of 68.2 μm or 23.7 μm has an aspect ratio of 1:7.3 or 1:21, respectively.

From the definition of the friction factor,

$$f = \frac{\tau_w}{\frac{1}{2}\rho U^2} = \frac{D_h}{2\rho U^2}\frac{dp}{dx} \tag{8-7}$$

where Dh is hydrodynamic diameter of the channel cross section and is equal to 2Hw/(H+w).

The product of friction factor, *f*, and the Reynolds number is defined as the Poiseuille number (Po) (Baviere, 2004) and can be written as follows for parallel plate channel:

$$Po = f\,\mathrm{Re} = 24 \tag{8-8}$$

The parallel plate channel has an infinite width in z direction, the velocity is a two dimensional parabolic profile. However, when the channel width in the z direction is finite the velocity close to the side wall becomes three dimensional. The side wall effect can produce greater pressure drop. The smaller the channel width in the z direction, the greater the aspect ratio, the greater the pressure drop becomes. For large scale channel, the greater Po value for channel with greater aspect ratios has been calculated and correlated in terms of aspect ratio, α, as follows Baviere et al. (2004):

$$Po(\alpha) = 24\,(1 - 1.3553\,\alpha + 1.9467\,\alpha^2 - 1.7012\,\alpha^3 + 0.9564\,\alpha^4 - 0.2537\,\alpha^5) \tag{8-9}$$

From Eqs. (8-7), (8-8) and (8-9) the pressure gradient can be re-calculated for the aspect ratio less than 1:21, as shown in Figure 33(a) and Figure 33(b) indicated by the red dash line. The pressure gradient predicted by Eq. (8-9) in the fully developed region agrees very well with the data for working fluid at the much higher concentration of KCl or when channel height is very large (68.2 μm) and EDL effect is negligible small. For clarity, the Po number for both the data and the prediction by Eq. (8-8) has been plotted as shown in Figure 33. The agreement between data and the prediction at channel height of 68.2 μm indicate that wall roughness in the current channel has a negligible effect on the pressure drop. Even at a smaller height of channel, 23.7 μm, the agreement between data at high concentration of KCl with negligible effect of EDL and the prediction further indicates the negligible effect of the wall roughness in the current channel.

In order to compare with the pressure drop data published in the literature, the normalized friction constant which is most frequently used is adopted and written as follows.

$$C_f^* = \frac{Po_{exp}}{Po_{theory}} \tag{8-10}$$

where the friction constant Poexp is found from the product of Re with the measured friction factor and Potheory is directly evaluated from Eq. (8-8). Large discrepancy among different works can be found as shown in Figure 34(a) and Figure 34(b). The data selected here for comparison are only limited to rectangular channels which have a channel height close to the current channel. More detailed sizes of the channels and the materials the channels used are also listed in the table underneath the figure. In general, the normalized friction constants for all the work in the past are not only much greater than unity which is a theoretical prediction for large scale channel, but also greater than the data of the current work. This is attributed to the use of the inlet and outlet pressure to calculate the pressure drop in the channel which is used to evaluate the friction constant. Most of the work in the past can not obtain the pressure inside the channel, but measure the pressure outside the channel at the inlet and outlet (Pfund et al., 2000; Xu et al., 2000; Papautsky et al.,1999; Peng et al., 1995). Since there is large pressure loss for flow entering or exiting the channel, the pressure drop measured outside the channel at the inlet and the outlet is much greater than the actual pressure drop in the channel. Therefore, much greater friction constants obtained from others are expected.

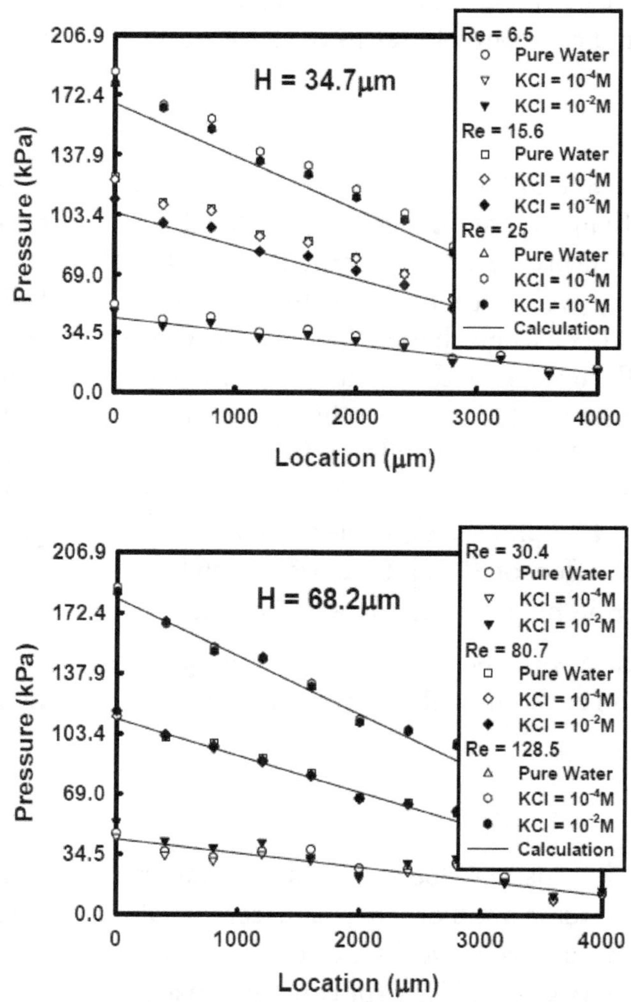

Fig. 31. local pressure distributions at different Reynolds numbers in a micro-channel with channel height of (a) 34.7 μm and (b) 68.2 μm (Ko, 2009).

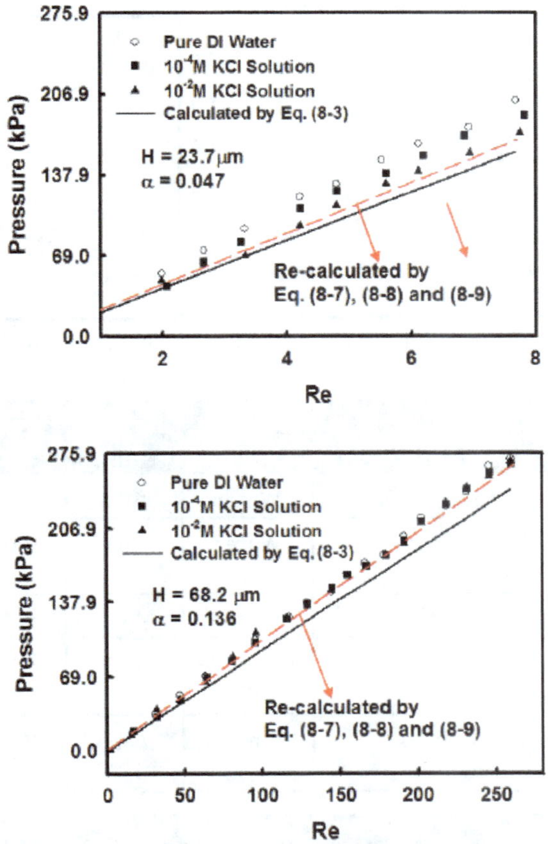

Fig. 32. Pressure drops at different Reynolds numbers in fully developed region a micro- channel with (a) channel height of 23.7 μm and (b) channel height of 68.2 μm (Ko, 2009).

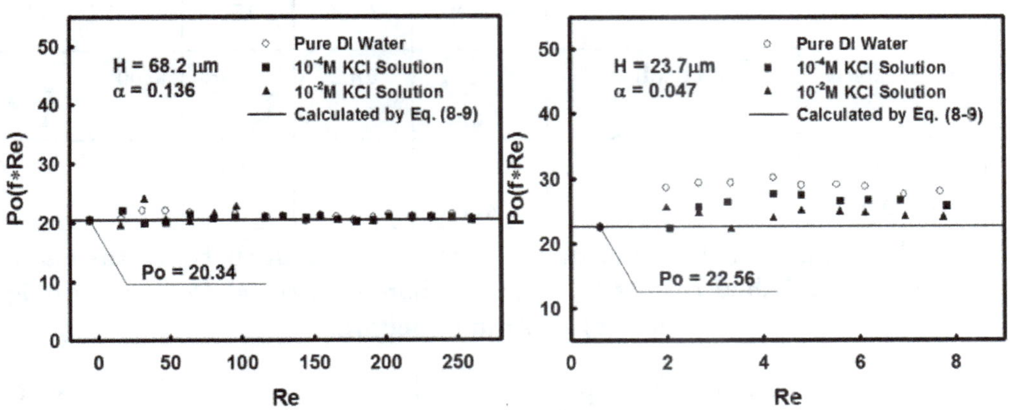

Fig. 33. The Poiseuille numbers at different Reynolds numbers in a micro-channel with (a) channel height of 68.2 μm and (b) channel height of 23.7 μm (Ko, 2009).

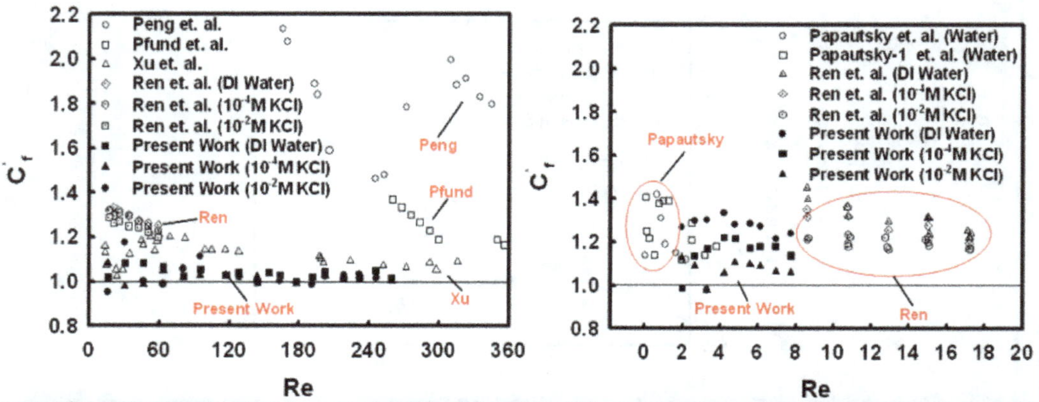

References	Channel Material	Working Fluid	Height (m)	Width (m)	Aspect Ratio	Dh (m)
Peng et al. (1995)	stainless steel	Water	100	200	0.5	133
Pfund et al.(2000)	Polycarbonate-Gasket-Polyimide	Water	263	10000	0.263	416.4
Xu et al. (2000)	Glass-Silicon	Water	43.7	415.3	0.106	79.08
Ren et al. (2001)	Silicon	Water/KCl	40.5	5000	0.0081	80.34
Present work (Ko, 2009)	Polymer	Water/KCl	68.2	500	0.1364	120.02

Reference	Channel Material	Working Fluid	Height (m)	Width (m)	Aspect Ratio	Dh (m)
Papautsky (1999)	Nickel	Water	24.65	600	0.0411	47.35
Papautsky-1(1999)	Nickel	Water	26.35	150	0.1757	44.82
Ren et al. (2001)	Silicon	Water / KCl	28.2	5000	0.0056	56.08
Present work (Ko, 2009)	Polymer	Water / KCl	23.7	500	0.0474	45.24

Fig. 34. (a) Comparison of the normalized friction constants between the current work and the published results for current channel with (a) 68.2 μm in height and (b) 23.7 μm in height.

In addition, data obtained from previous work scatter very much. The scatter in data among different works should not be attributed to the geometry of the channel which has a different aspect ratio since the comparison has already accounted for the aspect ratio effect. The scatter in data among different works is mostly attributed again to the use of the pressures at the inlet and the outlet outside the channel to evaluate the pressure drop inside the channel. This may lead to erroneous result. For example, the data from Xu et al., as shown in Figure 34(a) appears to be very good and is very close to the theoretical prediction. However, in a much smaller channel height, i.e. 15.4 μm, where the EDL effect can become significant and is expected to increase the pressure drop, the data from Xu et al. is identical to the theoretical prediction, as shown in Figure 34(a) of their paper (Xu et al., 2000). Most of the data did not discuss the effect of EDL except the work of Ren et al. (2001). Ren et al. has used a higher concentration of KCl solution to significantly reduce the thickness of the EDL and its effect on the pressure drop so that the friction constant in a smaller size of channel, as shown in Figure 34(b), with a higher concentration of KCl solution is smaller than the one with a lower concentration of KCl solution. However, all the data from Ren et al. are much greater than the theoretical prediction, as show in Figure 34(a) for a greater channel height and Figure 34(b) for a smaller channel height. The higher normalized friction constant obtained is attributed to the erroneous estimation of the entry length, which is used to estimate the pressure drop in the entry region, and the pressure loss at the exit. The normalized friction constant in Ren et al. is estimated from the pressure drop which is measured outside the channel at the inlet and the outlet and subtracted by the pressure drop in the entry region and the pressure loss at the outlet. Notwithstanding this limitation, the findings of Ren et al. consistently demonstrate that EDL can substantially influence the pressure drop when the channel height is minimal. The EDL effect may diminish with a high concentration of KCl solution or an increased channel height. Our measurements also corroborate these findings. The concordance between the observed pressure drop and the theoretical prediction indicates that many of the previously reported conflicting effects of microchannel size on pressure drop have been elucidated. The comparison further substantiates that the methodologies and fabrication of the micro-channel utilized in this work are appropriate for precise pressure drop measurement.

8.2 Flow characteristics for air

8.2.1 Theoretical analysis

Theoretical analysis of micro flow inside the channel is performed firstly, which is then used to compare with the data of pressure measured and analyze the characteristics of micro flow inside the channel. Since the channel has a large value in its width (500 μm) in comparison with its height (from 23 μm to 68 μm), the channel flow in the present study can be assumed as laminar flow between two parallel plates at steady state. The configuration with the coordinate system and flow velocity of micro-channel flow are defined in Figure 35. Therefore, the governing equations of the flow in this micro-channel system can be written in the following:

Continuity equation:

$$\frac{\partial \rho u}{\partial x}+\frac{\partial \rho v}{\partial y}=0 \tag{8-11}$$

Navier-Stokes Equation:

$$\rho(u\frac{\partial u}{\partial x}+v\frac{\partial u}{\partial y})=-\frac{\partial p}{\partial x}+\mu(\frac{\partial^2 u}{\partial x^2}+\frac{\partial^2 u}{\partial y^2}) \tag{8-12}$$

$$\rho(u\frac{\partial v}{\partial x}+v\frac{\partial v}{\partial y})=-\frac{\partial p}{\partial x}+\mu(\frac{\partial^2 v}{\partial x^2}+\frac{\partial^2 v}{\partial y^2}) \tag{8-13}$$

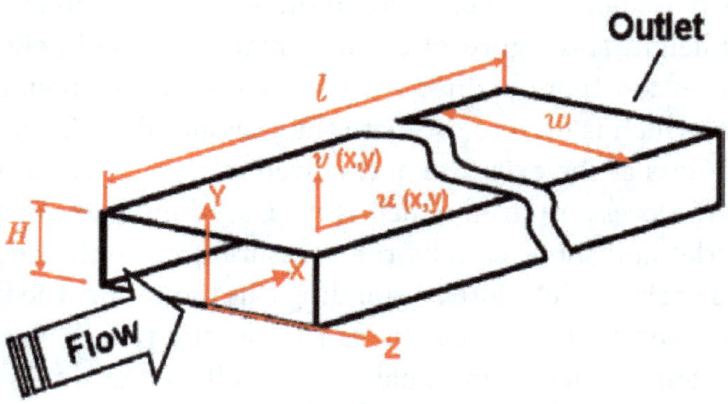

Fig. 35. Schematic diagram and the coordinate system used for the micro-channel flow system.

The micro-channel is actually fabricated by very low thermal conductivity materials which can be considered as well insulated channel. Therefore, the gas flow inside the channel can be assumed to be undergoing isentropic process as follows:

$$\frac{p}{\rho^k}=cons. \tag{8-14}$$

If non-slip boundary condition on the wall is used, then the boundary condition can be written as:

$$U(y)_w=0 \tag{8-15}$$

In Eq. (8-14), k is the ratio of specific heat Cp versus Cv. For micro flow in parallel plate channel, the height is much smaller than either the length or the width of the channel. One can assume that the gradient terms, normal to the surface of channel, in Navier-Stokes Equation are much larger than those along the channel. As a result, one can neglect terms that represent x-direction diffusion of momentum (along the channel) relative to those of y- direction (normal to the channel surface). That is

$$u >> v, \quad \frac{\partial}{\partial y} >> \frac{\partial}{\partial x} \; and \; \frac{\partial^2 u}{\partial y^2} >> \frac{\partial^2 u}{\partial x^2} \tag{8-16}$$

Therefore, the Navier-Stokes Equation for the micro-channel flow can be simplified as

$$\frac{\partial p}{\partial x} = \mu \frac{\partial^2 u}{\partial y^2} \tag{8-17}$$

$$\frac{\partial p}{\partial y} = 0 \tag{8-18}$$

Eq. (8-17) and Eq. (8-18) means the pressure varies only in the streamwise direction. Therefore, Eq. (8-17) can be readily solved analytically. This equation is the same as channel flow in the fully developed region. The pressure distribution measurements presented in the later section indicate that the entry length of flow in micro-channel could hardly be observed. It means that the flow in most part of the micro-channel is fully developed except in the very short entry region where x is not large as compared with the width of the channel. The solution of Eq. (8-17) is a parabolic velocity profile as follows:

$$u(y) = \frac{y^2}{2\mu}\frac{dp}{dx} - \frac{H}{2\mu}\frac{dp}{dx} y = \frac{1}{2\mu}\frac{dp}{dx}(y^2 - Hy) \tag{8-19}$$

where H is the height of the channel. To integrate the above equation along y direction of the channel, the average velocity can be obtained. Therefore, the mass flow rate (Qm) of in the channel can be shown as follows:

$$Q_m = -\frac{H^3 w \rho}{12\mu}\frac{dp}{dx} = -\frac{H^3 w p^{\frac{1}{k}}}{12\mu p_o^{\frac{1}{k}} \rho_o^{-1}}\frac{dp}{dx} \tag{8-20}$$

where w is width of the channel and the subscript o refers to the condition at the outlet location of the last pressure sensor in the micro-channel. The above equation can be re- arranged as follows:

$$Q_m dx = -\frac{H^3 w p^{\frac{1}{k}}}{12\mu p_o^{\frac{1}{k}} \rho_o^{-1}} dp \tag{8-21}$$

One can integrate above equation from the entrance to any location, x, in the channel and obtain the pressures distribution inside the channel as follows:

$$p_x = \left[p_i^{\frac{k+1}{k}} - \frac{12Q_m x(\frac{k}{k+1})\mu p_o^{\frac{1}{k}} \rho_o^{-1}}{H^3 w} \right]^{\frac{k}{k+1}} \tag{8-22}$$

Where the subscript i refers to the condition at the inlet location of the first pressure sensor in the micro-channel. To integrate Eq. (8-21) from the entrance to the end of the channel, the mass flow rate can be obtained explicitly as follows:

$$Q_m = \frac{H^3 w p_0^{\frac{k+1}{k}}}{12(\frac{k}{k+1})\mu p_o^{\frac{1}{k}} \rho_o^{-1} l} \left[\left(\frac{p_i}{p_o} \right)^{\frac{k+1}{k}} - 1 \right] \tag{8-23}$$

where l is the distance from the first pressure sensor to the last pressure sensor in the micro- channel and is equal to 4000 μm. During the calculation of pressure distribution, the dynamic viscosity (μ) in the equation is based on the entrance temperature of the channel flow.

For the case when the heat transfer occurs (with a given heat flux boundary condition on the wall), the wall is no longer adiabatic. The above equation can be modified with k replaced by n and is written as follows:

$$p_x = \left(p_i^{\frac{n+1}{n}} - \frac{12Q_m x(\frac{n}{n+1})\mu p_o^{\frac{1}{n}} \rho_o^{-1}}{H^3 w} \right)^{\frac{n}{n+1}} \tag{8-24}$$

where n can be obtained by fitting the above equation to the pressure distribution data. In this way, the pressure variation with either the density or the temperature can be found by a polytropic process as follows:

$$\frac{p}{\rho^n} = const. \tag{8-25}$$

8.2.2 Experimental results of the pressure drop for air flow

The wall roughness has been shown to play very important role on the pressure gradient, which contributed the scattering of friction factor data reported in others (Guo & Li, 2003; Kleinstreuer & Koo, 2004), in micro-channel flow. However, the wall information in most of the micro-channels reported in the literature is absent. Therefore, the wall roughness in the current channel is measured and studied. During etching and removing of the silicon substrate, the wall roughness of the channel has been minimized by proper etching process as described previously. The roughness of the current channel wall is measured by the profile meter using a small probe, scanning along axial location of the channel wall. The values of roughness are less than ± 5 nm to 15 nm. The relative roughness is much less than 1 % as compared with the height of the channel (the smallest height of the channel is 23 μm). Therefore, accurate results of pressure drop were obtained based on the small roughness of the wall surface in the micro-channel in this study. Typical results of the pressure distributions, both by theoretical prediction and experiment, for height of channel at 23 μm are shown with good agreement in Figure 36.

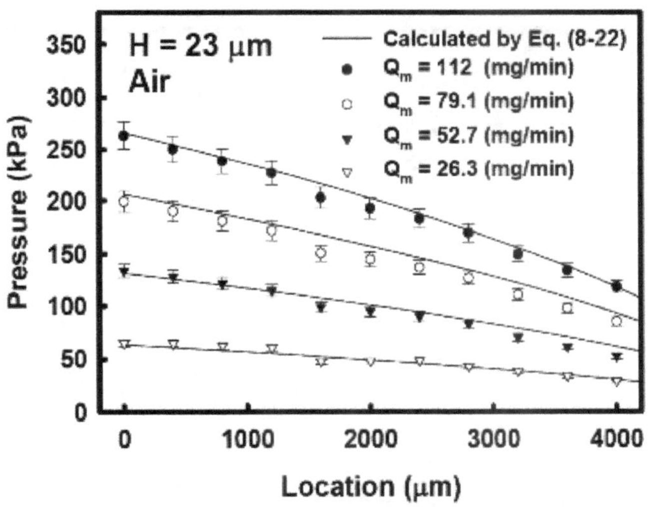

Fig. 36. Comparisons of the pressure distributions between the current data and the analytical results for the airflow along the micro-channel.

Both experimental data and the theoretical prediction indicate that the pressure distribution inside is not linear, due to the compressibility effect of the air flow in the micro-channel, except when the flow speed in the channel is lower than 0.3 Mach, which can be assumed as incompressible condition. The agreement between experimental data and theoretical prediction is very good even in the upstream region close to the entrance. Therefore, this means that the pressure distribution in the region close to the entrance has reached a fully developed flow. This clearly indicates that the hydrodynamic entrance length in the micro-channel flow is very short as mentioned previously in the derivation of the pressure distribution equation. Therefore, the assumptions used in the analysis are thus confirmed. In addition, all the pressure data drop smoothly from the entrance to the exit. There is no any sudden change in the pressure gradient where flow may undergo an earlier onset of transition from laminar to turbulent flow as mentioned by others (Peng et al., 1994, 1996), which causes greater friction factor in micro-channel.

The variation of the pressure difference between the inlet and the outlet with the volume flow rate is also measured and compared with theoretical prediction, as shown in Figure 37 for the channel height varied from 23 μm to 55 μm. The agreement remains highly advantageous for all micro-channels. This data further suggests that there is neither a reduction nor an augmentation in the friction on the wall, potentially attributable to a decrease in viscosity or the effects of surface roughness. There is once more no abrupt alteration in the pressure drop, suggesting that the flow may experience an earlier transition from laminar to turbulent flow, as noted by others. The Reynolds number seen in Figure 37 is calculated by dividing the mean flow velocity in the hydraulic diameter of the channel by the kinematic viscosity of the airflow.

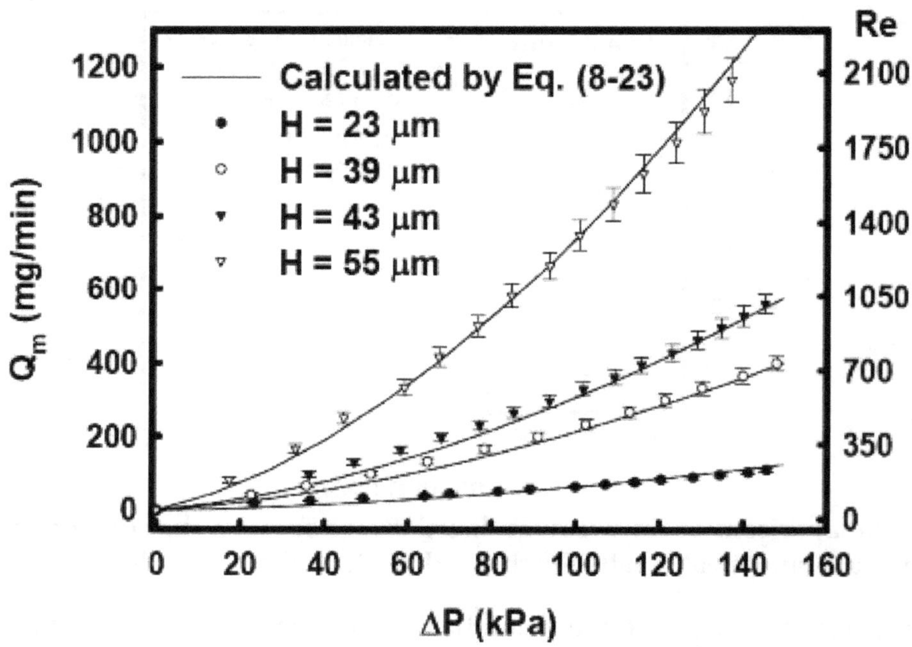

Fig. 37. Variation of the pressure difference between the inlet and the outlet with the mass flow rate or the Reynolds number.

It is noted that the friction factor in micro-channel of air flow is not a constant if one converts the pressure distribution in Figure 36 into the friction factor. This result is contrary to many of other experimental work which obtains a friction factor at a constant value (Takuto et al., 2000; Liu et al., 2004; Kohl et al., 2005; Bayraktar & Pidugu, 2006). That means that air flow in their micro-channels has been unconsciously assumed as incompressible. To calculate Reynolds number for the amount of mass flow rate in the current channel, the largest Reynolds number is very close to 2000, as shown in Figure 5.4, which is in the laminar regime. In the large scale channel, this is an extremely low Reynolds number flow condition and the flow usually can be assumed as incompressible. However, in the microchannel flow, the flow speed for Re = 2000 is 280 m/s for channel height of 55 μm. In this case, the March number for Re = 2000 is equal to 0.82, and the flow is highly compressible and can not be assumed as incompressible. Assuming incompressible flow will result in an inaccurate skin friction coefficient. Assuming incompressible flow results in a significantly lesser friction factor compared to that observed in large-scale channels, as indicated in various earlier studies.

With the constant heat flux imposed on the wall, the pressure distribution is also measured as shown in Figure 38. The pressure variation along the channel is very similar to the case when the channel wall is insulated except that the magnitude of the pressure is higher. Eq. (8-24) is used to fit the pressure distribution data by varying the value of n in this equation. The value of n obtained from the fit can be utilized to ascertain the correlation between pressure and density in the microchannel under constant heat flow conditions. The link between pressure and density is essential for calculating the heat transfer coefficient, as outlined in the subsequent section.

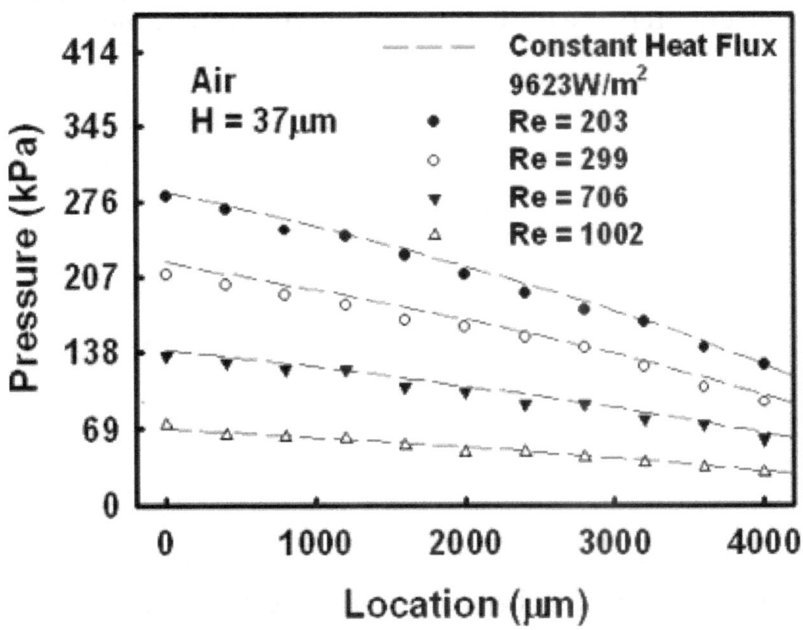

Fig. 38. Comparisons of the pressure distributions between the current data and the analytical results for the airflow along the microchannel for the case under constant wall heat flux, Eq. (8-24) is used to fit the data by varying the value of n.

8.3 Local heat transfer for air flow

In order to further study the heat transfer characteristics in the micro-channel, the current micro-channel is integrated with array of temperature sensors, pressure sensors and a set of heater. In the current experiment, the heat transfer coefficient, hb, is defined by the equation as follows:

$$h_b = \frac{\dot{q}}{(T_w - T_b)} \tag{8-26}$$

where q is the heat flux imposed along the wall by the electric heater and is equal to the product of the voltage and the current passing through the heater divided by the total area of the heater. Tw is the wall temperatures measured by the temperature sensors along the channel at particular locations x. Tb is the bulk temperature of the flow which can be estimated from the amount of heat flux imposed on the wall which heats up the air flow and rises the temperature of the bulk flow. By applying a control volume and conservation of energy for the flow in the channel as discussed by Incropera et al. (2007), one can obtain a conservation equation of energy as follows:

$$\dot{q}\pi D_h dx = \dot{m}c_p dT_b \tag{8-27}$$

where m is the mass flow rate of the air in the channel. Rearranging and integrating the above equation from the inlet of the channel, one can obtain:

$$\int_{T_{in}}^{T_b} dT_b \int_0^x \frac{\dot{q}\pi D_h}{\dot{m}c_p} dx \tag{8-28}$$

Therefore,

$$T_b = T_i + \frac{\dot{q}\pi D_h}{\dot{m}c_p} x \tag{8-29}$$

The local Nusselt number, Nub, can be defined as:

$$Nu_b = \frac{h_b \times D_h}{k} \tag{8-30}$$

where k is thermal conductivity of the gas flow. Once heat transfer coefficient is obtained, the local Nusselt number can be found from the above equations.

8.3.1 Direct measurements of the heat transfer for air flow

The heat transfer experiments have been performed with air flow in a channel with channel height of 37 μm. The velocity of air flow is controlled by mass flow controller and is assumed in the laminar regime where the Re from 203 to 1006. The thermal conductivity of the air used in the Nusselt number is evaluated at the mean bulk temperature, i.e. an average of the bulk temperature of the flow at the inlet and the outlet of the channel. As shown in Figure 39 (a), the heat flux is varied from 5413 to 21652 W/m^2 and the Reynolds number is kept at 500. The Nusselt numbers in the micro-channel with channel height at 37 μm is significantly higher than that of theoretical prediction by the parallel plates channel (Rohsenow et al., 1985), and become closer as the heat flux on the wall decreases. For the case of heat flux equal or lower than 9623 W/m^2, all the Nusselt number results collapse into a single line, as shown in Figure 39 (b), in the downstream of the channel where flow has become fully developed. The Nusselt number results collapsing into a single curve in the fully developed region agrees with theoretical results for prediction of large scale channel except that the Nusselt number is higher. This indicates that other effects, such as property variation with temperature, compressibility of flow, should be taken into account in order to obtain a good correlation of the heat transfer data.

8.3.2 Correction for the bulk temperature

The reason for this deviation may be partially attributed to property variation, such as compressible gas flow and leads to increase in the heat transfer. In order to account for the compressibility of the gas which can expand and reduce the kinetic energy of the flow, and change the bulk temperature of the flow, the theoretical calculation equation for the bulk temperature should be re-derived from concept of energy balance, as follows:

$$dq = \dot{m}(e_2 - e_1) + (\frac{U_2^2}{2} - \frac{U_1^2}{2}) \quad (8\text{-}31)$$

where e is enthalpy and the subscript indicates properties at different locations. For a differential control volume, the above equation can be written as:

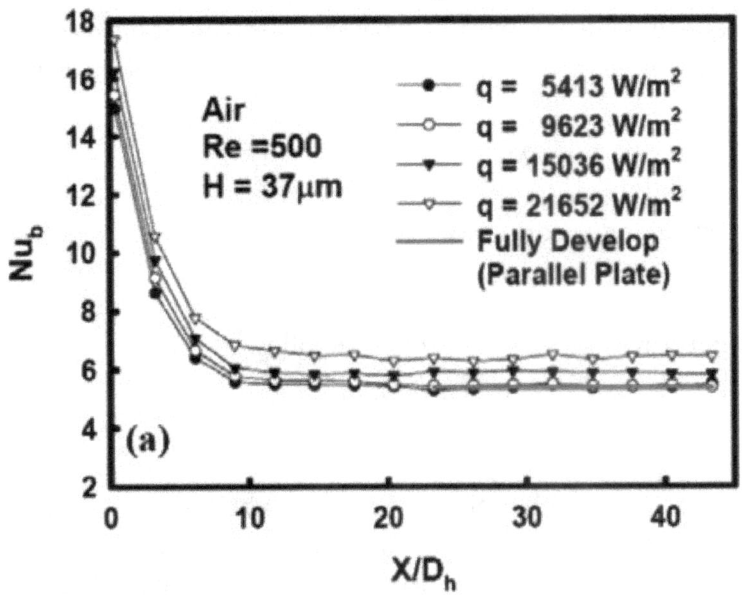

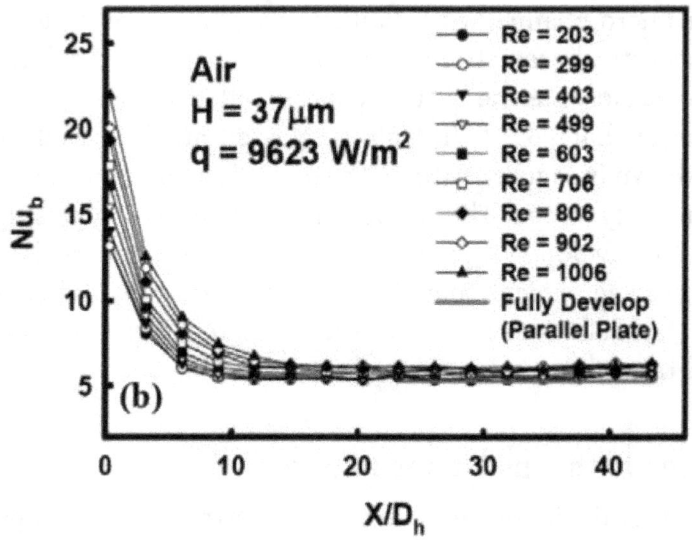

Fig. 39. Experimental results for local Nusselt number distributions under (a) different heating flux conditions and (b) different Reynolds number.

$$dq = \dot{m}\, de + d(\frac{U^2}{2}) \tag{8-32}$$

where $dq = \dot{q}_w \pi D_h$, and $de = c_p\, dT$, re-arranging the above equation, one obtains

$$\frac{dT_b}{dx} = \frac{\dot{q}_w \pi D_h}{\dot{m} c_p} - \frac{d(\frac{U^2}{2})}{c_p dx} \tag{8-33}$$

To integrate above equation from the entrance to the desired location, x, in the channel, one can obtain the bulk temperature of the flow as follows:

$$T_b = T_i + \frac{\dot{q}_w \pi D_h}{\dot{m} c_p} x - (\frac{U_x^2}{2} - \frac{U_i^2}{2}) \frac{1}{c_p} \tag{8-34}$$

From the continuity equation,

$$\rho_1 A_1 U_1 = \rho_2 A_2 U_2 \tag{8-35}$$

The bulk temperature can be re-arranged as follows:

$$T_b = T_i + \frac{\dot{q}_w \pi D_h}{\dot{m} c_p} x - \frac{U_i^2}{2c_p} \left[(\frac{\rho_i}{\rho_x})^2 - 1 \right] \tag{8-36}$$

Since the current micro-channel is actually fabricated by very low thermal conductivity materials, the channel wall can be considered as well insulated. Without heat transfer into the channel, the gas flow inside the channel can be assumed to be undergoing isentropic process. For the case when a desired heat flux is imposed on the bottom wall, the gas is assumed under polytropic process and the pressure density relationship can be written as follows:

$$\frac{p}{\rho^n} = cons. \tag{8-37}$$

where n is a constant and is determined from the heating condition on the wall of the channel. Therefore, the density in Eq. (8-36) can be replaced by the local pressure as follows:

$$T_m = T_i + \frac{\dot{q}_w \pi D_h}{\dot{m} c_p} x - \frac{U_i^2}{2c_p}\left[\left(\frac{p_i}{p_x}\right)^{\frac{2}{n}} - 1 \right] \tag{8-38}$$

8.3.3 Heat transfer data

The bulk temperature can be recalculated from the aforementioned equation using the local pressures recorded by the current pressure sensors along the micro-channel. Consequently, the local Nusselt numbers along the channel can be derived and are illustrated in Figure 40 under varying Reynolds number conditions. All Nusselt number findings converge into a singular curve in the fully developed region, as illustrated in Figure 40 (a) under varying heat flux circumstances. Again, for the case of heat flux at 9623 W/m^2 the Nusselt numbers at different Reynolds numbers also collapse into a single curve in the fully developed region, as shown in Figure 40 (b), but the curve is slightly lower than the prediction in large scale channel. Theoretical calculation for the heat transfer in large scale rectangular channel at different aspect ratios in the fully developed region has been performed and reported (Incropera et al., 2007). It is found that in large scale channel the Nusselt number is 4.55 for rectangular channel with aspect ratio of 1:10, but is 5.38 for ideal parallel plate channel. In fact, the current micro- channel is not an ideal parallel plates channel and is a channel with aspect ratio of 1:13. Therefore, the lower Nusselt numbers of experimental results in the current channel than the prediction for parallel plates channel is expected and is attributed to the effect of aspect ratio.

In order to compare the heat transfer in the entrance region, the flow is assumed to be hydrodynamically fully developed before entering into the channel and the problem becomes a thermal entry length problem because the hydrodynamic entry length is very short as defined as Eq. (8-6). In the thermal entry length region, the heat transfer process can be correlated in terms of the inverse Graetz number, where

$$G_z = \frac{\mathrm{Re\,Pr}}{x/D_h} \tag{8-39}$$

The thermal entry length in the large scale tube can be usually expressed as (Incropera et al., 2007),

$$Gz^{-1} \sim 0.05 \tag{8-40}$$

This expression or the entry length is found at the point when the Nusselt number results approach a constant value. From our heat transfer data, as shown in Figure 41, the entry length in the current micro-channel can be found and expressed as,

$$Gz^{-1} \sim 0.023 \tag{8-41}$$

It appears that the current heat transfer data does not have the same trend as in the large scale tube. This is attributed to the fact that the current micro-channel is not an ideal parallel plate channel. The correlation equation for parallel plates channel in the entry region (Naito, 1984) can be expressed as follows:

$$Nu_x = 0.461043Gz^{1/2}(1 + 6.35257Gz^{-1/2} - 33.4079Gz^{-1} + 419.158Gz^{-3/2} - 1537.84Gz^{-2}) \tag{8-42}$$

Comparison of the correlation equation for the local Nusselt numbers between the prediction for large scale channel and the experimental data for micro-channel with a channel height of 37 μm is made in Figure 41. The agreement in entrance region is very good except in the fully develop region the current data is slightly low due to the aspect ratio effect.

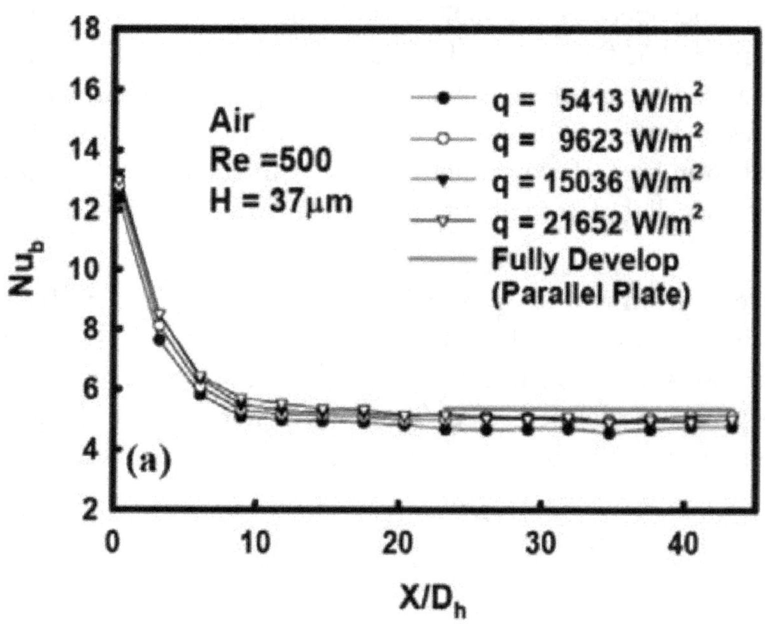

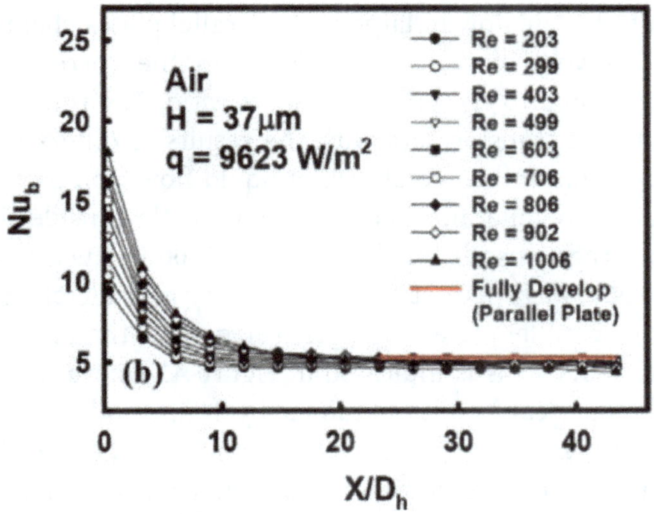

Fig. 40. Re-calculated results for local Nusselt number distributions under (a) different heating flux conditions and (b) different Reynolds number.

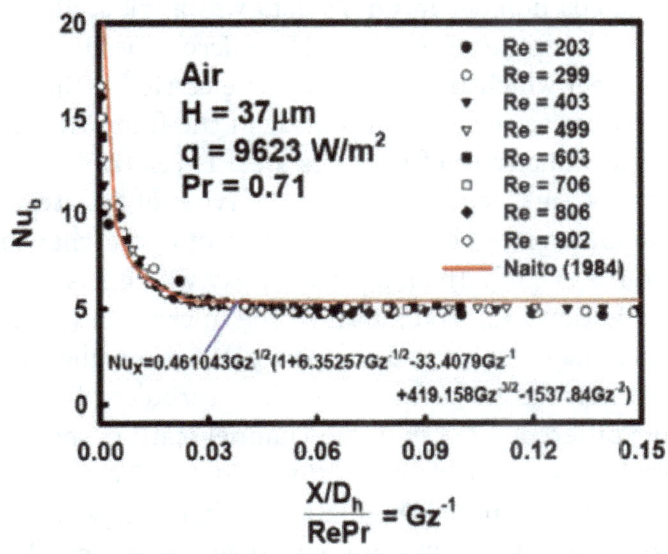

Fig. 41. Comparison of the local Nusselt numbers between the prediction for large scale channel and the experimental data.

Since some of the heat transfer data reported in the literature is the average heat transfer coefficient or the average Nusselt number, comparison with these data can be made when the average Nusselt number is defined as follows:

$$Nu_{ave} = \frac{1}{x}\int_0^x Nu_b \, dx \tag{8-43}$$

The average Nusselt numbers in the current micro-channel are obtained in the range of laminar flow and compared, as shown in Figure 42, with the correlation predicted by Naito for air flow in large scale parallel plates channel. The average Nusselt number is very close, but smaller than the correlation obtained by Naito. This is attributed to the effect of aspect ratio. The result is of interest to compare current heat transfer data with the results of others in micro or large scale channel. To avoid influence of EDL in liquid flow, comparison is limited to air flow in channels. In general, when Re > 2300, the flow inside channel becomes turbulent. The average Nusselt number for Re > 2300 is much larger than for Re < 2300 (Morini, 2004; Incropera et al., 2007). For large scale channel, the Nusselt number in the two parallel plates is, in general, greater than that in circular tube (Incropera et al., 2007). This is indicated in Figure 43 where the average Nusselt number of the numerical results obtained by Naito and experimental results obtained by this study for two parallel plates, and is higher than the correlation obtained by Hausen for circular tube. The heat transfer correlations obtained by Choi (1991) is for air flow in mcirotubes. Their data is much smaller than the correlation of Hausen. The much lower value of Nusselt number and its large increase with Reynolds number in Choi's data was attributed to the streamwise conduction of heat along the micro-tube. The micro-tubes used in the past were made of stainless steel which is convenient for electric heating, and is expected to cause large streamwise conduction of heat in the finite thickness of the tube since the thermal conductivity of stainless steel is relatively high and is in the range from 13 to 15 W/mK depending upon the types of the steel used. However, the streamwise conduction of heat along the current micro-channel wall has been minimized due to the use of a thick layer of epoxy which has a very low thermal conductivity. In addition, wall roughness in the micro-tubes of Choi is expected large enough which may result the discrepancy. However, the roughness of their micro-tubes is not reported. When one goes into turbulent flow region, the heat transfer data from different work in micro-channel scatters very much, but are all higher than those in large scale channel (Wu & Little, 1983; Yu et al., 1995; Morini, 2004; Kays et al., 2005). It is noted that data from both the work of Dittus-Boelter and Gnielinski (Rohsenow et al., 1985) is for air flow in large scale tubes. The higher heat transfer data in micro-channel can be attributed to wall roughness effect in the micro-tubes. It has been realized (Kays et al., 2005; Incropera et al., 2007) that wall roughness can significantly enhance turbulent heat transfer in the large scale tube. Consequently, wall roughness is anticipated to markedly improve turbulent heat transmission in micro-channels. Regrettably, the wall roughness of each microtube was not documented and is anticipated to be significantly elevated in the micro stainless steel tubes unless specific surface treatment is applied.

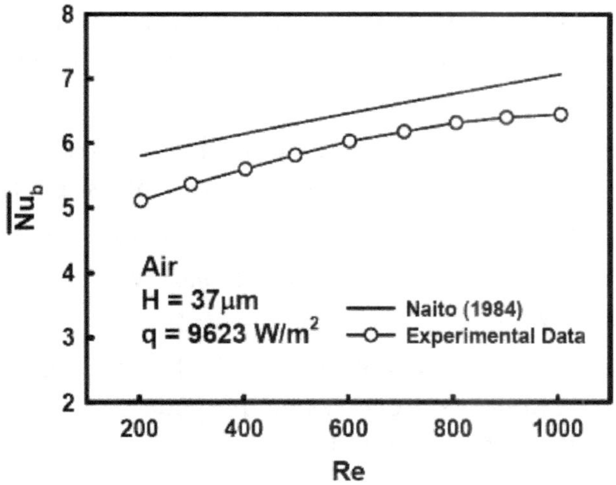

Fig. 42. Comparison of the average Nusselt numbers in the current micro-channel with the correlation predicted by Naito for air flow in large scale parallel plates channel.

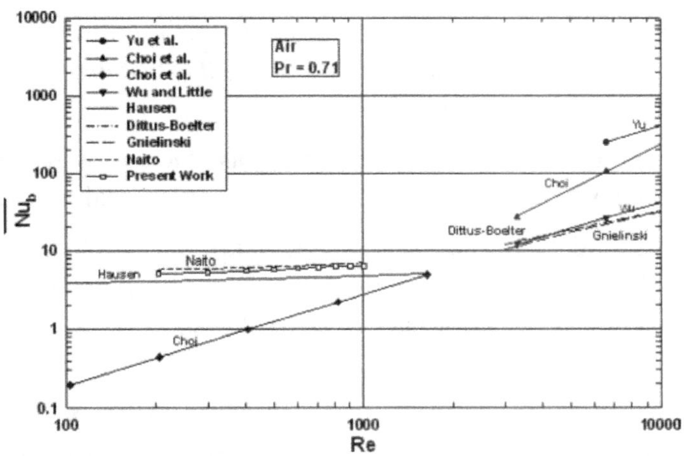

Fig. 43. Comparison of average Nusselt numbers between the current work and published data.

REFERENCES

Acosta, R. E., Muller, R.H., and Tobias, W. C. (1985). Transport processes in narrow (Capillary) channels, *AIChE Journal*, Vol. 31, pp.473–482.

Arkilic, E. B. and Schmidt, M. A. (1997). Gaseous slip flow in long microchannels, *Journal of Microelectromechanical System*, Vol. 6(2), pp.167-178.

Arshak, K., Morris, D., Arshak, A., Korostysnska, O., Jafer, E., Waldron, D., and Harris, J. (2006). Development of polymer-based sensors for integration into a wireless data aquisition system suitable for monitoring environmental and physiological processes, *Biomolecular Engineering*, Vol. 23, pp.253-257.

Arulanandam, S., and Li, D. (2000). Liquid transport in rectangular microchannels by electroosmotic pumping, *Colloids and Surfaces A: PhysicoChemical and Engineering Aspects*, Vol. 161, pp.89-102.

Bayraktar, T. and Pidugu, S. B. (2006). Review: Characterization of liquid flows in microfluidic systems, *Int. J. Heat Mass Transfer*, Vol. 49, pp.815-824.

Beskok, A., and Karniadakis, G. E. (1994) Simulation of heat and momentum transfer in complex microgeometries, *Journal of Thermophysics and Heat Transfer*, Vol. 8, No. 4, pp. 246-259.

Bhatnagar, P. L., Gross, E. P. and Krook, M. (1954) A Model for Collision processes in Gases. I. Small Amplitude processes in Charged and Neutral One-Component Systems, *Phys. Rev.*, Vol. 94, pp.511–525.

Bird, G. A. (1976a). *Molecular gas dynamics and the direct simulation of gas flows*, Clarendon Press, Oxford, pp. 183-207.

Bird, G. A. (1976b). *Molecular gas dynamics*, Clarendon Press, Oxford, pp. 110-117.

Brutin, D., Topin, F., and Tadrist, L. (2003) Experimental study of unsteady convective boiling in heated minichannels, *International Journal of Heat and Mass Transfer*, Vol. 46, pp. 2957-2965.

Chen, H. R. (2001). *Design and fabrication of the microchannel system for thermal fluid study*, PhD. Thesis, National Cheng Kung University, Tainan, Taiwan.

Chen, H. R., Gau, C., Dai, B. T. and Tsai, M. S. (2003a). A Novel Planarization process for Sacrificial Polysilicon Layers in a Micro-Thermal System, *Sensors and Actuators A: Physical*, Vol. 108/1-3, Nov. pp.86-90.

Chen, H. R., Gau, C., Dai, B. T. and Tsai, M. S. (2003b). A Monolithic Fabrication process for a Micro-Flow Heat Transfer Channel Suspended over an Air Layer with Arrays of Micro Sensors and Heaters, *Sensors and Actuators A: Physical*, Vol. 108/1-3, Nov. pp.81-85.

Chen, X. (1996). *Gas kinetics and its applications in heat transfer and flow*, TsingHua University Press, Beijing.

Cheng, P., Wang, G., and Quan, X. (2009). Recent work on boiling and condensation in microchannels, *ASME Journal of Heat Transfer*, Vol. 131, pp. 043211-043225.

Choi, S. B., Barron, R. F., and Warrington, R. O. (1991). Fluid flow and heat transfer in microtubes, Micromechanical Sensors, Actuators, and Systems, *ASME, DSC*, Vol. 32, pp. 123-134.

Choondal, B. S. and Suresh, V. G. (2000). A comparative analysis of studies on heat transfer and fluid flow in microchannels, *ProC. of the Micro-Scale Heat Transfer Conf.* pp.80-92.

Chu, W. K. H., Hsu, C. T., Wong, M., and Zohar, Y. (1994). Heat transfer in a microchannel flow, *ProCeedings of the 1994 IEEE Electron Devices Meeting*, Hong Kong, pp. 38-43.

Clark, S. K., and Wise, K. D. (1979). Pressure sensitivity in anisotropically etched thin- diaphragm pressure sensors, *IEEE Trans. Electron Devices*, ED-26, pp. 1887-1896.

Core, T. A., Tsang, W. K., and Sherman, S. J. (1993). Fabrication technology for an integrated surface micromachined sensor, *Solid State Technolog*, pp. 39-47.

Craighead, H. (2006). Future lab-on-chip technologies for interrogating individual molecules, *Nature,* Vol. 442 (27), pp.387-393.

Eckert, E. R. G., and Drake, R. M. (1972). *Analysis of heat and mass transfer*, McGraw-Hill, New York, pp.467-495.

Fan, Q., and Xue, H. (1998). Compressible effects in microchannel flows, *ProCeedings of the IEEE /CPMT Electronics Packaging Technology Conference*, pp.224-228.

George, M. (2006). The origins and the future of microfluidics, *Nature* Vol. 442, pp.368-373.

Giordani. N., Camberlein, L., Gaviot, E., Polet, F., Pelletier, N., and Bêche, B. (2007). Fast psychrometers as new SU-8 based microsystems, *IEEE Transactions on Instrumentation and Measurement*, Vol. 56, No. 1, pp.102-106, 2007.

Guckel, H. (1991). Surface micromachined pressure transducer, *Sens. Actuators A*, Vol. 28, pp. 133-146.

Guckel, H. et al. (1990). Microstructure sensors, *International Electron Devices Meeting*, pp. 613-616.

Guo, Z. and Li, Z. (2003). Size effect on microscale single-phase flow and heat transfer, *Int. J. Heat Mass Transfer* Vol. 46, pp.149–59.

Harley, J. C., Huang, Y. H., Bau, H., and Zemel, J. N. (1995). Gas flow in micro channels, *Journal of Fluid Mechanics*, Vol. 284, pp257-274.

Harms, T. M., Kazmierczak, M. J. and Gerner, F. M. (1999). Developing convective heat transfer in deep rectangular microchannels, *International Journal of Heat Fluid Flow*, Vol. 20, pp. 149–157.

Harms, T.M., Kazmierczak, M.J., Gerner, F.M., Holke, A., Henderson, H.T., Pilchowski, J. and Baker, K. (1997). Experimental investigation of heat transfer and pressure drop through deep microchannels in a (110) silicon substrate, *ProCeedings of ASME Heat Transfer Division*, ASME HTD, Vol. 351-1, pp. 347–357.

Hetsroni, G., Mosyak, A., Segal, Z., and Ziskind, G. (2002). A uniform temperature heat sink for cooling of electronic devices, *International Journal of Heat and Mass Transfer*, Vol. 45, pp.3275-3286.

Hetsroni, G., Mosyak, A. and Segal, Z. (2001). Nonuniform temperature distribution in electronic devices cooled by flow in parallel microchannels, *IEEE Transactions on components and packaging technologies*, Vol. 24, pp.17-23.

Hu, H. Y., Peterson, G. P., Peng, X. F. and Wang, B. X. (1998). Interface fluctuation propagation and superposition model for boiling nucleation, *International Journal of Heat and Mass Transfer*, Vol. 41, pp.3483-3490.

Incropera F. P., DeWitt D. P., Bergman T. L. and Lavine A. S. (2007). *Fundamentals of Heat and Mass Transfer* 6th Ed., New York, John Wiley & Sons Inc.

Jiang, P. X., Fan, M. H., Si, G. S. and Ren, Z. P. (1997). Thermal-hydraulic performance of small scale micro-channel and porous-media heat exchangers, *International Journal of Heat and Mass Transfer*, Vol. 44, pp.1039–1051.

Jiang, L., Wong, M. and Zohar, Y. (1999). Phase change in microchannel heat sinks with integrated temperature sensors, *J. MEMS,* Vol., 8, pp.358-365.

Jiang, L., Wong, M., and Zohar, Y. (2001). Forced convection boiling in a microchannel heat sink, *Journal of Microelectromechanical System*, Vol. 10, pp.80-87.

Jiang, X. N., Zhou, Z. Y., Yao, J., Li, Y. and Ye, X. Y. (1995). Micro-fluid flow in micro channel, *ProCeedings of the 8th International Conference on Solid-State Sensors and Actuators, and Eurosensors IX (Transducers '95)*, StoCkholm, Sweden, pp.317-320, June 25-29.

Kanda, Y. (1982). A graphical representation of the piezoresistive coefficients in silicon, *IEEE Transactions on electron devices*, Vol. ED-29, No. 1, pp. 64-70.

Kandlikar, S. G. (2006). Nucleation characteristics and stability considerations during flow boiling in microchannels, *Experimental Thermal and Fluid Science*, Vol. 30, pp.441-447.

Kays, W. M., Crawford, M. E. and Weigand, B. (2005). *Convective Heat and Mass Transfer*, 4th ed., McGraw-Hill, New York.

Kennard, E. H. (1938) Kinetic theory of gases-with an introduction to statistical mechanics, McGraw- Hill, Inc., New York, pp.291-337.

Khantuleva, T. A. (1982). Nonlocal hydrodynamical models of gas flows in the transition regime. In: *Rarefied gas dynamics 1*, Belotserkovskii, O. M., Kogan, M. N., Kutateladze, S. S. and Rebrov, A. K. Ed., Plenum Press, New York, pp.229-235.

Khrustalev, D., and Faghri, A. (1993). Thermal analysis of a micro heat pipe, *Heat Pipes and Capillary Pumped Loops*, ASME, HTD, Vol. 236, pp.19-30.

Kim, C. J., Kim, J. Y., and Sridharan, B. (1998). Comparative evaluation of drying techniques for surface micromachining, *Sens. Actuators A*, Vol. 64, pp.17-26.

Kleinstreuer, C. and Koo, J. (2004). Computational Analysis of Wall Roughness Effects for Liquid Flow in Micro-Conduits, *J. Fluid Engineering,* Vol. 126, pp.1-9.

Knudsen, M. (1934). *The kinetic theory of gases*, Methuen, London.

Ko, H. S. and Gau, C. (2011). Micro scale thermal fluid transport process in a microchannel integrated with arrays of temperature and pressure sensors, *Microfluidics and Nanofluidics*, (in press).

Ko, H. S. and Gau, C. (2009) Bonding of a complicated polymer microchannel system for study of pressurized liquid flow characteristics with electric double layer effect, *Journal of Micromechanics and Microengineering*, Vol. 19, art no. 115024.

Ko, H. S. (2009). *Fabrication and development of micro fluidic system with embedded micro pressure and temperature sensors for study of thermal and fluid flow properties*, Ph D. thesis, National Cheng Kung University, Tainan, Taiwan.

Ko, H. S., Liu, C. W., Gau, C. and Jeng, D. Z. (2008). Flow characteristics in a microchannel system integrated with arrays of micro pressure sensors using polymer material, *Journal of Micromechanics and Microengineering*, Vol. 18, art no. 075016.

Ko, H. S., Liu, C. W., and Gau, C. (2007). Micro pressure sensor fabrication without problem of stiction for a wider range of measurement, *Sensors and Actuators A: Physical,* Vol. 138, pp.261–267.

Ko, H. S., Liu, C. W., and Gau, C. (2007). A novel fabrication for pressure sensor with polymer material and evaluation of Its performance, *Journal of Micromechanics and Microengineering*, Vol. 17, pp.1640-1648.

Kohl, M.J., Abdel-Khalik, S. I., Jeter, S. M. And Sadowski, D. L. (2005). An experimental investigation of microchannel flow with internal pressure measurements, *International Journal of Heat and Mass Transfer*, Vol. 48, pp.1518-1533.

Komvopoulos, K. (1996). Surface engineering and microtribology for micro-electromechanical systems, *Wear*, Vol. 200(1-2), pp.305-327.

Koo, J. and Kleinstreuer, C. (2003). Liquid flow in microchannels: experimental observations and computational analyses of microfluidics effects, *J. Micromech. Microeng.* Vol. 13 pp.568-579.

Kundt, A. and Warburg, E. (1875). On the friction and thermal conductivity in rerefied gases, *Phil. Mag.*, Vol. 50, pp.53.

Lalonde, P., Colin, S., and Caen, R. (2001). Mesure de debit de gaz dans les microsystemes, *Mechanical Industry*, Vol. 2, pp.355–362.

Lee, G. B., Lin, C. H. and Chang, G. L. (2003). Micro flow cytometers with buried SU-8/ SOG optical waveguides, *Sensors and Actuators: Physical A*, Vol. 103, pp.165-170.

Lee, M., Wong, Y. Y., Wong, M. and Zohar, Y. (2003). Size and shape effects on two-phase flow patterns in microchannel forced convection boiling, *Journal of Micromechanics and Microengineering*, Vol. 13, pp.155-164.

Lee, W. Y., Wong, M. and Zohar, Y. (2002). Pressure loss in constriction microchannels, *Journal of MicroElectroMechanicalSystem*, Vol. 11, No. 3, pp.236-244.

Legtenberg, R., Elders, J. and Elwenspoek, M. (1993). Stiction of surface microstructure after rinsing and drying: model and investigation of adhesion mechanisms, *ProC. 7th Int. Conf. Solid-State Sensors and Actuators (Transducer'93)*, Yogohama, Japan, pp.198-201.

Li, D. (2001). Electro-viscous effects on pressure-driven liquid flow in microchannels, *Colloids and Surface A*, Vol. 195, pp.35-57.

Li, X., Lee, W. Y., Wong, M. and Zohar, Y. (2000) Gas flow in constriction microdevices, *Sensor and Actrators A*, Vol. 83, pp.277-283.

Li, Z. X., Du, D. X. and Guo, Z. Y. (2003). Experimental study on flow characteristics of liquid in circular microtubes, *Microscale Thermophysicals Engineering*, Vol. 7, pp.253–265.

Li, Z. X., Du, D. X. and Guo, Z. Y. (2000). Experimental study on flow characteristics of liquid in circular microtubes, *ProCeedings of International Conference on Heat Transfer and Transport Phenomena in Microscale*, Begell House, New York, USA, pp.162–168.

Lin, L. and Pisano, A. P. (1991). Bubble forming on a micro line heater, *Micromechanical Sensors, Actuators, and Systems*, ASME, DSC, Vol. 32, pp.147-164.

Lin, L., Udell, K. S. and Pisano, A. P. (1994). Liquid-vapor phase transition and bubble formation in micro structures, *Thermal Science Engineering*, Vol. 2, pp.52-59.

Liu, C. W. (2004). *Fabrication development for micro channel system by MEMS technology with measurements of the inside thermal transport* process, Ph D. thesis, National Cheng Kung University, Tainan, Taiwan.

Liu, C. W., Gau, C. and Dai, B. T. (2004). Design and Fabrication Development of a Micro Flow Heated Channel with Measurements of the inside Micro-Scale Flow and Heat Transfer process, *Biosensors and Bioelectronics*, Vol. 20, No. 1, pp.91-101.

Liu, J. and Tai, Y. (1995). MEMS for pressure distribution studies of gaseous flows in microchannels, *An Investigation of Micro Structures, Sensors, Actuators, Machines, and Systems*, pp. 209-215.

Liu, T. Y., Li, P.L., Liu, C.W., Gau, C. (2011). Boiling flow characteristics in microchannels with very hydrophobic surface to super-hydrophilic surface, *International Journal of Heat and Mass Transfer*, Vol. 54, pp.126-134.

Madou, M. J. (2002). *Fundamentals of microfabrication*, 2nd ed., CRC Press, pp.276-277.

Mala, G. M. and Li, D. (1999). Flow characteristics of water in microtubes, *International Journal of Heat and Fluid Flow*, Vol. 20, pp.142-148.

Mala, G. M., Yang, C. and Li, D. (1998). Electrical double layer potential distribution in a rectangular microchannel, *Colloids and Surfaces A: PhysicoChemical and Engineering Aspects*, Vol. 135, pp.109-116.

Mala, G. M., Li, D. and Dale, J. D. (1997). Heat transfer and fluid flow in microchannels, *International Journal of Heat and Mass Transfer*, Vol. 40, No. 13, pp.3079-3088.

Mala, G. M., Li, D., Werner, C., Jacobasch, H. J. and Ning, Y. B. (1997). Flow characteristics of water through a microchannel between two parallel plates with electrokinetic effects, *International Journal of Heat and Fluid Flow*, Vol. 18, No. 5, pp.489-496.

Mala, G. M., Li, D. and Dale, J. D. (1996). Heat transfer and fluid flow in microchannels, *Microelectromechanical Systems (MEMS)*, ASME, DSC, Vol. 59, pp.127-136.

Martin, B. W., Hagena, O. F., and Schomburg, W. K. (1998). Strain gauge pressure and volume-flow transducers made by thermoplastic molding and membrane transfer, *ProC. MEMS '98*, Jan. 25-29, Heidelberg, Germany (1998), pp. 361-366.

Mason, W. P. (1969). Use of solid-state transducers in mechanics and acoustics, *J. Aud. Eng. SoC.*, Vol. 17, pp.506-511.

Maxwell, J. C. (1890). On the condition to be satisfied by a gas at the surface of a solid body, *The Scientific Papers of James Clerk Maxwell*, Vol. 2, pp.704, Cambridge University Press, London.

Morini, G. L. (2004). Single-phase convective heat transfer in microchannels: a review of experimental results, *International Journal of Thermal Science,* Vol. 43, pp.631-651.

Monat, C., Domachuk, P., and Eggleton, B. J. (2007). Integrated optofluidics: A new river of light, *Nature*, Vol. 1, pp. 106-114.

Naito, E. (1984). Laminar heat transfer in the entrance region between parallel plates - the case of uniform heat flux, *Heat transfer, Japanese research*, Vol. 13(3), pp.92-106.

Obermeier, E. and Kopystynski, P. (1992). Polysilicon as a material for microsensor applications, *Sensors and Actuators A*, Vol. 30, pp.149-155.

Obot, N. T. (2002). Toward a better understanding of friction and heat/mass transfer in microchannel – A literatures review, *Microscale Thermophysical Engineering*, Vol. 6, pp.155-173.

Papautsky, I., Gale, B. K., Mohanty, S., Ameel, T. A. and Frazier, A. B. (1999). Effect of rectangular microchannel aspect ratio on laminar friction constant, SPIE, Vol. 3877, pp.147-158.

Peake, E. R., Zias, A. R. and Egan, J. V. (1969). Solid-state digital pressure transducer, *IEEE Trans. on Electron Devices*, Vol. 16, pp.870-876.

Pelletier, N., Bêche, B., Tahani, N., Zyss, J., Camberlein, L. and Gaviot, E. (2007). SU-8 waveguiding interferometric micro-sensor for gage pressure measurement, *Sensors and Actuators: Physical A*, Vol. 135, pp.179-184.

Peng, X. F., Hu, H. Y. and Wang, B. X. (1998a). Boiling nucleation during liquid flow in microchannels, *International Journal of Heat and Mass Transfer*, Vol. 41, pp.101-106.

Peng, X. F., Hu, H. Y. and Wang, B. X. (1998b), Flow boiling through V-shape microchannels, *Experimental Heat Transfer*, Vol. 11, pp.87-90.

Peng, X. F., Liu, D., Lee, D. J. and Yan, Y. (2000). Cluster dynamics and fictitious boiling in microchannels, *International Journal of Heat and Mass Transfer*, Vol. 43, pp.4259-4265.

Peng, X. F., and Peterson, G. P. (1996). Convective heat transfer and flow friction for water flow in microchannel structures, *International Journal of Heat and Mass Transfer*, Vol. 39, No. 12, pp.2599-2608.

Peng, X. F., Peterson, G. P. and Wang, B. X. (1994). Heat transfer characteristic of water flowing through microchannels, *Experimental Heat Transfer*, Vol. 7, pp.265-283.

Peng, X. F., Peterson, G. P. and Wang, B. X. (1996). Flow boiling of binary mixtures in microchanneled plates, *International Journal of Heat and Mass Transfer*, Vol. 39, No. 6, pp.1257-1264.

Peng, X. F., Tien, Y. and Lee, D. J. (2001). Bubble nucleation in microchannels: statistical mechanics approach, *International Journal of Heat and Mass Transfer*, Vol. 44, pp. 2957-2964.

Peng, X. F. and Wang, B. X. (1993). Forced convection and flow boiling heat transfer for liquid flowing through micro channel, *International Journal of Heat and Mass Transfer*, Vol. 36, No. 14, pp.3421-3427.

Peng, X. F. and Wang, B. X. (1994). Evaporation space and fictitious boiling for internal evaporation of liquid, *Science Foundation in China*, Vol. 2, pp.55-5.

Peng, X. F. and Wang, B. X. (1998a), Forced-convection and boiling characteristics in microchannels, *ProCeedings of the 11th International Heat Transfer*, Vol. 1, pp.371-390.

Peng, X. F. and Wang, B. X. (1998b). Boiling characteristics in microchannels/micro structures, *The 11th Int. Symposium on Transport Phenomena*, ISTP-11, Hsinchu, Taiwan, pp 485-491.

Peng, X. F., Wang, B. X., Peterson, G. P. and Ma, H. B. (1995). Experimental investigation of heat transfer in flat plates with rectangular microchannels, *International Journal of Heat and Mass Transfer*, Vol. 38, No. 1, pp.127-137.

Pong, K. C., Ho, C. M., Liu, J. and Tai, Y. C. (1994). Nonlinear pressure distribution in uniform microchannel, *Application of Microfabrication to Fluid Mechanics*, ASME FED, Vol. 197, pp.51-56.

Pfahler, J., Harley, J. and Bau, H. (1991). Gas and liquid flow in small channels, *Micromechanical Sensors Actuators, and Systems*, ASME, DSC, Vol. 32, pp.49-59.

Pfahler, J., Harley, J., Bau, H. and Zemel, J. (1990a). Liquid transport in micron and submicron channels, *Sensors and Actuators A*, Vol. 21/23, pp.431-434.

Pfahler, J., Harley, J., Bau, H. and Zemel, J. (1990b). Liquid and gas transport in small channels, *Microstructures, Sensors, and Actuators*, ASME, DSC, Vol. 19, pp.149-157.

Pfund, D., Rector, D., Shekarriz, A., Popescu, A. and Welty, J. (2000). Pressure drop measurements in a microchannel, *Fluid Mechanics and Transport Phenomena*, Vol. 46, pp.1496-1507.

Pfund, D. A., Shekarriz, A., Popescu, A. and Welty J. R. (1998). Pressure drops measurements in microchannels, *ProCeedings of MEMS*, ASME DSC, Vol. 66, pp.193- 198.

Phan, H. T., Caney, N., Marty, P., Colasson, S. and Gavillet, J. (2009). Surface wettability control by nanoCoating: The effects on pool boiling heat transfer and nucleation mechanism, *International Journal of Heat and Mass Transfer*, Vol. 52, pp.5459–5471.

Qu, W., Mala, G. M. and Li, D. (2000a). Pressure-driven water flows in trapezoidal silicon microchannels, *International Journal of Heat and Mass Transfer*, Vol. 43, pp.353-364.

Qu, W., Mala, G. M. and Li, D. (2000b). Heat transfer for water flow in trapezoidal silicon microchannels, *International Journal of Heat and Mass Transfer*, Vol. 43, pp.3925-3936.

Qu, W. and Mudawar, I. (2002). Experimental and numerical study of pressure drop and heat transfer in a single-phase micro-channel heat sink, *International Journal of Heat and Mass Transfer*, Vol. 45, pp.2549-2565.

Qu, W. and Mudawar, I. (2003a). Flow boiling heat transfer in two-phase micro-channel heat sinks-I. Experimental investigation and assessment of correlation methods, *International Journal of Heat and Mass Transfer*, Vol. 46, pp.2755-2771.

Qu, W. L. and Mudarwar, I. (2003b). Measurement and prediction of pressure drop in two- phase micro-channel heat sinks, *International Journal of Heat and Mass Transfer*, Vol. 46, pp. 2737-2753.

Ren, L. and Li, D. (2001). Electroosmotic flow in heterogeneous microchannels, *Journal of Colloid and Interface Science*, Vol. 243, pp.255-261.

Ren, L., Li, D. and Qu, W. (2001a). Electro-viscous effects on liquid flow in microchannels, *Journal of Colloid and Interface Science*, Vol. 233, pp.12-22.

Ren, L., Qu, W. and Li, D. (2001b). Interfacial electrokinetic effects on liquid flow in microchannels, *International Journal of Heat and Mass Transfer*, Vol. 44, pp.3125-3134.

Ribeiro, J. C., Minas, G., Turmezei, P., Wolffenbuttel, R. F. and Correia, J. H. (2005). A SU-8 fluidic microsystem for biological fluids analysis, *Sensors and Actuators: Physical A*, Vol. 123-124, pp.77-81.

Rohsenow, W. M., Hartnett, J. P. and Ganic, E. N. (1985). *Handbook of Heat Transfer Fundamentals* 2nd ed., McGraw-Hill, New York, pp.7-49.

Rostami, A. A., Mujumdar, A. S. and Saniei, N. (2002). Flow and heat transfer for gas flowing in microchannel: a review, *Heat and Mass Transfer,* Vol. 38, pp.359-367.

Sharp, K. V., Adrian, R. J. and Beebe, D. J. (2000). Anomalous transition to turbulence in microtubes, *ProCeedings of International Mechanical Engineering Cong. Expo.*, 5th Micro-Fluidic Symp., Orlando, FL, pp. 150–158.

Shen, C. H. and Gau, C. (2004). Thermal chip fabrication with arrays of sensors and heaters for micro-scale impingement cooling heat transfer analysis and measurements, *Biosensors and Bioelectronics*, Vol. 20, No. 1, pp.103-114.

Shen, C. H. (2003). *Thermal chip fabrication with arrays of sensors and heaters with analysis and measurements of micro scale impingement cooling flow and heat transfer*, Ph D. Thesis, National Cheng Kung University, Tainan, Taiwan.

Shih, J. C., Ho, C. M., Liu, J. and Tai, Y. C. (1996). Monatomic and polyatomic gas flow through uniform microchannels, *ASME DSC*, Vol.59, pp.197–203.

Shirinov, A. V. and Schomburg, W. K. (1996). Polymer pressure sensor from PVDF, *ProC. Eurosensor*, Göteborg, Schweden Sep. 17th-20th, pp.84-85.

Smoluchowski, M. Von (1910). Uber den temperaturesprung bei warmeleitung in gasen, *Anz. Wiss. Krakau ser. A*, Vol. 5, pp.129.

Sugiyama, S., Shimaoka, K. and Tabata, O. (1991). Surface micromachined micro-diaphragm pressure sensors, *ProC. 6th Int. Conf. Solid-State Sensors and Actuators (Transducer'91)*, pp.188-191.

Sze, M. (1994). *Semiconductor Sensors*, John Wiley and Sons, Inc., New York, pp.17-80.

Sze, M. (2002). *Semiconductor Devices, Physics and Tecnology*, 2nd ed., John Wiley and Sons, Inc., New York, pp.171-200.

Takuto, A., Soo, K. M., Hiroshi, I. and Kenjiro, S. (2000). An experimental investigation of gaseous flow characteristics in microchannels, *ProCeedings of the International Conference on Heat Transfer and Transport Phenomena in Microscale*, Banff, Canada, pp.155-161.

Takao, K. (1961). Rarefied gas flow between two parallel plates, in: *Rarefied gas dynamics*, Talbot, L., Ed., Academic Press, New York, Section 1, pp.465-473.

Tao, C. P. and Mahulikar, S. P. (1998). Laminar convection behavior in microchannels in conventional thermal entry length and beyond, *ProCeedings of the IEEE/CPMT Electronics Packaging Technology Conference*, pp.126-132.

Touloukina, Y. S. (1970-1979). *Thermophysical properties of matter: the TPRC data series; a compressive compilation of data*, IFI/Plenum, New-York.

Trung, N., Nguyen, N. and Wu, Z. (2005). Micromixer-a review, *J. Micromech. Microeng.*, Vol. 15, pp.R1-R16.

Tsien, H. S. (1948). Superaerodynamics, the mechanics of rarefied gases, *Journal of Aeronautic SoCiety*, Vol. 13, pp.653-664.

Tufte, O. N., Chapman, P. W. and Long, D. (1962). Silicon diffused-element piezoresistive diaphragms, *Journal of Applied Physics*, Vol. 33, pp.3322-3327.

Turner, S. E., Sun, H., Faghri, M. and Gregory, O. J. (2001). Compressible gas flow through smooth and rough microchannels, *ProCeedings of IMECE*, New York, USA, HTD- 24145.

Vilkner, T., Janasek, D. and Manz, A. (2004). Micro total analysis systems: recent developments, *Anal. Chem.*, Vol. 76, pp.3373-3386.

Wadsworth, D. C., Erwin, D. A. and Muntz, E. P. (1993). Transient motion of a confined rarefied gas due to wall heating or cooling, Journal of Fluid Mechanics, Vol. 248, pp.219-235.

Wang, G. D., Cheng, P. and Bergles, A. E. (2008). Effects of inlet/outlet configurations on flow boiling instability in parallel microchannels, *International Journal of Heat and Mass Transfer*, Vol. 51, pp.2267-2281.

Wang, G. D., Cheng, P. and Wu, H. Y. (2007). Unstable and stable flow boiling in parallel microchannels and in a single microchannel, *International Journal of Heat and Mass Transfer*, Vol. 50, pp.4297-4310.

Webb, R. L. and Zhang, M. (1998). Heat transfer and friction in small diameters channels, *Microscale Thermo physics Engineering*, Vol. 2, pp. 189–202. White, F. M. (1991). Viscous fluid flow, McGraw-Hill, New York.

Wu, H. Y. and Cheng, P. (2003a). Friction factors in smooth trapezoidal silicon microchannels with different aspect ratios, *International Journal of Heat and Mass Transfer*, Vol. 46, pp. 2519–2525.

Wu, H. Y. and Cheng, P. (2003b). Visualization and measurements of periodic boiling in silicon microchannels, *International Journal of Heat and Mass Transfer*, Vol. 46, pp.2603-2614.

Wu, H. Y. and Cheng, P. (2003c). Liquid/two-phase/vapor alternating flow during boiling in microchannels at high heat flux, *International Communications in Heat and Mass Transfer*, Vol. 30, pp.295-302.

Wu, H. Y. and Cheng, P. (2004). Boiling instability in parallel silicon microchannels at different heat flux, *International Journal of Heat and Mass Transfer*, Vol. 47, pp.3631- 3641.

Wu, P. and Little, W. H. (1983). Measurement of friction factor for the flow of gases in very fine channels used for micro-miniature Joule-Thompson refrigerators, *Cryogenics*, pp 272-277.

Xu, B., Ooi, K. T., Wong, N. T. and Choi, W. K. (2000). Experimental investigation of flow friction of liquid flow in microchannels, *Int Comm. Heat Mass Transfer*, Vol. 27, pp.1165-1176.

Xu, B., Ooi, K. T., Wong, N. T., Liu, C. Y. and Choi, W. K. (1999). Liquid flow in microchannels, *ProCeedings of the 5th ASME/JSME Joint Thermal Engineering Conference*, San Diego, CA, pp.150–158.

Yang, C. and Li, D. (1997). Electrokinetic effects on pressure-driven liquid flows in rectangular microchannels, *Journal of Colloid and Interface Science*, Vol. 194, pp.95- 107.

Yang, C., and Li, D. (1998). Analysis of electrokinetic effects on the liquid flow in rectangular microchannels, *Colloids and Surface A: PhysicoChemical and Engineering Aspects*, Vol. 143, pp.339-353.

Yang, C., Li, D. and Masliyah, J. H. (1998). Modeling forced liquid convection in rectangular microchannels with electrokinetic effects, *International Journal of Heat and Mass Transfer*, Vol. 41, pp.4229-4249.

Yen, S. M. (1971). Monte carlo solutions of nonlinear Boltzmann equation for problems of heat transfer in rarefied gases, *International Journal of Heat and Mass Transfer*, Vol. 14, pp. 1865-1869.

Yu, D., Warrington, R.O., Barron, R. and Ameel, T. (1995). An experimental and theoretical investigation of fluid flow and heat transfer in microtubes, *ProCeedings of ASME/JSME Thermal Engineering Joint Conference*, Maui, pp.523–530.

Ziering, S. (1961). Plane Poiseuille Flow, in: *Rarefied gas dynamics*," Talbot, L., Ed., Academic Press, New York, Section 1, pp.451-463.

Zohar, Y., Lee, Y. K., Lee, W. Y., Jiang, L. and Tong, P. (2002). Subsonic gas flow in a straight and uniform microchannel, *J. Fluid Mech.* Vol. 472, pp.125-151.

6 Chapter THERMAL CONDUCTION IN MOLECULAR CRYSTALS

1. INTRODUCTION: AN OVERVIEW

This brief overview does not aim to encompass all existing information regarding the thermal conductivity of molecular crystals, particularly at low temperatures (below 20K). For such kind of information see, for example, Batchelder, 1977; Gorodilov et al., 2000; Jezowski et al., 1997; Ross et al., 1974; Stachowiak et al., 1994. The goal of this paper consists in presentation and generalization of the new experimental results and theoretical models, accumulated over the past 2-3 decades, to a certain extent changing existing view about the heat transfer in crystals. Quite recently, it was not doubted that high-temperature thermal conductivity of molecular crystals is proportional to the inverse temperature, $\Lambda \propto 1/T$. It was based on both the experimental data and assumptions being evident at first sight from which this dependence followed. In simple kinetic model, the phonon thermal conductivity can be represented as $\Lambda = 1/3 Cvl$, where C (the heat capacity) and v (the sound velocity) can be considered to be constant at $T \geq \Theta_D$, and averaged phonon mean-free path l is inversely proportional to the temperature. More precise expression (see, for example, Berman, 1976; Slack, 1979) can be written in the form:

$$\Lambda = K \frac{ma\Theta_D^3}{\gamma^2 T}, \tag{1}$$

where m is the average atomic (molecular) mass; a^3 is the volume per atom (molecule); $\gamma = -(\partial\, ln\Theta_D / \partial\, lnV)_T$ is the Grüneisen parameter, and K is a structure factor. In time, data on the deviation from $1/T$ dependence has accumulated, and in a number of cases some ideas qualitatively explaining the observed behaviour of thermal conductivity have been proposed. The problem has been, however, that the theory predicts the $1/T$ law at the constant volume of the sample, whereas the measurements were carried out at constant pressure. In this case, thermal expansion, been usually rather essential at high temperatures (the molar volume of molecular crystals may change up to 10-20% in the temperature interval from zero and up to the melting temperature) leads, as a rule, to additional decrease of Λ with rise of temperature. Moreover, in many cases, the phonons are not the only excitations determining the heat transfer and scattering process. The dependence of the thermal conductivity on the molar volume can be described using Bridgman's coefficient:

$$g = -(\partial \ln \Lambda / \partial \ln V)_T , \qquad (2)$$

It follows from Eqs. (1) and (2) that for crystals:

$$g = 3\gamma + 2q - 1/3 , \qquad (3)$$

were $q=(\partial \ln\gamma/\partial \ln V)_T$. Ordinarily, it is assumed that $\gamma \propto V$ and the second Grüneisen coefficient $q \cong 1$ (Slack, 1979; Ross et al., 1984). Taking into account that $\gamma \cong 2 \div 3$ for a number of simple molecular crystals (Manzhelii et al., 1997) it is expected that $g \cong 8 \div 11$ and $\Lambda \propto V8 \div 11$. It means that 1% change in volume may result in 8-11% change in thermal conductivity. Data measured at saturated vapour and atmosphere pressures can be considered as equivalent because the difference between them is much smaller than accuracy of experiment and they will be further denoted as isobaric ($P \cong 0$, MPa) data.

Constant-volume investigations are possible for molecular solids having a comparatively large compressibility coefficient. Using a high-pressure cell, it is possible to grow a solid sample of sufficient density. In subsequent experiments it can be cooled with practically unchanged volume, while the pressure in the cell decreases. In samples of moderate densities the pressure drops to zero at a certain characteristic temperature $T0$ and the isochoric condition is then broken; on further cooling, the sample can separate from the walls of the cell. In the case of a fixed volume, melting of the sample occurs in a certain temperature interval and its onset shifts towards higher temperatures as density of samples increases (For more experimental details see Konstantinov et al., 1999).

As the temperature increases, phonon scattering processes intensify, the mean-free path length l decreases and it may approach to the lattice parameter. The question of what occurs when the phonon mean-free path becomes comparable to the lattice parameter or its own wavelength is one of the most intriguing problems in the thermal conductivity of solids (see, for example, Auerbach & Allen, 1984; Feldman et al., 1993; Sheng et al., 1994). According to preferably accepted standpoint, in this case the vibrational modes assume a "diffusive" character, but the basic features of the kinetic approach retain their validity. Some progress in the description of the heat transport in strongly disordered materials has come about through the concept of the minimum thermal conductivity Λmin (Slack, 1979; Cahill et al., 1992), which is based on the picture where the lower limit of the thermal conductivity is reached when the heat is being transported through a random walk of the thermal energy between the neighboring atoms or molecules vibrating with random phases. In this case Λmin can be written as the following sum of three Debye integrals:

$$\Lambda_{min} = \left(\frac{\pi}{6}\right)^{1/3} k_B n^{2/3} \sum_i v_i \left\{ \left(\frac{T}{\Theta_i}\right)^2 \int_0^{\Theta_i/T} \frac{x^3 e^x}{\left(e^x - 1\right)^2} dx \right\}, \tag{4}$$

The summation is taken over three (two transverse and one longitudinal) sound modes with the sound speeds υ_i; Θ_i is the Debye cutoff frequency for each polarization expressed in degrees *K*; $\Theta_i = v_i(\hbar/k_B)\left(6\pi^2 n\right)^{1/3}$; *n* is the number density of atoms or molecules. Although no theoretical justification exists as yet for this picture of the heat transport, the evidence for its validity has been obtained on a number of amorphous solids in which the high temperature thermal conductivity has been found to agree with the value predicted by this model. Indirect evidence has also been obtained in measurements of the thermal conductivity of highly disordered crystalline solids, in which no thermal conductivity smaller than that predicted by this model seems to have ever been observed (Cahill et al., 1992). It is evident, that thermal conductivity approaches its lower limit Λ_{min} in amorphous solids and strongly disordered crystals (Auerbach & Allen, 1984; Cahill et al., 1992; Sheng et al., 1994). This raises the question whether or not the three-phonon scattering processes in themselves may result in Λ_{min} in perfect crystals with rise of temperature.

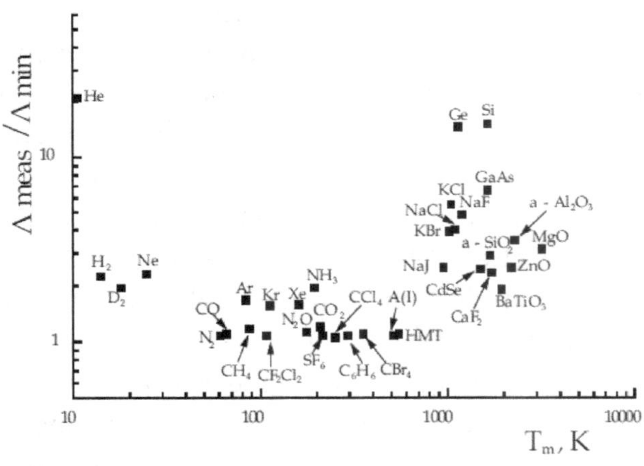

Fig. 1. The ratio $\Lambda_{meas}/\Lambda_{min}$ immediately below the corresponding melting temperatures (T_m) versus T_m for crystals with different types of chemical bonds.

To find an answer let us compare the measured thermal conductivity Λ_{meas} of a number of crystals with different types of chemical bonds and the lower limit to thermal conductivity Λ_{min} at the corresponding melting temperatures *Tm* (see Fig. 1). It is evident that the ratio $\Lambda_{meas}/\Lambda_{min}$ increases as the crystal bond becomes stronger. In van-der-Waals-type crystals $\Lambda_{meas}/\Lambda_{min} \cong 1.5 \div 2$, while in the crystals with diamond-type structure it is of the order of $10 \div 20$, i.e. van-der-Waals-type crystals are the most suitable objects for observing the thermal conductivity "minimum" due to umklapp processes only and this will be demonstrated further.

Crystals containing molecules or molecular ions are more complicated than crystals containing only atoms and ions, since the former possess translational, orientational and intramolecular degrees of freedom. An important common feature of simple molecular crystals is that in the condensed phases the intermolecular forces are much weaker than the intramolecular ones, so that the molecular parameters remain close to those in the gas. As a rule, the intramolecular forces and associated intramolecular vibration frequencies ($\sim 1000 cm^{-1}$) exceed by an order of magnitude the intermolecular ones (the corresponding lattice-mode frequencies are below $\sim 1000 cm^{-1}$). Such a large difference between the two types of frequencies makes it possible to safely regard the respective types of motion as independent. So as far as the lattice vibrations are concerned, the molecules can be treated as rigid bodies. In such an approximation each molecule participates in two types of motion: translational, when the molecular center of mass shifts, and rotational, when the center of mass rests. Many features in the dynamics of the simple molecular solids are related to the rotational motion of the molecules (Parsonage & Stavaley, 1972). At very low temperature, the structure of a crystal is perfectly ordered and the molecules can perform only small amplitude translational vibrations at the lattice sites and oscillations around selected axes (libration) in a manner so that the motion of neighboring molecules is correlated and collective translational and orientational excitations (phonons and librons) propagate throw the crystal. Calculation of anharmonic effects show that translational vibrations are characterized by relatively small amplitudes while the amplitude of librational vibrations in molecular crystals is sufficiently large even at $T \cong 0$, so that the harmonic theory can hardly claim to give more than qualitative picture of the librational motion (Briels et al., 1985; Manzhelii et al., 1997). As the temperature increases, the rotational motion may, in principle, pass through the following stages depending on the relationship between the central and anisotropic forces: enlargement of the libration amplitudes, the appearance of jump-like reorientations of the molecules, increase of the frequency of reorientations, hindered rotation of the molecules, and, finally, nearly free rotation of the molecules. In the last two cases a phase transition takes place, as a rule, before the crystal melts, giving rise to a structure in which translational long-range order is preserved while the orientational order is lost. It is a characteristic property of crystals consisting of high-symmetry "globular" molecules like CH_4, N_2, adamantane ($C_{10}H_{16}$) or, in some degree, of "cylindrical" molecules like benzene, C_2F_6 or long-chain n-alkanes. They form high-temperature "plastic" or orientationally-disordered (commonly called ODIC: Orientational Disorder In Crystals) phases in which the rotational motion of molecules resembles their motion in the liquid state (Parsonage & Stavaley, 1972). In crystals consisting of molecules of a lower symmetry the long-range orientational order is preserved, as a rule, up to the melting points.

In harmonic approximation, phonons and librons (rotational excitations) are treated as independent entities. Real phonons are, however, coupled together and with rotational excitations by anharmonic terms of the crystal Hamiltonian (Manzhelii et al., 1997; Lynden-Bell & Michel, 1994). Therefore, translational and orientational types of motions in molecular crystals are not independent of one another, but rather they occur as coupled translational-orientational vibrations. That involves considerable difficulties to describe this case with analytical expressions. As a consequence, a simplified model where the translational and orientational subsystems are described independently is usually used (Manzhelii et al., 1997; Kokshenev et al., 1997). The coupling produces a shift of phonons frequency with respect to the harmonic value as well as a broadening of bands, associated to the finite phonon lifetime. In such approximation the TO coupling results in an additional contribution to the thermal resistance of the crystal $W=1/\Lambda$. This additional thermal resistance may decrease if the frequency of reorientations becomes sufficiently large, so that the TO coupling reduces. The relative simplicity of the investigated molecular crystals made possible an appropriate theoretical interpretation and provided establishing of the general relationships in heat transfer that result from the presence of rotational degrees of freedom of the molecules. In the experimental part of the paper the results of study of isochoric thermal conductivity of solidified inert gases, simple molecular crystals and their solutions at $T \geq \Theta_D$ will be considered then the models intended to explain temperature and volume dependences of thermal conductivity will be discussed.

2. EXPERIMENTAL RESULTS

2.1 The solidified inert gases

The solidified inert gases *Ar*, *Kr* and *Xe* are convenient object for comparison of experimental data with theoretical calculations of thermal conductivity of a lattice since they are simplest solids closely conformable to theoretical models. This fact stimulated a considerable number of experimental and theoretical works (see, for example, the review of Batchelder, 1977). At $T \geq \Theta_D$ the phonon-phonon interaction is the only mechanism, which determines the magnitude and temperature dependence of the thermal conductivity Λ in perfect crystals. If the scattering is not too strong and the picture of elastic waves can be used, theory predicts the thermal conductivity $\Lambda \propto 1/T$ at fixed volume of the sample (Berman, 1976). The more rapid decrease of the thermal conductivity as $\Lambda \propto 1/T^2$ observed in these inert gases under saturated vapour pressure was originally ascribed to scattering process with participation of four or more phonons (Krupski et al., 1968). Later, Slack, 1972 suggested that the change of crystal volume with temperature by itself may lead to considerable deviations from the dependence $\Lambda \propto 1/T$ because of a "quasi-harmonic" change of the spectrum of vibrational modes, and Ecsedy &

Klemens, 1977 showed theoretically that four-phonon processes are expected to be weak even at premelting temperatures. Subsequent thermal conductivity studies of solid *Ar*, *Kr* and *Xe* at fixed density (Clayton & Batchelder, 1973; Bondarenko et al., 1982; Konstantinov et al., 1988) confirmed that roughly the dependence $\Lambda \propto 1/T$ is valid when $T \geq \Theta_D$. Fig. 2 shows both isochoric for samples of different densities (Clayton & Batchelder, 1973) and measured under saturated vapour pressure $P \cong 0$, *MPa*, (Krupski et al., 1968) experimental data for solid argon in the *W* versus *T* coordinates, where $W=1/\Lambda$ is the thermal resistance of crystal. It is seen that appreciable deviations are observed at the highest temperatures, with the isochoric thermal conductivity varying markedly more slowly than $\Lambda \propto 1/T$ or $W \propto T$ dependence. A similar behaviour was also observed for krypton (Bondarenko et al., 1982) and xenon (Konstantinov et al., 1988). It was found that Λ can be described by the expression where *A* and *B* are constants independent on the temperature; *V0* is molar volume at *T=0K*, and *V* is actual molar volume *V=V(T)*. The Bridgman's coefficients *g* for *Ar*, *Kr* and *Xe* were found in good agreement with calculated by Eq. 3 and they are equal to 9.7, 9.4 and 9.2 correspondingly.

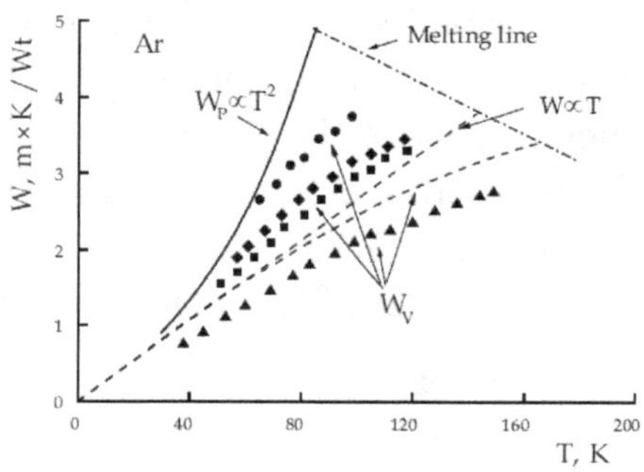

Fig. 2. Isobaric WP (Krupski et al., 1968) and isochoric WV (Clayton et al., 1973) thermal resistance W=1/Λ of crystalline argon for samples of different densities.

The deviations observed were primary attributed to anharmonic renormalisation of the law of phonon dispersion at a fixed volume (Konstantinov et al., 1988). The quantitative calculation has not yet been carried out because of the complexity of model proposed. Later on the thermal conductivity of solid *Ar*, *Kr* and *Xe* was calculated within framework of the Debye model which allows for the fact that the mean-free path of phonons cannot become smaller than half the phonon wavelength (Konstantinov, 2001a).

2.2 Nitrogen-type crystals and oxygen

The N_2-type crystals (N_2, *CO*, N^2o and CO_2) consisting of linear molecules have rather simple and largely similar physical properties. In these crystals the anisotropic part of the molecular interaction is determined mostly by the electric-quadrupole forces. At low temperatures and pressures, these crystals have a cubic lattice with four molecules per unit cell. The axes of the molecules are along the body diagonals of the cube. In N_2 and CO_2 having equivalent diagonal directions the crystal symmetry is Pa_3, for the noncentrosymmetrical molecules *CO* and N^2o the crystal symmetry is $P2_13$. In CO_2 and N^2o the anisotropic interaction is so strong that the crystals melt before the complete orientational disorder occurs. In N_2 and *CO* the barriers impeding the rotation of the molecules are an order of magnitude lower; as a result, the orientational disordering phase transitions occur at 35.7 and 68.13*K*, respectively. In the high-temperature phases, N_2 and *CO* molecules occupy the sites of the *HCP* lattice of the spatial group $P6_3/mmc$.

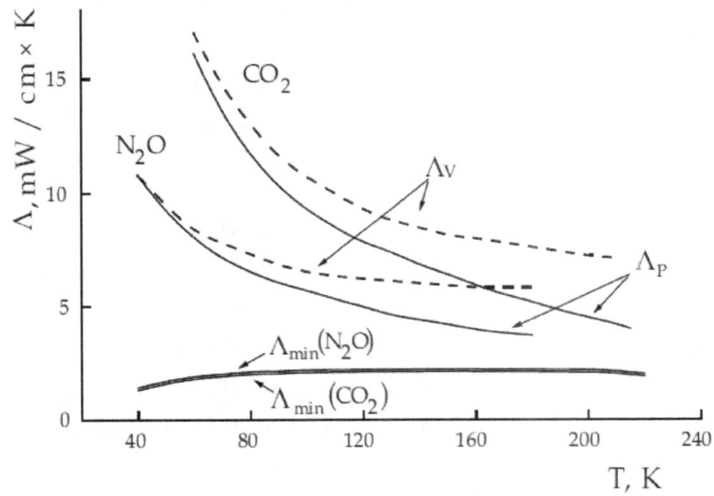

Fig. 3. Isochoric (dashed lines) and isobaric (solid lines) thermal conductivities Λ of CO_2 and N^2o.

The thermal conductivity of solid CO_2, N^2o, N_2 and *CO* was studied at saturated vapor pressure by Manzhelii et al., 1975. Isochoric thermal conductivity of CO_2 and N^2o was studied by Konstantinov et al., 1988b, and N_2 and *CO* by Konstantinov et al., 2005b; 2006a. The data for CO_2 and N^2o is shown in Fig. 3, solid and dashed lines depict isobaric and isochoric thermal conductivity respectively. The isochoric thermal conductivity is recalculated to the molar volumes which CO_2 and N^2o have at zero temperature and pressure. The lower limits to thermal conductivity, calculated accordingly to Eq. 4 are shown at the bottom. The Bridgman's coefficients *g* for nitrogen-type crystals were found in poor agreement with calculated by Eq. 3. The reason for it will be discussed later.

Whereas isobaric thermal conductivity of CO_2 roughly follows *1/T* dependence, isochoric one deviates rather more strongly from the above dependence than in solidified inert gases. In N^2o both isochoric and isobaric thermal conductivities deviates strongly from *1/T* dependence. To reveal the features to be associated with the anisotropic component of the molecular interaction it is necessary to compare molecular crystals with rare-gas solids in the reduced coordinates (de Bour, 1948). Such a comparison is of interest for the following reasons: the thermal resistance $W_{ph\text{-}ph}$ of an ideal crystal of an inert gas is due solely to phonon-phonon scattering. In CO_2 an additional phonon thermal resistance $W_{ph\text{-}lib}$ (or $W_{ph\text{-}rot}$) appears due to interaction phonons with librons (rotational excitations). In the case of N^2o the scattering resulting from dipole disordering is added to these phonon scattering mechanisms (W_{dip}). To a first approximation the total thermal resistance W is:

$$W = W_{ph-ph} + W_{ph-rot} + W_{dip} \tag{6}$$

It is convenient to make a comparison between the crystals mentioned because all of them have a *FCC* lattice. A modified version of the method of reduced coordinates was used. It is important to note that in this case there is no need to resort to some approximate model or other. As a rule, the reducing parameters used are the values of $T_{mol}=\varepsilon\sqrt{k}$, $W_{mol}=\sigma^2/k\sqrt{(m/\varepsilon)}$, and $V_{mol}=N\sigma^3$, were ε and σ are the parameters of the Lennard-Jones potential, *m* is molecular weight and *k* is the Boltzmann constant. It is reasonable to use as an alternative to this the values of the temperature and molar volume of abovementioned substances at the critical points T_{cr} and V_{cr}. The choice of the given parameters is explained as follows. For simple molecular substances T_{cr} and V_{cr} are proportional to ε and σ^3, respectively. However, the accuracy of determination is much higher for the critical parameters than for the parameters of the binomial potential. It should be mentioned that the quantities ε and σ depend substantially on the choice of binomial potential and the method used to determine it. Temperature dependence of isochoric thermal resistances of *Xe*, CO_2 and N^2o in the reduced coordinates is shown in Fig. 4. It is seen that all the contributions to the total thermal resistance are of the same order of magnitude and the deviations from the $W \propto T$ dependence increase from *Xe* to N^2o. The reason has to do with increasing of the phonon scattering and approaching of Λ to its lower limit Λ_{min} as it is seen in Fig. 3.

The behavior of thermal conductivity in the orientationally-ordered phases of N_2 and *CO* is very similar to CO_2 and N^2o. In the orientationally-disordered β-phases isochoric thermal conductivity of all samples of different density increases with rise of temperature, whereas isobaric one is nearly temperature independent (see in Fig. 5 experimental data for N_2; data for *CO* is very similar). In the framework of simple kinetic model, an increase of thermal conductivity with rise of temperature may be explained by an increase of the phonon mean- free path because of the weakening of the effect of some scattering mechanism. It is logically to assume that the interaction of phonons with rotational excitations provides such a mechanism. At the α→β transition and over the orientationally-disordered β-phase a gradual transition from librations to hindered rotation takes place. In contrast to libration, free molecule rotation does not lead to phonon scattering. From the above it follows that there is a temperature interval where phonon scattering by the rotational excitations weakens with rise of temperature or, in other words, TO coupling decreases. The isobaric thermal conductivity is determined by partial compensation of this effect as a result of decreasing of thermal conductivity with rise of temperature due to thermal expansion.

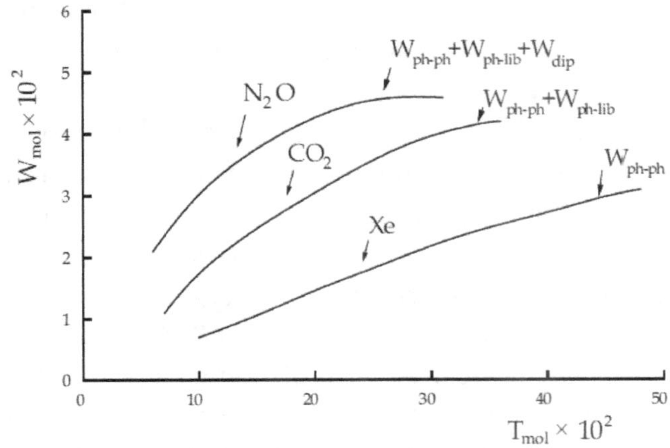

Fig. 4. Temperature dependences of isochoric thermal resistance of Xe, CO_2 and N^2o in the reduced coordinates.

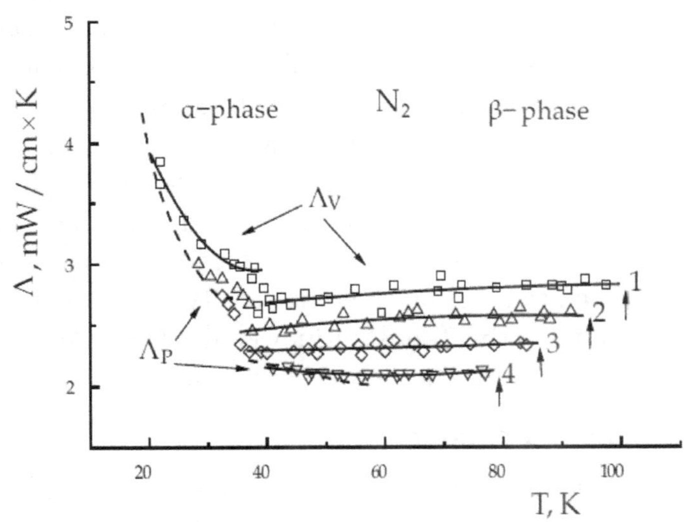

Fig. 5. Thermal conductivity of solid nitrogen. Solid lines are isochoric data for samples of different densities numbered from 1 to 4. Dashed lines are isobaric data.

Solid oxygen belongs to a small group of molecular crystals consisting of linear molecules. In contrast to the N_2-class crystals having the orientationally ordered *Pa3*-type structure, solid oxygen, much like halogens, has a collinear orientational packing because the valence rather than quadrupole forces predominate in the anisotropic interaction. Besides, in the ground state the O_2 molecule has the electron spin *S*=1 which determines the magnetic properties of oxygen. Another specific feature of solid O_2 is the fact that the energy of the magnetic interaction makes up a considerable portion of the total binding energy. This unique combination of molecular parameters has stimulated much interest in the physical properties of O_2, in particular its thermal conductivity. The thermal conductivity of solid O_2 was investigated under saturated vapor pressure in α, β and γ-phases over a temperature interval 1-52*K* (Ježowski, et al., 1993). The low-temperature α-O_2 phase is orientationally and magnetically ordered. On heating to 23.9K, the structure changes into the rhombohedral magnetically-ordered β-phase of the symmetry *R3m*. This is the simplest orientational structure, similar to α-O_2. On a further heating, orientational disordering (cubic cell, *Pm3m* symmetry with *Z*=8) occurs at *T*= 43.8*K*. Under atmospheric pressure oxygen melts at 54.4K. The thermal conductivity has a maximum in the α-phase at T≈6*K*, drops sharply on a change to the β-phase, where it is practically constant, jumps again at the β→γ transition and increases in γ-O_2. The experimental results were interpreted as follows. In the magnetically ordered α-phase the heat is transferred by both phonons and magnons, and their contributions are close in magnitude: $\Lambda_{ph} \approx \Lambda_m$. On the α→β transition the thermal conductivity decreases sharply (~60%) because the magnon component disappears during magnetic disordering. The weak temperature dependence of the thermal conductivity in the β-phase was attributed to the anomalous temperature dependence of the sound velocity in β-O_2 which is practically constant for the longitudinal modes and increases for the transverse ones. The growth of the thermal conductivity in γ-O_2 was attributed to the decay of the phonon scattering at the rotational excitations of the molecules and at the short-range magnetic order fluctuations at rising temperature. The isochoric thermal conductivity of γ-O_2 has been studied on samples of different density in the temperature interval from 44*K* to the onset of melting (Konstantinov et al., 1998b). More sharp increase of isochoric thermal conductivity was observed in γ-O_2 than in the isobaric case.

2.3 Methane and halogenated methanes

The solid halogenated methanes, composed of tetrahedral molecules, serve as suitable subjects for examining the relationship between molecular rotational motion and thermal conductivity behavior. Methane (CH_4), and carbon tetrahalogenides (CF_4, CCl_4, CBr_4 and CJ_4) form high-temperature "plastic" or orientationally-disordered phases in which the rotational motion of molecules is similar to their motion in the liquid state. In crystals consisting of low-symmetry molecules such as chloroform ($CHCl_3$), methylene chloride (CH_2Cl_2) or dichlorodifluoromethane (CCl_2F_2) the anisotropic forces are much stronger and the long- range order persists in them up to the melting points. A special case is trifluoromethane CHF_3, where the second

NMR momentum decreases sharply above *T*=80*K* from 11.5*G2* to 3.0*G2* immediately prior to melting at T_m=118*K*, which suggests enhancement of the molecule rotation about the three-fold axes.

The molecule of methane can be presented as a regular tetrahedron with hydrogen atoms at the vertex positions and carbon atom in the center. The symmetry causes the molecule to exhibit permanent octupole electrostatic moment. At the equilibrium vapor pressure CH_4 solidifies at 90.7*K* and displays unchanged crystallographic structure called phase *I* down to 20.4*K* which is the temperature of phase transition to phase *II*. In both phases, the carbon atoms at the center of the tetrahedral molecule occupy sites of the face-centered cubic lattice. In the low-temperature phase *II* the orientation dependent octupole-octupole interaction leads to a partial orientational ordering. The crystal structure with six orientationally- ordered and two disordered sublattices belongs to the space group *Fm3c*. The orientationally-ordered molecules at D_2d site symmetry positions perform collective librations, while those at *Oh* positions rotate almost freely down to the lowest temperatures. In phase *I* all the tetrahedral molecules are orientationally-disordered, performing rotations which do not show any long-range correlation. $CH_4(I)$ is unique between ODIC molecular crystals since its molecular rotation is virtually free at premelting temperatures. The isobaric thermal conductivity of solid methane was measured within the temperature range of 21- 90*K* in phase *I* (Manzhelii & Krupski, 1968) and within the temperature interval of 1.2–25*K* in phase *II* (Jeżowski et al., 1997). The results obtained revealed an existence of the strong phonon scattering mechanisms connected with rotational excitation of the methane molecules. The isochoric thermal conductivity was studied by (Konstantinov et al., 1999) on samples with molar volumes 30.5 and 31.1 cm³/*mole*. The experimental data for the orientationally-disordered phase of $CH_4(I)$ is shown in Fig. 6.

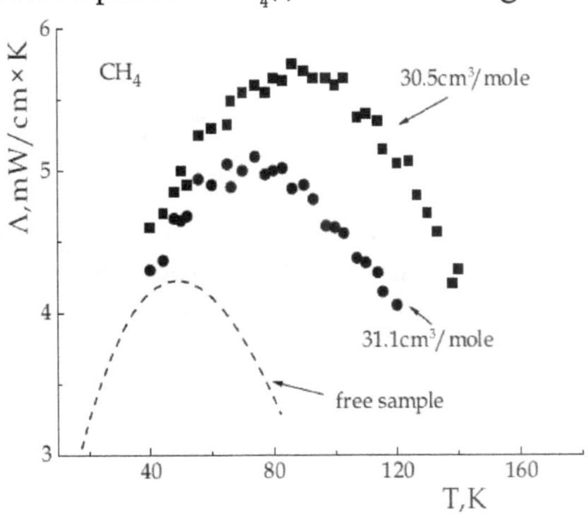

Fig. 6. The isochoric thermal conductivity of solid methane for samples having molar volumes 30.5 (■) and 31.1 (●) cm³/mole together with the isobaric data (dashed line).

It is seen that both isobaric and isochoric thermal conductivities first increase with rise of temperature, pass through a maximum and then decrease up to melting. Note, than regular „kinetic" maximum of thermal conductivity is observed at considerably lower temperature in phase *II* (Jeżowski et al., 1997). Maximum shifts towards higher temperatures as the density of the sample increases. The Bridgman coefficient is equal to 8.8±0.4.

The origin of such behavior of thermal conductivity is the same that for orientationally- disordered phases of other molecular crystals, it is decrease of phonon scattering on rotational excitations of molecules. However, in contrast to "plastic" phases of other molecular crystals where rotation is hindered, methane molecules rotate almost freely at premelting temperatures. Above the maximum, phonon-rotation contribution $W_{ph\text{-}rot}$ to the total thermal resistance W of methane tend to zero, and behavior of thermal conductivity is determined solely by increase of phonon-phonon scattering. It is clearly seen in Fig. 7, where the appropriate contributions were calculated using the method of reduced coordinates. The theoretical models proposed to describe thermal conductivity of solid methane will be discussed later.

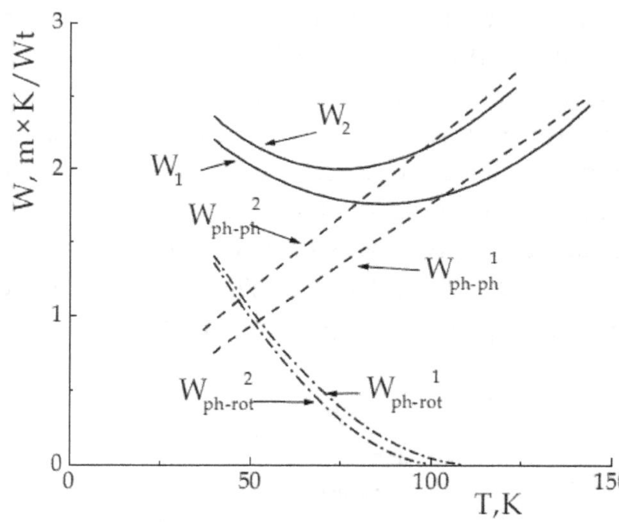

Fig. 7. Contributions of phonon-phonon scattering $W_{ph\text{-}ph}$ and phonon scattering by rotational molecule excitations $W_{ph\text{-}rot}$ to the total thermal resistance W of solid methane samples having molar volumes 30.5 (1) and 31.1 (2) cm^3/mole.

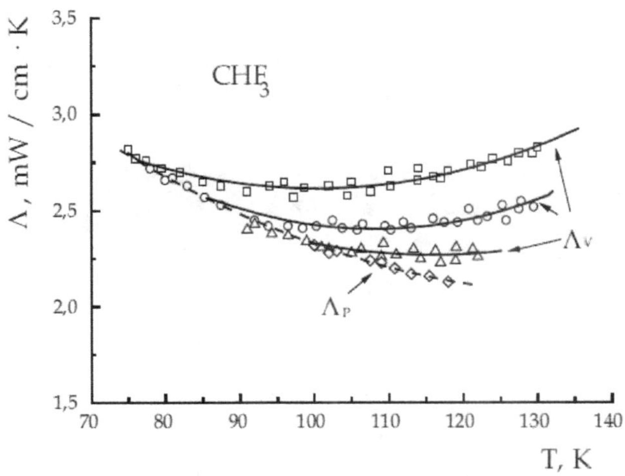

Fig. 8. The isochoric thermal conductivity of three solid CHF_3 samples of different densities. Solid lines show smoothed values of isochoric thermal conductivity. Dashed line and rhombus are for the thermal conductivity of a free sample.

Slow increase of isochoric thermal conductivity was also observed in the orientationally- disordered phases of CCl_4 (Konstantinov et al., 1991a) and CBr_4 (Ross et al., 1984) at recalculation the last experimental data to constant volume. The isochoric thermal conductivity in orientationally-ordered phases of halogenated methanes ($CHCl_3$, CH_2Cl_2 and CCl_2F_2) decreases with rise of temperature deviating markedly from $1/T$ dependence like case of CO_2 and N^2o (Konstantinov et al., 1991b; 1994; 1995). An interesting behavior of the isochoric thermal conductivity was found in trifluoromethane CHF_3 (Konstantinov et al., 2009a). CHF_3 melts at T_m=118K, the melting entropy being $\Delta S_f/R$ = 4.14. Neutron scattering investigations of the crystallographic structure of CHF_3 revealed only one crystalline phase of the spatial symmetry $P2_1/c$ with four differently oriented molecules in the monoclinic cell. The Debye temperature of CHF_3 is Θ_D=88±5K. Fig. 8 shows isochoric thermal conductivity of CHF_3 for three samples of different densities in the interval from 75K to the onset of melting. The isochoric thermal conductivity first decreases with increasing temperature, passes through a minimum at T~100K, and then starts to increase slowly. The weak growth of isochoric thermal conductivity with temperature in solid CHF_3 suggests that the translational-orientational coupling becomes weaker in this crystal at premelting temperatures owing to the intensive molecule reorientations about the three-fold axes.

Some parameters of the halogenated methanes discussed are presented in Table 1.

Table 1. Melting temperature Tm; phase transition temperature TI-II; structure and the number of molecules per unit cell z; melting entropy ΔSf /R; Debye temperature ΘD; Bridgman coefficient g = -(∂ ln Λ / ∂ lnV)T ; dipole momentum of molecule μ.

Substance		T_m, $T_{I\text{-}II}$	Structure	z	$\Delta S_f/R$	Θ_D, K	g	μ, D
CH_4	(I)	90.6	F_m3_m	4	1.24	96	8.8	0
	(II)	20.5	P43m	32		141		
CCl_4	(I)	250.3	F_m3_m	4	1.21		6.0	0
	(II)	225.5	C2/c	32		92	6.5	
CBr_4	(I)	363	F_m3_m	4	1.3	62	3.8	0
	(II)	320	C2/c	32			3.4	
CHF_2Cl	(I)	115.7	$P4_2/n$	8	4.25		4.5	1.41
	(II)	59	P112/n	8		70		
$CHCl_3$		210	Pnma	4	5.4	86*	3.9	1.01
CH_2Cl_2		176	P_{bcn}	4	3.13	115*	4.6	1.6
CF_2Cl_2		115	Fdd2	8	4.2	80*	5.0	0.51
CHF_3		118	$P2_1/c$	4	4.14	88*	4.6	1.6

* - Estimates obtained from *IR* and Raman spectra.

2.4 Some special cases: SF_6 and C_6H_6

Sulphur haxafluoride SF_6 is often assigned to substances that have a plastic crystalline phase. Indeed, the relative entropy of melting $\Delta S_f/R$ of SF_6 is 2.61, which is close to the Timmermanns criterion. However, the nature of orientational disorder in the high- temperature phase of SF_6 is somewhat different from that of plastic phases in other molecular crystals, where the symmetries of the molecule and its surroundings do not coincide. The interaction between the nearest neighbors in the *bcc* phase is favorable for molecule ordering caused by the *S-F* bonds along the *{100}* direction, and the interaction with the next nearest neighbors is dominated by repulsion between the *F* atoms. According to *X*-ray and neutron diffraction data a strict order is observed in SF_6 *(I)* just above the phase transition point. The structural dynamical factor $\mathfrak{R}$ characterizing the degree of the orientational order is close to unity in the interval 95-130*K*. This feature sets off SF_6 from other plastic crystals, such as methane, carbon tetrachloride, adamantane and so on, where the long-range orientational order becomes disturbed immediately after the phase transition. Orientational disordering in SF_6 starts to intensify only above 140*K*. As follows from the analysis of the terms of the Debye-Waller factor derived from neutron- diffractometric data for the high-temperature phase of SF_6, the *F* atoms have large effective libration amplitudes. As the temperature rises, the amplitudes increase to 20° and higher, but the *F* localization is still appreciable near *{100}* direction. This implies that the orientational structure of SF_6 *(I)* does not become completely disordered even at rather high temperatures. The disordering itself is dynamic by nature. The increasing amplitudes of librations are not the only factor responsible for the increasing orientational disordering at rising temperature. It is rather connected with dynamic reorientations, which become more intensive due to frustrations of the molecular interactions. Owing to these features, SF_6 offers a considerable possibility for investigating the influence of wide-range rotational states of the molecules on the thermal conductivity in a monophase one-component system, where such states can vary from nearly complete orientational ordering to frozen rotation.

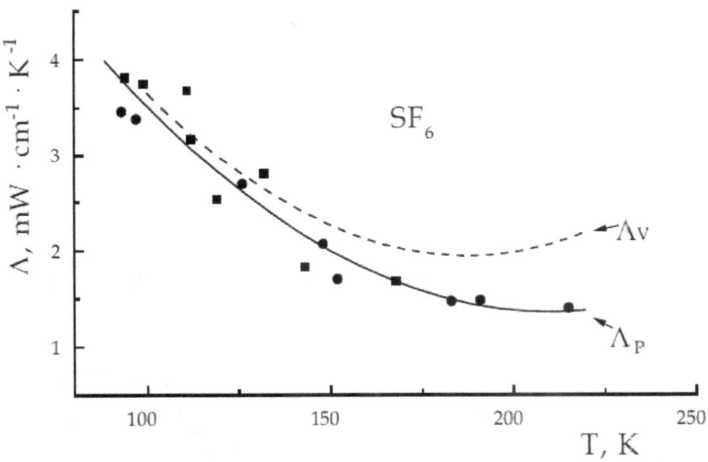

Fig. 9. Isochoric (V_m=58.25 cm^3/mole) and isobaric thermal conductivity of solid SF_6.

Isochoric thermal conductivity of SF_6 was studied by Konstantinov et al., 1992b, while isobaric one by Purski et al., 2003. The data are shown in Fig. 9. The isobaric thermal conductivity first decreases with increasing temperature and flattens out at premelting temperatures. The isochoric thermal conductivity first decreases with increasing temperature, passes through a minimum at *T*~180*K*, and then starts to increase. Such behavior was attributed to decreasing of phonon scattering on rotational excitations of molecules with rise of temperature.

Solid benzene has only one crystallographic modification under the saturated vapor pressure: it has the orthorhombic spatial symmetry P*bca* (D_{2h}^{15}) with four molecules per unit cell. Benzene melts at 278.5K with entropy of melting $\Delta S_f /R$ = 4.22, which is much higher that the Timmermanns criterion for *ODIC* phases. The Debye temperature of C_6H_6 is 120K. In the interval 90-120K the second *NMR* moment of C_6H_6 drops considerably as a result of the molecule reorientations in the plane of the ring around the six-fold axis. The activation energy of the reorientational motion estimated from the spin-lattice relaxation time is 0.88 *kJ/mole*. The frequency of molecule reorientations at 85*K* is 104 sec^{-1}. On a further heating it increases considerably reaching 10^{11} sec^{-1} near T_m. The basic frequency of the benzene molecule oscillations about the six-fold axis at 273*K* is 1.05×10^{12} sec^{-1}. Isochoric thermal conductivity of C6H6 was studied by Konstantinov et al., 1992a, while isobaric one by Purski et al., 2003. The data are shown in Fig. 10.

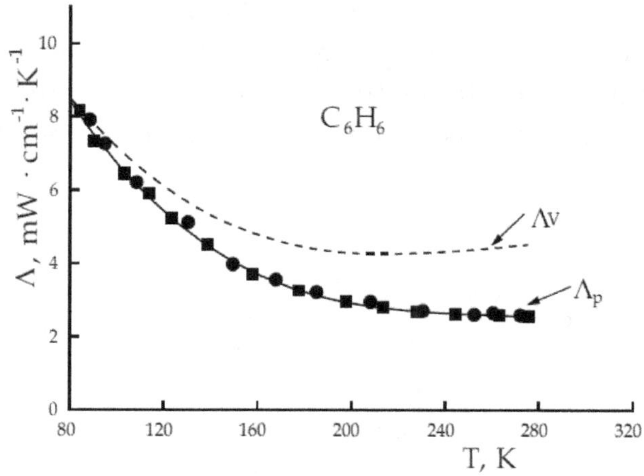

Fig. 10. Isochoric (V_m=70.5 cm^3/mole) and isobaric thermal conductivity of solid C_6H_6.

Like solid SF_6 the isobaric thermal conductivity of benzene first decreases with increasing temperature and flattens out at premelting temperatures. The isochoric thermal conductivity first decreases with increasing temperature, passes through a minimum at $T\sim210K$, and then starts to increase slowly. In contrast to SF_6 where the rotation is multi-axial, molecule of benzene rotates about released six-fold axis. The increase of thermal conductivity with rise of temperature was attributed like to previous cases to weakening of phonon scattering on rotational excitations of molecules.

2.5 Solid n-alkanes

Normal alkanes (*n*-paraffins) of the C_n*sssH2n+2*-type form a class of substances that are intermediate in changing-over to long-chain polymers. *N*-alkanes have a comparatively simple structure and a molecular packing: in the solid state the axes of all molecules are always parallel to one another irrespective of a particular crystalline modification. Owing to their relative simplicity, normal alkanes are naturally considered as the starting point for understanding the structural and thermophysical properties of more complex long- chain compounds. *N*-alkanes exhibit an extremely diverse dynamic behavior both in the solid and liquid states. The melting temperature increases in this series of compounds with the length of the chain and its behavior is nonmonotonic: the *n*-alkanes with an odd number of carbon atoms (odd *n*–alkanes) melt at relatively lower temperatures than those with an even number of *C* atoms (even *n*–alkanes). An interesting effect is observed when the orthorhombic, monoclinic and triclinic structures alternate with the even and odd members of the series. The even *n*–alkanes with *n*=6–24 (*n*=the number of *C* atoms) crystallize at low temperatures forming a triclinic cell. Heptane (*n*=7) and nonane (*n*=9) have an orthorhombic structure at low temperatures. Glass–like cylindrical "rotational" phases with a hexagonal symmetry were found in rather narrow temperature interval below the melting points of odd *n*–alkanes starting with *n*=9. The region of existence of the "rotational" phase increases with the length of the chain. Hexagonal modifications also occur in even *n*–alkanes starting with *n*=22. Some parameters of *n*-alkanes discussed above are submitted in Table 2.

Table 2. The structure of n-alkanes; the temperature $T\alpha\_\beta$ and the entropy $\Delta S\alpha\_\beta/R$ of the transition to the "rotational" phase; the temperature T_m and the melting entropy $\Delta S_m/R$; a complete change of the entropy and variations of thermal conductivity $\Lambda\alpha$ / ΛL during the ordered-phase – liquid transition.

Substance	Structure	$T\alpha\_\beta$,K	$\Delta S\alpha\_\beta/R$	T_m,K	$\Delta S_m/R$	$\Delta S\alpha_{-L}/R$	$\Lambda\alpha / \Lambda_L$
C_2H_6	$P2_1/n$, z=2	89.8	2.74	90.3	0.77	3.6	1.3
C_3H_8	$P2_1/n$, z=4	-	-	85.5	4.95	4.95	2.2
C_6H_{14}	P ī, z=1	-	-	177.8	8.85	8.85	1.9
C_9H_{20}	P ī, z=1	217.2	3.48	219.7	8.47	12.0	2.4
$C_{11}H_{24}$	P_{bcn}, z=4	236.6	2.9	247.6	10.8	13.7	2.4
$C_{13}H_{28}$	P_{bcn}, z=4	255.0	3.6	267.8	12.8	16.4	2.3
C15 H_{32}	P_{bcn}, z=4	270.9	4.1	283.1	14.7	18.8	2.1
$C_{17}H_{36}$	P_{bcn}, z=4	284.3	4.8	295.1	16.4	21.2	2.0
$C_{19}H_{40}$	P_{bcn}, z=4	296.0	5.6	304.0	18.8	24.3	2.0

The isochoric thermal conductivity of methane, ethane, propane and hexane was studied by Konstantinov et. al., 1999a; 2006c; 2009b and 2010 correspondingly. Isobaric thermal conductivity of "odd" numbered *n*-alkanes with *n*=9-19 was investigated by Forshman & Andersson, 1984. The isobaric thermal conductivity of n-alkanes discussed above is shown in Fig. 11 along with the thermal conductivities of the liquid phases of these compounds measured immediately after melting. It was noted (Forshman & Andersson, 1984) that the thermal conductivity of long-chain odd *n*-alkanes has some features in common. As the "rotational" phase melts, the thermal conductivity changes by about 35% and is independent of the chain length. The jump of the thermal conductivity on changing to the low temperature ordered phase decreases with the increasing length of the chain and makes ~85% for *n*-undecane and ~40% for *n*-nonadecane. The absolute value of thermal conductivity increases in the "rotational" phase with increasing of the chain length.

According to our studies the isobaric thermal conductivity exhibits closely similar behavior in short and long-chain *n*-alkanes. On the transition from the ordered phase to a liquid the thermal conductivity of the *n*-alkanes starting with propane changes nearly twice and is independent of the total transitions entropy and the chain length. This change is much smaller in the case of spherical and elliptic molecules: for example, $\Delta\Lambda/\Lambda_L$ is only 20–30% in methane and ethane. This can be related to the higher degree of orientational order in solid long-chain *n*-alkanes as compared to spherical molecules. The isochoric thermal conductivity of solid *n*-alkanes decreases with rise of temperature following a dependences weaker than $\Lambda \propto 1/T$. The deviation of the isochoric thermal conductivity from the dependence $\Lambda \propto 1/T$ in solid *n*-alkanes was explained proceeding from the concept of the lower limit to thermal conductivity (Konstantinov et. al., 2009b).

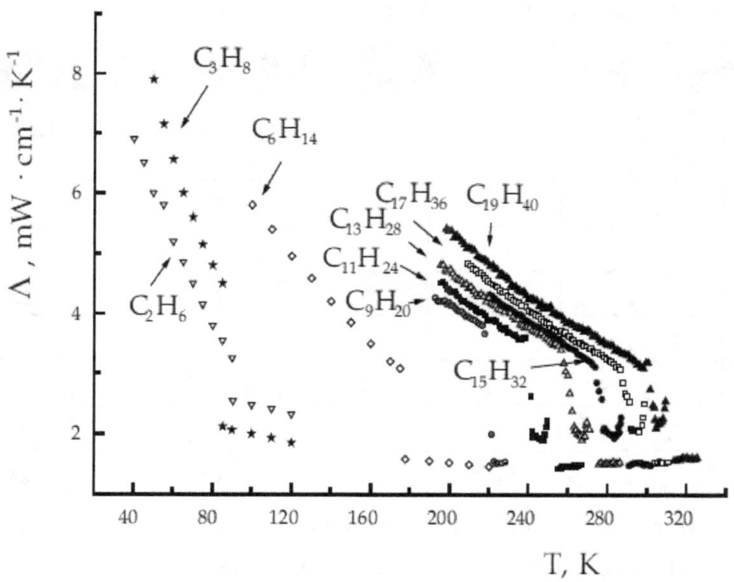

Fig. 11. Isobaric thermal conductivity of some n-alkanes.

2.6 Mixed molecular crystals

Mixed molecular crystals are convenient object for testing of the concept of the lower limit to thermal conductivity. In heavy solidified inert gases *Ar*, *Kr* and *Xe* the thermal conductivity approaches its lower limit at premelting temperatures. In the case when the thermal conductivity approaches Λ_{min} with rise of temperature, the effect of impurities should manifest itself in a specific manner. Impurities cannot considerably decrease the thermal conductivity at premelting temperatures at which Λ has already been close to the minimum value. As temperature decreases the contribution of impurities to the thermal resistance of crystal $W=1/\Lambda$ should increase. The isochoric thermal conductivity of solid $(CH_4)1-_{\xi}Kr_{\xi}$, $Kr1-_{\xi}Xe_{\xi}$ and $(CO_2)1-_{\xi}Kr_{\xi}$, $(CO_2)1-_{\xi}Xe_{\xi}$ solutions ($\xi=0\div1$) has been studied by Konstantinov et al., 2000, 2001a; 2002b and 2006b, respectively. Fig. 12 shows the temperature dependence of isochoric thermal conductivity for pure *Kr* and $Kr_1-_{\xi}Xe_{\xi}$ solid solution reduced to samples for which condition of constant volume starts from 80*K*.

It can be seen that the thermal conductivity of $Kr_1-_{\xi}Xe_{\xi}$ solid solution decreases and its temperature dependence becomes weaker with an increase in *Xe* concentration. At $\xi=0.14$, the thermal conductivity virtually coincides with the lower limit to thermal conductivity calculated by Eq. 4.

Kr and CH_4 form a homogeneous solid solution with fcc structure above 30*K* at all ξ. They have similar molecule/atom radii and parameters of the pair potential. The Debye temperatures Θ_D of *Kr* and CH_4 are 72 and 143*K* respectively. At the same time the masses of the *Kr* atom and the CH_4 molecule are very different, 83.8 and 16 atomic units, respectively. The isochoric thermal conductivity of $(CH_4)1-_{\xi}Kr_{\xi}$ solid solutions has been studied between 40*K* and ~ 150*K* over the wide range of concentrations, (ξ=0.013, 0.032, 0.07, 0.115, 0.34, 0.71, 0.855, 0.937, and 0.97). A gradual transition from the thermal conductivity of a highly perfect crystal to the minimum thermal conductivity was observed as the crystal becomes increasingly more disordered (see Fig. 13).

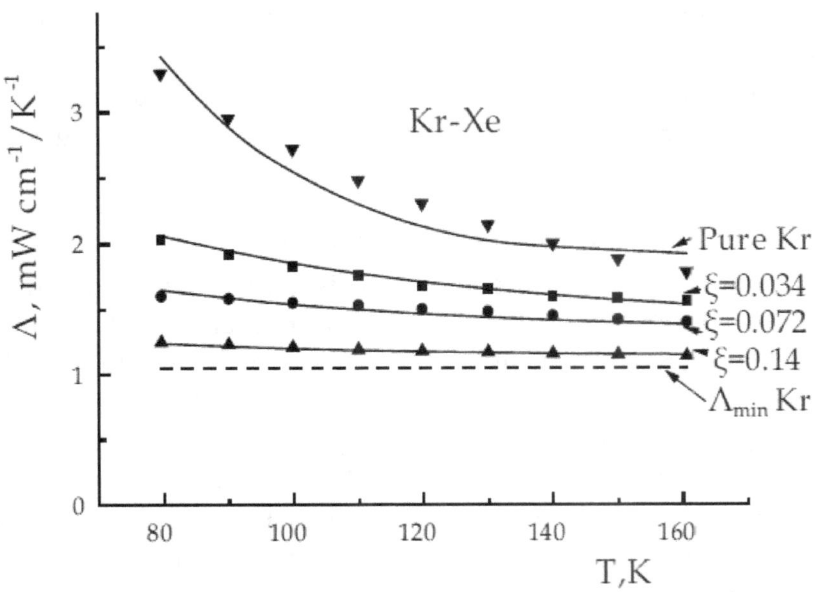

Fig. 12. Smoothed values of isochoric thermal conductivity of pure Kr and $Kr1-_{\xi}Xe_{\xi}$ solid solution for samples, whose volume is constant starting from 80K. The dashed line is Λmin of pure Kr, calculated according to Eq. 4.

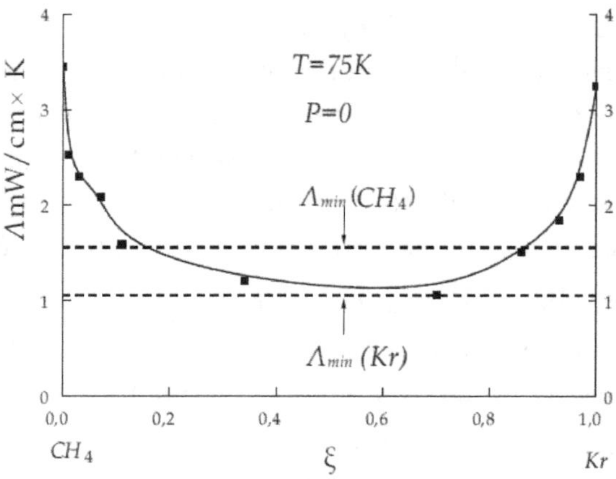

Fig. 13. The concentration dependence of the thermal conductivity of solid solution $(CH_4)_{1}$-ξKr_ξ at T=75K and P=0. The horizontal lines are Λ_{min} of pure Kr and CH4 under the same conditions calculated by Eq. 4.

A strong decrease of the Bridgman coefficient was observed, from a value $g\cong 9$ characteristic for pure *Kr* and CH_4 to $g\cong 4$ at only a tiny impurity concentration.

An unusual effect of point defects on the thermal conductivity has been detected in $(CO_2)_{1-\xi}Xe_\xi$ and $(CO_2)_{1-\xi}Kr_\xi$ solid solutions (Konstantinov et al., 2006b). In pure CO_2 at T >150K the isochoric thermal conductivity decreases smoothly with increasing temperature. In contrast, the thermal conductivity of CO_2 /*Kr* and CO_2 /*Xe* solid solutions first decreases, passing through a minimum at 200–210*K*, and then starts to increase with temperature up to the onset of melting. Such behavior of the isochoric thermal conductivity was attributed to the rotation of the CO_2 molecules, which gain more freedom as the spherically symmetrical inert gas atoms are dissolved in the CO_2.

2.7 Short Summary to the experimental part

It has been demonstrated that the $\Lambda \propto 1/T$ law is not obeyed in molecular crystals at all. Because of the strong translational-orientational (TO) coupling in the orientationally-ordered phases molecular librations contribute considerably to the thermal resistance $W=1/\Lambda$ of the crystal. As a result, the isochoric thermal conductivity approaches its lower limit Λ_{min} at premelting temperatures and shows significantly more slow dependence than $1/T$. In the case of hindered or almost free rotation of molecules in crystal the isochoric thermal conductivity increases with rise of temperature. It is originated due to weakening of TO coupling as the rotational motion of molecules attains more freedom. The isobaric thermal conductivity is determined by the partial compensation of these effects as a result of decreasing of thermal conductivity with rise of temperature due to thermal expansion.

3. THEORETICAL MODELS.

3.1 The Debye model of thermal conductivity

In simple Debye model (see, for example, Berman, 1976) thermal conductivity Λ can be expressed as

$$\Lambda = \frac{\hbar^2}{2\pi^2 v^2 k_B T^2} \int_0^{\omega_D} l_\Sigma(\omega)\omega^4 \frac{e^{\frac{\hbar\omega}{k_B T}}}{\left(e^{\frac{\hbar\omega}{k_B T}} - 1\right)^2} d\omega, \tag{7}$$

where v is the sound velocity; ω_D is the Debye frequency $\omega_D = v(6\pi^2 n)1/3$; $l\Sigma(\omega)$ is the combined phonon mean-free path determined by the package of all scattering mechanisms

$$l_\Sigma(\omega) = \left(\sum_i l_i(\omega)^{-1}\right)^{-1}, \tag{8}$$

If the scattering governed by *U*-processes only the mean-free path of phonons is

$$l_u(\omega) = v/A\omega^2 T, \quad A = \frac{18\pi^3}{\sqrt{2}} \frac{k_B \gamma^2}{m a^2 \omega_D^3}; \tag{9}$$

were m is the average atomic (molecular) mass; *a3* is the volume per atom (molecule); $\gamma = -(\partial \ln\Theta_D / \partial \ln V)_T$ is the Grüneisen parameter. It is easy to check that in high-temperature limit $T \geq \Theta_D$ Eqs. 7 and 9 reduce to Eq. 1. In the case of point defect scattering the mean-free path of phonons is

$$l_i(\omega) = v/B\omega^4; \quad B = \frac{3\pi\Gamma}{2\omega_D^3}; \quad \Gamma = \xi(1-\xi)\left(\frac{\Delta M}{\bar{M}} + 6\gamma\frac{\Delta a}{\bar{a}}\right)^2 \tag{10}$$

where ΔM is the mass difference of atoms (molecules) of the impurity and host; Δa is the change of the lattice parameter upon introduction of impurity.

3.2 Phonon-libron scattering

The contribution of phonon-libron scattering was considered by (Manzhelii et al., 1975), and (Kokshenev et al., 1997). In harmonic approximation decomposition of the interaction energy with respect to translational and angular displacements of molecules near equilibrium positions can be written as

$$H_{harm} = \sum_{\bar{q},\lambda} \hbar\omega_{\bar{q},\lambda}\left(b^+_{\bar{q},\lambda} b_{\bar{q},\lambda} + \frac{1}{2}\right) + \sum_{\bar{k},j} \varepsilon_j \left(a^+_{\bar{k},j} a_{\bar{k},j} + \frac{1}{2}\right), \tag{11}$$

where $b_{\bar{q},\lambda}$, $a_{\bar{k},j}$ are the operators of deletion of phonons and librons; λ and j are the numbers of phonon and libron branches. So long as the anharmonicity of molecular librations is substantial even at low temperatures a self-consistent theory was used for description of the libron subsystem (see, for example, Manzhelii et al., 1997). Only the interaction processes leading to the linear dependence of thermal resistance in high temperature $T \geq \Theta_D$ limit were taken into consideration

$$V = \frac{\gamma_1 a G}{\sqrt{N}} \sum_{q,q',k} \sqrt{qq'} a_k^+ b_q b_{q'} \Delta(k - q - q') + \gamma_2 G \sqrt{a/N} \sum_{q,k,k'} \sqrt{q} a_k a_{k'}^+ b_a \Delta(k - k' + q), \quad (12)$$

where $k \equiv (\vec{k}, j)$, $q \equiv (\vec{q}, \lambda)$; N is number of lattice points; $G = U\eta$ (U is a constant of molecular field, η is a parameter of the long-range orientational order). In Eq. 12 only the members were taken into consideration which allow for conservation law of energy and quasi- momentum. Three types of interactions were considered:

$$Ph_A + Ph_A \leftrightarrow Lib, \quad (13a)$$

$$Ph_A + (Ph_O \text{ or } Lib) \leftrightarrow (Lib \text{ or } Ph_O), \quad (13b)$$

$$Ph_A + Lib \leftrightarrow Lib', \quad (13c)$$

where Ph_A and Ph_O designate acoustic and optical phonons. The constants of interactions γ_1 and γ_2 can be expressed via characteristic temperatures Θ and Θ_1 of translational and librational spectrum of crystal:

$$\gamma_1 = C_1 \hbar^3 / M a^2 \Theta (J \Theta_1)^{1/2} \, k_B^{3/2}, \quad (14a)$$

$$\gamma_2 = C_2 \hbar^3 / J \Theta_1 (M a^2 \Theta)^{1/2} \, k_B^{3/2}, \quad (14b)$$

where J is a momentum of inertia of molecule. In quasi-harmonic approach the characteristic temperature is associated with a value of barrier G hindering of molecular rotation by the relation

$$\Theta_1 = (3\hbar^2 G / J \, k_B^2)^{1/2}, \quad (15)$$

An additional thermal resistance of crystal determined by scattering processes of the type (13a) and (13b) can be presented as

$$W_1(T) = C_1 T \left(\frac{h}{k_B} \right)^3 \frac{\gamma_{lib}^2}{m a \Theta^3} \frac{m a^2}{J} S_1; \quad S_1 = \sum_{\nu} g_\nu \varphi_\nu, \quad (16a)$$

$$W_2(T) = C_2 T \left(\frac{h}{k_B} \right)^3 \frac{\gamma_{lib}^2}{m a \Theta^3} \frac{m a^2}{J} S_2; \quad S_2 = \sum_{\nu,\nu'} g_\nu g_{\nu'} f_{\nu,\nu'}, \quad (16b)$$

where γ_{lib} is Grüneisen parameter; g_ν and g_ν , are degrees of degeneration of libron and optical branches; m and J are the mass and momentum of inertia of molecule; a is inter-atomic distance; φ_ν and $f_{\nu,\nu}$, are the surfaces of energy conservations in scattering processes like (13a) and (13b). In the simple Ziman model they can be written as:

$$\varphi_\nu = (\alpha_\nu^2 - 1)(6\alpha_\nu - 4 - \alpha_\nu^3)\,; \quad \alpha_\nu = \theta_\nu/\theta, \tag{17a}$$

$$f_{\nu,\nu'} = \sigma_{\nu,\nu'}^3(12 - \sigma_{\nu,\nu'}^2)\,; \quad \sigma_{\nu,\nu'} = \left|\Theta_\nu - \Theta_{\nu'}\right|/\Theta, \tag{17b}$$

where θ_ν and $\Theta_{\nu'}$ are the frequencies of optical and phonon modes.

The total thermal resistance of crystal at presence of phonon-phonon and two phonon-libron mechanisms of scattering in the reduced coordinates is:

$$W^* = B^*T^* = B_0^*T^*\left[1 + (ma^2/J)(\alpha S_1 + \beta S_2)\right], \tag{18}$$

where coefficients B^*_0, α and β are equal for the group of one-type crystals. By this means that an additional phonon-libron scattering leads in the first approximation only to renormalization of coefficient *A* in Eq. 9.

3.3 Phonon scattering on rotational excitations of molecules

In the orientationally-disordered phases of molecular crystals there is no long-range order, what suggests that the distinct pure libration modes cannot propagate in the crystal. Nevertheless, the correlation effects are still strong immediately after the phase transition and the short-range orientational order persists. In this region there is an additional phonon scattering at the short-range orientational order fluctuations and it becomes weaker on a further temperature rise. To explain the behaviour of the thermal conductivity in the orientationally-disordered phases of solid methane and deuteromethane Manzhelii & Krupski, 1968 used the analogy between molecular and spin systems. In a number of magnetic crystals the thermal conductivity was observed to increase above the magnetic phase transition. Reason for these anomalies is the scattering of phonons by critical fluctuations of the short-range magnetic order above the Neel point (Kawasaki, 1963). In the orientationally-disordered phases of the molecular crystals an increase of the isochoric thermal conductivity with increasing temperature is due to weakening of phonon scattering by fluctuations of the short-range orientational order. By existing analogy, using the equations for one and two-phonon relaxation times of Kawasaki, 1963, the phonon mean- free path of each of the examined scattering mechanisms can be expressed as

$$l_I(\omega) = \rho\upsilon^5/B^2\Lambda_{rot}T\omega^2, \tag{19a}$$

$$l_{II}(\omega) = \pi\rho^2\upsilon^8/C^2k_BC_{rot}T^2\omega^4, \tag{19b}$$

where *B* and *C* are anisotropic molecular interaction constants; Λ_{rot} and C_{rot} are the thermal conductivity and the heat capacity of the rotational subsystem, respectively; ρ is the density. It is assumed that $B=C_2$ in the first approximation. The coefficient *B* can be roughly estimated from the dependence of the phase transition temperature T_f upon pressure $B=-\chi_T^{-1}\partial(\ln T_f)/\partial P$, where χ_T is isothermal compressibility. The thermal conductivity of the rotational subsystem can be calculated from the known gas-kinetic expression $\Lambda_{rot}=\frac{1}{3}C_{rot}a^2\tau^{-1}$, where τ is the characteristic time of the site-to-site transfer of the rotational 3 energy and can be estimated as a mean period of the librations. Taking into account Eqs. (8, 9, 19a, 19b), the phonon mean-free path in the orientationally-disordered phase can be written as

$$l_\Sigma(\omega)=\left(\frac{AT\omega^2}{\upsilon}+\frac{B^2\Lambda_{rot}T\omega^2}{\rho\upsilon^5}+\frac{C^2k_BC_{rot}T^2\omega^4}{\pi\rho^2\upsilon^8}\right)^{-1}, \tag{20}$$

The validity of such description was corroborated subsequently by the first principles calculations of the thermal conductivity of methane by Yasuda, 1978.

3.4 Phonons and "diffusive" modes

Eq. 8 is inapplicable if *l*(ω) becomes of the order of or less than half the phonon wavelength $\alpha\lambda/2=\alpha\pi v/\omega$, where α is a numerical factor of the order of unity (Roufosse & Klemens, 1974).

$$l(\omega)=\begin{cases} l_\Sigma(\omega), & 0\le\omega\le\omega_0, \\ \alpha\pi\upsilon/\omega=\alpha\lambda/2, & \omega_0<\omega\le\omega_D, \end{cases} \tag{21}$$

Taking into account Eqs. 9, 18, and 21 the phonon mobility edge ω_0 is equal for orientationally-ordered molecular crystals to

$$\omega_0=1/\alpha\pi AT\ . \tag{22}$$

Eq. 21 is the well-known Ioffe-Regel' criterion, which presumes localization, so that we shall assume excitations whose frequencies lie above the phonon mobility edge ω_0 to be "localized". Since completely localized states do not contribute to the thermal conductivity, localization is assumed to be weak and excitations can hop from site to site in a diffusion manner, as Cahill et al., 1992 supposed. This point of view is, on the hole, consistent with the results of Feldman et al., 1993, where the theory for intermediate case (where disorder is sufficient for oscillations not to propagate but insufficient for localization) was formulated for amorphous silicon. The idea was that the dominant scattering was correctly described by a harmonic Hamiltonian and reduced to a single-particle problem of decoupled oscillators. On this basis the thermal conductivity could be calculated exactly using a formula similar to the Kubo-Greenwood formula for the electric conductivity of disordered metals (Auerbach et al., 1984). The calculation performed by authors showed that in this case it is incorrect to indentify the Ioffe-Regel' edge with localization. Although the oscillations dominating in high-temperature heat transfer lie near the Anderson localization edge, with the exception of narrow band of states they are not completely localized. The Boltzmann theory assigns to each vibrational mode with wave vector k and propagation velocity $\vec{v} = \partial\omega / \partial\vec{k}$ a diffuseness $D_k = vl/3$, where l is the mean-free path length. The authors found that even though vibrations are not localized, a definite wave vector or velocity cannot be assigned to them. Nonetheless, they transfer heat and contribute to the thermal conductivity an amount of the order of $C_i(T)D_i/V$ for the ith mode, where the specific heat $C_i(T) = k_B$ for $T \geq \Theta_D$, D_i is the temperature-independent "diffuseness" of the mode. The numerical calculation is in many respects similar to the calculation performed according to Boltzmann's theory and it confirms the concept of a "minimum" thermal conductivity in the form it is discussed by Roufosse & Klemens, 1974; Slack, 1979 and Cahill et al., 1992.

If $\omega_0 \geq \omega_D$, then the mean-free path length of all modes is greater than $\lambda/2$ and Eqs. 7 and 22 come to Eq. 1 at $T \geq \Theta_D$. If $\omega_0 < \omega_D$, then the integral of thermal conductivity (7) splits into two parts which describe the contributions of low-frequency phonons and high-frequency "diffusive" modes to heat transfer:

$$\Lambda = \Lambda_{ph} + \Lambda_{dif} \qquad (23)$$

In the high-temperature limit $T \geq \Theta_D$ these contributions are

$$\Lambda_{ph} = \frac{k_B \omega_0}{2\pi^2 v A T}\,;\ \Lambda_{dif} = \frac{\alpha k_B}{4\pi v}(\omega_D^2 - \omega_0^2)\,, \qquad (24)$$

The experimental results for a number of crystals were computer-fitted by the least-square method (see for details Konstantinov et al., 2003b). Figs. 14 and 15 show the contributions of phonons and "diffusive" modes to the isochoric thermal conductivity of Kr and CO_2.

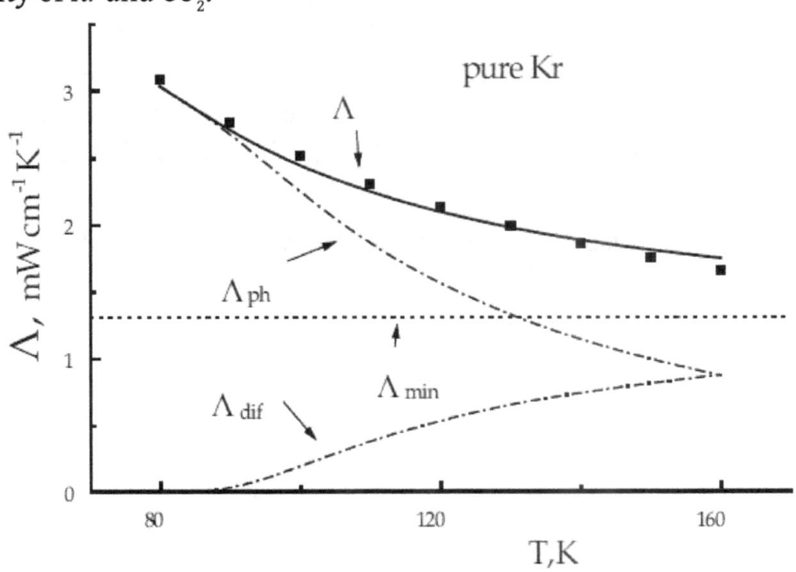

Fig. 14. Contributions Λ_{ph} and Λ_{dif} for Kr sample with molar volume 28.5 cm^3/mole.

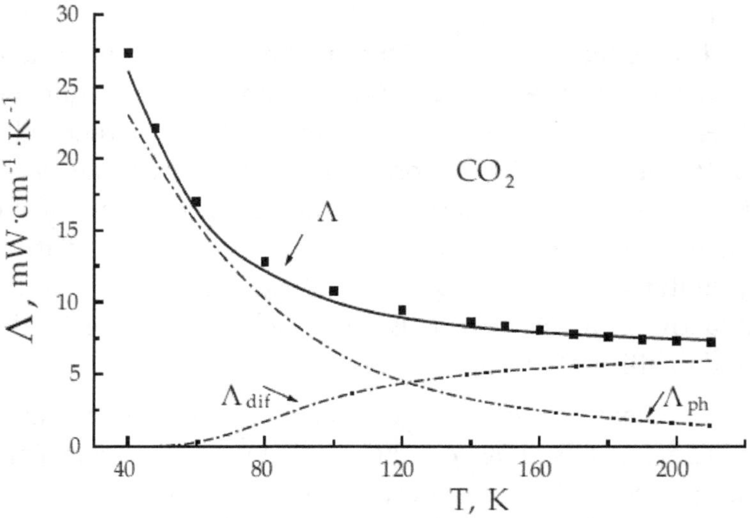

Fig. 15. Contributions Λ_{ph} and Λ_{dif} for CO_2 sample with molar volume 25.8 cm^3/mole.

It is seen that in *Kr* thermal conductivity possesses pure phonon character up to 80*K*. "Diffusive" modes arise above 80*K* and $\Lambda_{ph} \cong \Lambda_{dif}$ at 160*K*. In CO_2 where the strong additional phonon scattering arises due to TO coupling, "diffusive" modes arise at 50*K* and in the high temperature region most part of the heat is transferred by "diffusive" modes.

The applicability of such description is supported by the straightforward calculations of the thermal conductivity by the method of molecular dynamics using the Kubo-Greenwood formula. Recently, the thermal conductivity of solid argon with the Lennard-Jones potential has been described using two contributions made by low-frequency phonons with mean- free paths exceeding half the wavelength and high-frequency phonons with mean-free path of about half the wavelength (McGaughey & Kaviany, 2004).

In the orientationally-disordered phases ω_0 follows from Eqs. (20) and (21):

$$\omega_0 = -\frac{u}{\left(-\eta + \sqrt{u^3 + \eta^2}\right)^{1/3}} + \left(-\eta + \sqrt{u^3 + \eta^2}\right)^{1/3} \quad (25)$$

where the parameters *u* and η are

$$u = \frac{\pi \rho^2 \upsilon^7}{3C^2 k_B C_{rot} T}\left(A + \frac{B^2 \Lambda_{rot}}{\rho \upsilon^4}\right), \quad \eta = -\frac{\rho^2 \upsilon^7}{2\alpha C^2 k_B C_{rot} T^2} \quad (26)$$

The appropriate contributions to the thermal conductivity from the low-frequency phonons Λ_{ph} and the high-frequency "diffusive" modes Λ_{dif} are

$$\Lambda_{ph} = \frac{\hbar^2}{2\pi^2 v^2 k_B T^2} \int_0^{\omega_0} l_\Sigma(\omega)\omega^4 \frac{e^{\frac{\hbar\omega}{k_B T}}}{\left(e^{\frac{\hbar\omega}{k_B T}} - 1\right)^2} d\omega \quad (27a)$$

$$\Lambda_{dif} = \frac{\alpha\hbar^2}{2\pi v k_B T^2} \int_{\omega_0}^{\omega_D} \omega^3 \frac{e^{\frac{\hbar\omega}{k_B T}}}{\left(e^{\frac{\hbar\omega}{k_B T}} - 1\right)^2} d\omega \quad (27b)$$

Figs. 16 and 17 show the ratio ω_0/ω_D upon temperature and contributions Λ_{ph} and Λ_{dif} to the thermal conductivity of solid *CO*. The results were computer-fitted by the least-square method to the smoothed values of the thermal conductivity of the sample with V_m=27.93 cm³/*mole* individually in the α– and β–phases. The polarization–averaged sound velocity corresponding to this density was 1280 *m/s*. It was assumed that C_{rot} varies linearly from the value *2R* to *R* during β–phase (Manzhelii et al. 1997). The varied parameters were α and *A* in the α–phase and α, *A* and *B* in the β–phase. The best agreement with the experimental results was achieved with *α=1.55* and *A= 4.3 ×10⁻¹⁶s/K* in the α–phase and with *α=1.25*, *A= 1.0×10⁻¹⁷s/K*, *B=5.0* and *C=2.24* in the β–phase. As the temperature rises, the ratio ω_0/ω_D decreases in the α-phase and increases in the β-phase. This increase can be attributed to decreasing of rotational correlations between the neighboring molecules become weaker.

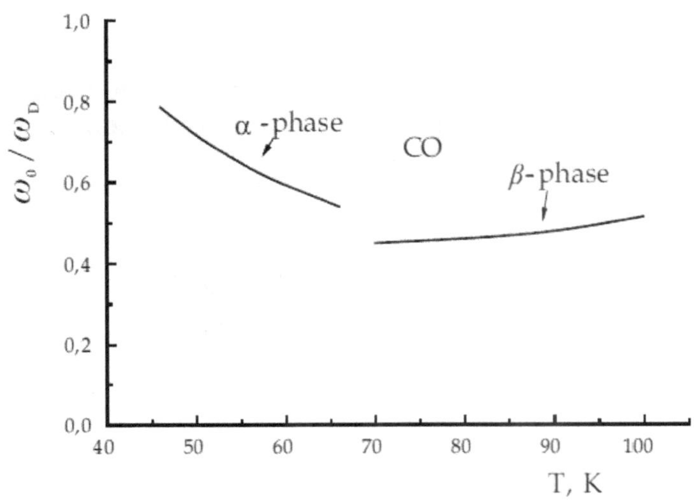

Fig. 16. The ω_0/ω_D ratio upon temperature.

It is seen that near *T=45K* most of the heat is transported by the phonons (the contribution of the "diffusive" modes is no more than *10%*). However, immediately before the $\alpha \rightarrow \beta$ transition over half of the heat is transported by the "diffusive" modes. In the orientationally-disordered phase the contribution of the "diffusive" modes immediately after phase transition is about two times larger than that of the phonons. As the temperature rises, the contribution of the "diffusive" modes decreases and that of the phonons increases because the scattering of the phonons by the short-range orientational order fluctuations becomes weaker due to their attenuation damping. The estimates show that both three- phonon scattering and one-phonon scattering became the dominant mechanisms. The lower limit of the thermal conductivity is reached when the mean-free paths of all the modes are $\alpha\lambda/2$. It absolutely agrees with Eq. (4) if the polarization-averaged sound velocity $v=(v_l+2v_t)/3$ and $\alpha=1$ are used.

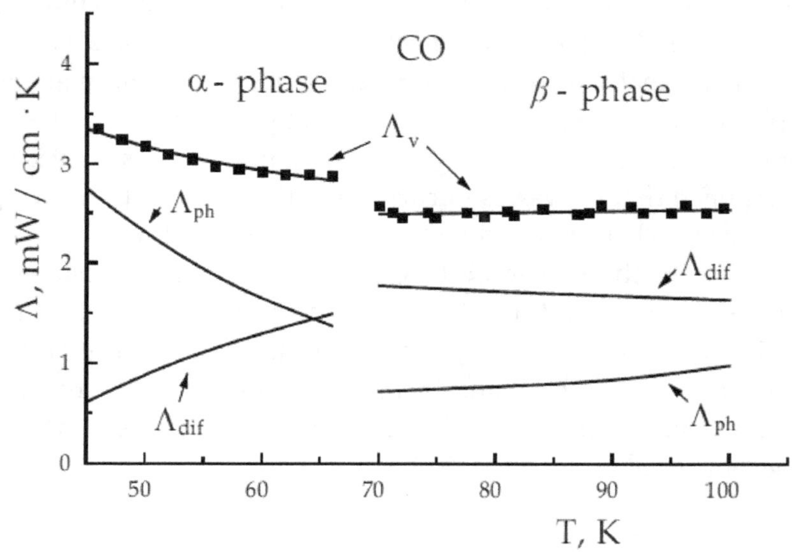

Fig. 17. Results of fitting Λ_V and the contributions to the thermal conductivity of CO sample with V_m=27.93 cm^3/mole from low-frequency phonons Λ_{ph} and "diffusive" modes Λ_{dif}.

3.5 Mixed crystals

In the case if thermal conductivity is governed by combined *U*-processes and point defect scattering it follows from Eqs. 8, 10 and 21 that

$$\omega_0 = \frac{1}{(2\alpha\pi B)^{\frac{1}{3}}}\left[\sqrt[3]{1+\sqrt{1+u}}+\sqrt[3]{1-\sqrt{1+u}}\right] \tag{28}$$

where

$$u = \frac{4\alpha^2\pi^2 A^3 T^3}{27B} \tag{29}$$

The appropriate contributions to the thermal conductivity from the low-frequency phonons Λ_{ph} and the high-frequency "diffusive" modes Λ_{dif} are

$$\Lambda_{ph} = \frac{k_B}{2\pi^2 v} \frac{1}{\sqrt{ATB}} arctg \sqrt{\frac{B}{AT}} \omega_0 \tag{30a}$$

$$\Lambda_{dif} = \frac{\alpha k_B}{4\pi v} (\omega_D^2 - \omega_0^2) \tag{30b}$$

Fig. 18 presents the relative low-frequency phonon contribution Λ_{ph}/Λ to the thermal conductivity of pure *Kr* and the $(CH_4)_{1-\xi}Kr_\xi$ solid solution. A gradual transition from the thermal conductivity of a highly perfect crystal to the "lower limit" to thermal conductivity Λ_{min} was observed at $T \geq \Theta_D$ in the $(CH_4)_{1-\xi}Kr_\xi$ solid solution as the crystal becomes increasingly more disordered. It is seen that in *Kr* the thermal conductivity is of pure phonon character up to about 90*K*. As the CH_4 concentration increases, progressively more heat is transferred by the "diffusive" modes, but even at the highest concentration (29% of CH_4 in *Kr*) and the highest temperatures (150*K*) an appreciable part of the heat (about 10%) is transferred by the low-frequency phonons. The values of α vary from 1 to 1.4. This supports the view about the vibrations localized in the λ/2 regions as the limiting case of the phonon picture (Cahill, 1992) and validity of Eq. (4) for prediction of the lower limit to the thermal conductivity of crystalline lattice. The ω_0/ω_D ratio decreases with rise of temperature and concentration of impurity CH_4 in *Kr*.

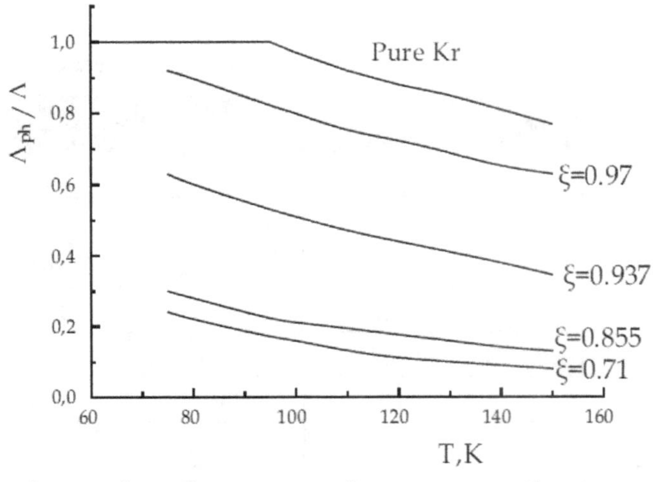

Fig. 18. The relative low-frequency phonon contribution to the thermal conductivity of pure Kr and $(CH_4)_{1-\xi}Kr_\xi$ solid solution.

3.6 The Bridgman coefficient g and molar volume dependence of thermal conductivity.

Subsequent studies of the thermal conductivity as a function of pressure (Ross et al., 1984) and as a function of density (Konstantinov et.al., 1988-2010) for a wide range of substances have shown that the values of the Bridgman coefficient vary, as a rule, in a range from 3–4 to 10–15. The general tendency is for *g* to decrease with increasing structural disorder; the weakest volume dependence of the thermal conductivity is found in glasses and polymers. In three cases—the *Ih* phase of ice, the phase NH_4F *(I)*, and *CuCl*—the Bridgman coefficient is negative. This is explained by anomalous behavior of the transverse modes, the velocity of which decreases with increasing pressure.

Values of the Bridgman coefficient *g*, measured at temperature *T* for some molecular substances are shown in Tab. 3. The problem of determining the Bridgman coefficient *g* in the present model reduces to finding the volume derivative of the expressions for Λ_{ph} and Λ_{dif}. Since $(\partial \ln A/\partial \ln V)_T = 3\gamma + 2q - 2/3$ we obtain

$$g = \frac{\Lambda_{ph}}{\Lambda} g_{ph} + \frac{\Lambda_{dif}}{\Lambda} g_{dif}, \tag{31}$$

where

$$g_{ph} = 5\gamma + 4q - 1, \tag{32a}$$

$$g_{dif} = -\left(\frac{\partial \ln \Lambda_{dif}}{\partial \ln V}\right)_T = -\gamma + \frac{1}{3} + \frac{2}{\omega_D^2 - \omega_0^2}(\omega_D^2\gamma - \omega_0^2\gamma_0), \tag{32b}$$

$$\gamma_0 = 3\gamma + 2g - 1/3, \tag{33}$$

Table 3. Values of the Bridgman coefficient g, measured at temperature T.

Substance	T, K	g	Substance	T, K	g
Ar	90	9.7	HMT	300	8.9
Kr	120	9.4	Adamantane	320	6.4
Xe	160	9.2	Adamantane	300	9.8
CO_2	220	6.2	Cyclohexane	273	5.5
N^2o	180	5.7	Naphthalene	300	8.5
α-N_2	35	6.0	Anthracene	300	8.9
β-N_2	60	4.3	Sulphure	300	6.0
α-*CO*	60	5.2	NH_4Cl *(II)*	298	-6.2
β-*CO*	70	4.0	NH_4Cl *(III)*	160	8.6

Substance	T, K	g	Substance	T, K	g
γ-O_2	55	3.8	NH_4F (I)	298	-6.2
CH_4 (I)	90	8.8	NH_4F (II)	380	7.5
CCl_4 (Ib)	250	5.8	NH_4F (III)	386	18
CCl_4 (II)	225	6.5	H^2o (Ih)	120	-3.9
CBr_4 (I)	360	3.4	H^2o (VII)	286	4.8
CBr_4 (II)	300	3.8	H^2o (VIII)	246	4.8
CHF_3	118	4.6	C_2H_6	88	5.5
$CHCl_3$	210	3.9	C_2F_6	170	4.5
CH_2Cl_2	175	4.6	C_3H_8	85	7.5
CF_2Cl_2	115	5.0	C_6H1_4	178	7.6
CHF_2Cl(I)	115	4.5			
C_6H_6	273	7.5			
SF_6 (I)	220	5.2			

In our opinion, Eq. 31 has a general character. The main idea pursued in this paper is that the contributions to the molar volume dependence of the thermal conductivity from acoustic phonons and "diffusive" modes are sharply different. If the heat is transferred mainly by acoustic phonons (perfect crystals), then the Bridgman coefficient is described by Eq. 3. In the opposite case, when the thermal conductivity has reached its lower limit Λ_{min} and all the heat is transferred by "diffusive" modes (amorphous solids and strongly disordered crystals), then for $T>\Theta_D$ the lower limit to the thermal conductivity is described by $\Lambda_{min} \propto v/a^{2/3}$, and

$$g=\gamma+1/3 \tag{34}$$

This also follows from Eq. 32b in the limit $\omega_0 \to 0$. In the general case the Bridgman coefficient g is a weighted average over the acoustic and "diffusive" modes. Fig. 18 shows the Bridgman coefficients calculated for CO_2 and N^2o according to Eqs. 31-33. The temperature dependence of g was not investigated experimentally. For CO_2 and N^2o the Bridgman coefficients were determined only at the triple-point temperature. It is seen in Fig. 19 that the agreement between the experimental and computed values of g is completely satisfactory. It was also found for a number of molecular crystals and their solutions (Konstantinov et al., 2002b, 2003b).

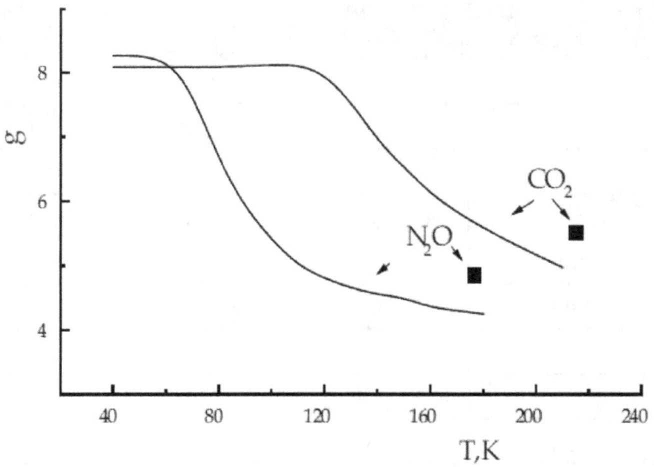

Fig. 19. Temperature dependence of the Bridgman coefficient g for solid CO_2 and N^2o calculated according to Eqs. 31-33. Dark squares are experimental values (Konstantinov et al., 1988b).

3.7 Lower limit to thermal conductivity and rotational energy transfer

It was found that the values of coefficient α for molecular crystals, which express the ratio of the lower limit to thermal conductivity obtained by fitting procedure to Λ_{min} calculated from Eq. 4 vary from 2 to 4 (Konstantinov et al., 2003). These values are much larger than for solidified inert gases, where α lies in the range 1.2-1.4. The most obvious reason for this difference is that the site-to-site rotational energy transfer must be taken into account. Eq. 4 for the lower limit of the thermal conductivity is valid for substances consisting of atoms other than molecules with rotational degrees of freedom. Slack, 1979 has taken into account the possibility of thermal energy transfer by optical phonons in crystals consisting of atoms of different kinds. In molecular crystals heat is transferred by mixed translational-orientational modes, whose specific heat for $T \geq \Theta_D$ saturates in proportion to the number of degrees of freedom. On this basis the following expression can be suggested for the lower limit of the thermal conductivity of molecular crystal Λ^*_{min} whose molecules have *z* rotational degrees of freedom.

$$\Lambda^*_{min} = \frac{1}{2}\left(\frac{\pi}{6}\right)^{1/3}\left(1+\frac{z}{3}\right)k_B n^{2/3}(v_l + 2v_t), \tag{35}$$

It should be noted that although the Eq. 4 for Λ_{min} describes well, over all, the thermal conductivity of amorphous substances and strongly disordered crystals, it is nonetheless semi-empirical. The assumption that the minimum phonon mean-free path length is equal to half the wavelength is only one of many possibilities. Thus, Slack, 1979 assumed that it is equal to the phonon wavelength. In addition, the expression was obtained on the basis of a very simple Debye model which neglects phonon dispersion and the real density of states. The coefficient α is in such an event an integral factor that effectively takes account of the imperfection of the model.

3.8 Additional issues

To date, no coherent theory exists that describes the thermal conductivity of molecular crystals from first principles based on the characteristics of crystalline potential. This description is exclusively applicable to solid methane and deuteromethane (Yasuda, 1978). There is no definitive assurance that the semi-empirical technique of Manzhelii et al., 1968, is entirely sufficient for characterizing the enhancement of thermal conductivity with increasing temperature in scenarios involving impeded molecular rotation, especially in the context of uniaxial rotation. Is there a distinct correlation between the frequency of molecule reorientations and the characteristics of thermal conductivity? What is the minimum value of thermal conductivity in molecular crystals? Is it necessary to permit the movement of rotational energy across sites? The reasons supporting the suitability of Equation 35 in delineating the lower limit of heat conductivity in molecular crystals are insufficient. There is insufficient evidence regarding the temperature dependency of the Bridgman coefficient g across a broad temperature range to provide comparison with estimated values inside the proposed model.

REFERENCES

Auerbach, A. & Allen P.B., (1984). Universal high-temperature saturation in phonon and electron transport. *Physical Review B.*, Vol. 29, No. 6, pp. 2884-2890.

Batchelder, D.N. (1977). In *Rare Gas Solids.* Eds. Klein, M.L.; Venables, J.A., London, New York, San Francisco: Academic Press, Vol. II, 1977, pp. 883-920.

Berman, R. (1976). *Thermal Conduction in Solids*, Clarendon Press, Oxford (1976). Bondarenko, A.I.; Manzhelii, V.G.; Popov, V.A.; Strzhemechnyi, M.A. & Gavrilko V.G. (1982). Isochoric thermal conductivity of crystalline *Kr* and *Xe*. Heat transfer by vacancies. *Soviet Journal Fizika Nizkikh Temperatur*, Vol. 8, No. 11, pp. 1215-1224.

Bour, J. (1948). Quantum theory of condensed permanent gases I. Quantum-mechanical low of corresponding states. *Physica*, Vol. 14, No 1, pp. 139–148.

Briels, W.J.; Jansen, A.P.J. & van-der Avoird A. (1985). Translational-rotational coupling in strongly anharmonic molecular crystals with orientational disorder, *Chemie Physics et de Physico-Chemical Biologique*, Vol. 82, No. 2-3, pp. 125-136.

Cahill, D.G.; Watson, S.K. & Pohl R.O., (1992). Lower limit to thermal conductivity of disordered crystals. *Physical Review B.*, Vol. 46, No. 10, pp. 6131-6140.

Clayton, F. & Batchelder, D.N. (1973). Temperature and volume dependence of the thermal conductivity of solid argon. *Journal of Physics. C.*, Vol. 6, No. 7, pp. 1213-1228.

Ecsedy, D.J. & Klemens, P.G. (1977).Thermal resistivity of dielectric crystals due to four- phonon processes and optical modes. *Physical Review B.*, Vol. 15, No. 12, pp. 5957– 5962.

Feldman, J.L.; Kluge, M.D.; Allen, P.B. & Wooten F. (1993). Thermal conductivity and localization in glasses: Numerical study of a model of amorphous silicon. *Physical Review B.*, Vol. 48, No. 17, pp. 12589-12602.

Forsman, H. & Andersson, P. (1984). Thermal conductivity at high pressure of solid odd numbered *n*-alkanes ranging from *C9H20* to *C19H40*. *Journal of Chemical Physics*, Vol. 80, No. 6, pp. 2804-2807.

Gorodilov, B.Ya.; Korolyuk O.A.; Krivchikov, A.I. & Manzhelii V.G. (2000). Heat Transfer in Solid Solutions Hydrogen-Deuterium. *Journal of Low Temperature Physics*, Vol. 119, No. 3/4, pp. 497-505.

Jeżowski, A.; Stachowiak, P.; Sumarokov V.V., Mucha J. & Freiman, Yu.A. (1993). Thermal conductivity of solid oxygen. *Physical Review Letters*, Vol. 71, pp. 97–100.

Jeżowski A.; Misiorek H.; Sumarokov V.V. & Gorodilov B.Ya. (1997). Thermal conductivity of solid methane. *Physical Review B.*, Vol. 55, No. 9, pp. 5578-5580.

Kawasaki K. (1963). On the behavior of the thermal conductivity near the magnetic transition point. *Progress of Theoretical Physics,* Vol. 29, No 6, pp. 801- 816.

Konstantinov, V.A.; Manzhelii, V.G.; Strzhemechnyi, M.A. & Smirnov, S.A. (1988a). The $\Lambda \propto 1/T$ law and isochoric thermal conductivity of rare gas crystals. *Soviet JournalFizika Nizkikh Temperatur*, Vol. 14, No. 1, pp. 48-54.

Konstantinov, V.A.; Manzhelii, V.G.; Tolkachev, A.M. & Smirnov, S.A. (1988b). Heat transfer in solid CO_2 and N^2o: dependence on the temperature and volume. *Soviet Journal Fizika Nizkikh Temperatur* Vol. 14, No. 2, pp. 412-417.

Konstantinov, V.A.; Manzhelii,V.G. & Smirnov, S.A. (1988c). Isochoric thermal conductivity of solid CO_2 with N^2o and *Xe* impurities. *Soviet Journal Fizika Nizkikh Temperatur*, Vol. 14, No. 7, pp. 412-413.

Konstantinov, V.A.; Manzhelii, V.G. & Smirnov, S.A. (1991a). Isochoric thermal conductivity and thermal pressure of solid *CCl4*. *Physica Status Solidi (b).* Vol. 163, pp. 368-374.

Konstantinov, V.A.; Manzhelii,V.G. & Smirnov, S.A. (1991b). Isochoric thermal conductivity of solid *CHCl3* and *CH2Cl2*. The role of rotational motion of molecules. *Low Temperature Physics*, Vol. 17, No. 7, pp. 462-464.

Konstantinov, V.A.; Manzhelii,V.G. & Smirnov, S.A. (1992a). Temperature dependence of isochoric thermal conductivity of crystalline benzene. *Ukrainian Physical Journal*, Vol. 37, No. 5, pp. 757-760.

Konstantinov, V.A.; Manzhelii,V.G. & Smirnov, S.A. (1992b). Influence of rotational motion of molecules on heat transfer in solid *SF_6*. *Low Temperature Physics*, Vol. 18, No.11 pp. 902-904.

Konstantinov, V.A. & Manzhelii, V.G. (1994). Phonon scattering and heat transfer in simple molecular crystals. In "*Die Kunst of Phonons*" ed. by T. Paskiewizc, T. Rapsewizc., New York, London: Plenum Press. 1994. pp. 321-332.

Konstantinov, V.A.; Manzhelii, V.G. & Smirnov, S.A. (1995). Isochoric thermal conductivity of solid freons of methane series: *CF_2Cl_2* and *CHF_2Cl*. *Low Temperature Physics*, Vol. 21, No. 1. pp. 78-81.

Konstantinov, V.A.; Manzhelii, V.G.; Revyakin, V.P. & Smirnov, S.A. (1999a). Heat transfer in the orientationally disordered phase of solid methane. *Physica B*. Vol. 262. pp.421-425.

Konstantinov, V.G.; Revyakin, V.P. & Smirnov, S.A. (1999b). A setup for measuring of isochoric thermal conductivity of solidified gases and liquids. *Instruments and Experimental Techniques*. Vol. 42, No. 1, pp. 132-135.

Konstantinov V.A. (1999c). Heat transfer in simple molecular crystals at temperatures of the order of Debye ones and above them. *Functional Materials*,Vol. 6, No. 2, pp. 335–339.

Konstantinov, V.A.; Manzhelii, V.G.; Revyakin, V.P. & Smirnov, S.A. (2000). Manifestation of lower limit to thermal conductivity of solid *Kr* with *CH_4* impurities. *Physica B.*, Vol. 291, pp. 59-65.

Konstantinov, V.A. Manifestation of the lower limit to thermal conductivity in the solidified inert gases. (2001a). *Journal of Low Temperature Physics*. Vol. 122, No. 3/4, pp. 459-465.

Konstantinov, V.A.; Manzhelii, V.G.; Pohl, R.O. & Revyakin V.P. (2001b). Search for the minimum thermal conductivity in mixed cryocrystals $(CH)_{1-\xi}Kr_{\xi}$. *Low Temperature Physics*, Vol. 27, No. 1. pp. 1159-1169.

Konstantinov, V.A.; Pohl, R.O. & Revyakin V.P. (2002a). Heat transfer in solid *$Kr_{1-\xi}Xe_{\xi}$* solution. *Soviet Physics of Solid State*, Vol. 44, No. 5, pp. 824-829.

Konstantinov, V.A.; Revyakin, V.P. & Orel, E.S. (2002b). Molar volume dependence of thermal conductivity in mixed cryocrystals. *Low Temperature Physics*, Vol. 28, No. 2. pp. 136-139.

Konstantinov, V.A.; Revyakin, V.P. & Orel, E.S. (2003a). Heat transfer by low-frequency phonons and „diffusive" modes in cryocrystal solutions: the *Kr-Xe* system. *Low Temperature Physics*, Vol. 29, No. 9-10. pp. 759-762.

Konstantinov, V.A. (2003b). Heat transfer by low-frequency phonons and „diffusive" modes in molecular crystals. *Low Temperature Physics*, Vol. 29, No. 5. pp. 422-428.

Konstantinov, V.A.; Manzhelii, V.G.; Revyakin, V.P. & Sagan V.V. (2005a). Heat transfer in γ-phase of oxygen. *Journal of Low Temperature Physics*, Vol. 139, No. 5-6. pp. 703-709.

Konstantinov, V.A.; Revyakin, V.P.; Manzhelii V.G. & Sagan V.V. (2005b). Isochoric thermal conductivity of solid nitrogen. *Low Temperature Physics*, Vol. 31, No. 5. pp. 419-422.

Konstantinov, V.A.; Manzhelii, V.G.; Revyakin, V.P. & Sagan V.V. (2006a). Isochoric thermal conductivity of solid carbon oxide: the role of phonons and "diffusive" modes. *Journal of Physics: Condensed Matter*, Vol. 18, pp. 9901-9909.

Konstantinov, V.A.; Manzhelii, V.G.; Revyakin, V.P. & Sagan V.V. (2006b). Extraordinary temperature dependence of isochoric thermal conductivity of crystalline CO_2 doped with inert gases. *Low Temperature Physics*, Vol. 32, No. 11, pp. 1076-1077.

Konstantinov, V.A.; Revyakin, V.P. & Sagan V.V. (2006c). Rotation of the methyl groups and thermal conductivity of molecular crystals: ethane. *Low Temperature Physics*, Vol. 32, No. 7. pp. 689-694.

Konstantinov, V.A.; Revyakin, V.P. & Sagan V.V. (2009a). Heat transfer in solid halogenated methanes: trifluoromethane. *Low Temperature Physics*, Vol. 35, No. 4. pp. 286-289.

Konstantinov, V.A.; Revyakin, V.P. & Sagan V.V. (2009b).The isochoric thermal conductivity of solid *n*-alkanes: propane C_3H_8. *Low Temperature Physics*, Vol. 35, No. 7. pp. 577-579.

Konstantinov, V.A.; Revyakin, V.P. & Sagan V.V. (2010). Isochoric thermal conductivity of solid *n*-alkanes: hexane *C6H14*. *Low Temperature Physics*, Vol. 36, (in press).

Kokshenev, V.B.; Krupskii I.N. & Kravchenko Yu.G. (1997). High-temperature thermoconductivity of nitrogen-type crystals. *Brazilian Journal of Physics*, Vol. 27, No. 4, pp. 510-514.

Krupski, I.N. & Manzhelii V.G. (1968). Multi phonon interaction and thermal conductivity of crystalline argon, krypton and xenon. *Soviet Journal for Experimental and Theoretical Physics*. Vol. 55, No. 6, pp. 2075-2082.

Krupskii, I.N.; Koloskova L.A. & Manzhelii V.G. (1974). Thermal conductivity of deuteromethane. *Journal of Low Temperature Physics*. Vol. 14, No. 3-4, pp. 403-410.

Lynden-Bell, R.M. & Michel K.H. (1994). Translation-rotation coupling, phase transitions, and elastic phenomena in orientationally disordered crystals. *Review of Modern Physics*, Vol. 66, No. 3, pp. 721-762.

McGaughey, A.J.H. & Kaviany, M. (2004). Thermal conductivity decomposition and analysis using molecular dynamics simulations. Part I. Lennard-Jones argon. *International Journal of Heat and Mass Transfer*, Vol. 47, pp. 1783–1798.

Manzhelii, V.G. & Krupski, I.N. (1968). Thermal conductivity of solid methane. *Soviet Journal Physics of Solids*. Vol. 10, No 1, pp. 284-286.

Manzhelii V.G.; Kokshenev V.B.; Koloskova L.A. & Krupski I.N. (1975). Phonon-libron interaction and thermal conductivity of simplest molecular crystals. *Soviet Journal Fizika Nizkich Temperetur*, Vol. 1, No. 10, pp. 1302-1310.

Manzhelii V.G. & Freiman Yu.A. (1997). In: *Physics of cryocrystals*. Eds. Manzhelii, V.G. & Freiman Yu. A., Woodbury, New York: AIP PRESS. pp. 1-514.

Parsonage, N. & Stavaley, L. (1972). *Disorder in crystals*, Clarendon Press, Oxford.

Pursky, O.I.; Zholonko N.N. & Konstantinov, V.A. (2003). Influence of the rotational motion of molecules on the thermal conductivity of solid *SF6, CHCl3, C6H6*, and *CCl4*. *Low Temperature Physics*, Vol. 29, No. 9/10, pp. 771-775.

Pursky, O.I. & Konstantinov V.A. (2008). Thermal conductivity of solid CHF2Cl. *Physica B.*, Vol. 403, pp. 190-194.

Purski, O.I.; Zholonko, N.N. & Konstantinov V.A. (2000). Heat transfer in the orientationally disordered phase of *SF6*. *Low Temperature Physics*, Vol. 21, No. 1. pp. 278- 281.

Ross, R.G.; Andersson, P.; Sundqvist, B. & Bäckström G. (1984). Thermal conductivity of solids and liquids under pressure. *Report of Progress in Physics*, Vol. 47, pp. 1347–1355.

Roufosse, M.C. & Klemens P.G. (1974). Lattice thermal conductivity of minerals at high temperatures. *Journal of Geophysical Researches*, Vol. 79, No. 5, pp. 703-705.

Sheng, P.; Zhou M. & Zhang Z. (1994). Phonon transport in strong-scattering media. *Physical Review Letters*, Vol. 72, No. 2, pp. 234-237.

Slack, G.A. (1972). In: *Proceedings of International Conference on phonon scattering in solids*, Eds. Albany H.J., Service de documentation du CEN, Saclay, pp. 24-32.

Slack, G.A. (1979). The thermal conductivity of nonmetallic crystals. *Solid State Physics*. Vol. 34, pp. 1-71.

Stachowiak, P.; Sumarokov V.V.; Mucha J. & Jezowski A. (1994). Thermal conductivity of solid nitrogen. *Physical Review B.*, Vol. 50, No. 1, pp. 543-546.

Yasuda H. (1978). Thermal conductivity of solid *CH4* and *CD4*. *Journal of Low Temperature Physics,* Vol. 31, No. 1/2, pp. 223-256.

7 Chapter NONLINEAR BUBBLE DYNAMICS INDUCED BY THERMAL TRANSFER

1. INTRODUCTION: AN OVERVIEW

Prior investigations into the forced oscillation of a spherical bubble in a solution have employed the Rayleigh equation to derive the time-dependent bubble radius and a polytropic relation to ascertain the internal gas pressure as a function of bubble volume (Lauterborn, 1976). The polytropic approximation, with appropriate index values, has been extensively employed for gases undergoing quasi-equilibrium processes including heat transfer. However, the polytropic pressure-volume relationship fails to account the thermal damping due to heat transfer through the bubble wall because $P_b\, dV_b$ is a perfect differential and consequently its integral over a cycle vanishes (Prosperiretti et al., 1988) where P_b is the gas pressure inside the bubble and V_b is the bubble volume. Furthermore, the polytropic approximation assumes the uniform temperature for the gas intrinsically, which is valid only for a particular case and it is hard to tell whether the gas inside the bubble oscillating under ultrasound behaves isothermally or adiabatically (Loefstedt et al., 1993).

In this study, we have formulated a general bubble dynamics model, which is as follows.

The density, velocity and pressure distributions for the gas inside a spherical bubble were obtained by solving the continuity and momentum equations analytically (Kwak et al., 1995, Kwak and Yang, 1995). With the set of analytical solutions for the conservation equations, the temperature distribution for the gas inside the bubble was also obtained by solving the energy equation for the gas. The heat transfer through the bubble wall was considered to obtain the instantaneous thermal boundary layer thickness from the mass and energy conservations for the liquid layer adjacent to the bubble wall by the integral method. The mass and momentum equations for the liquid outside the bubble wall provided the well known equation of motion for the bubble wall, the Rayleigh-Plesset equation in an incompressible limit or the Keller-Miksis equation in a compressible limit. The bubble dynamics model was applied to an evolving bubble formed form the fully evaporated droplet at the superheat limit (Kwak et al., 1995) and phenomena of sonoluminescence which is light emission associated with the catastrophic collapse of a micro-bubble oscillation under ultrasound (Young, 2005).

With uniform density, temperature and pressure approximations which are valid for the characteristic time scale of ms, the calculated values of the far field pressure signal from the evolving the bubble formed form the fully evaporated droplet at its superheat limit (Kwak et al., 1995) are in good agreement with the experimental results (Shepherd and Sturetevant, 1982). With uniform pressure approximation which is valid for the characteristic time scale of *μs*, the calculated values of the minimum velocity of the bubble wall, the peak temperature and pressure are excellent agreement with the observed ones for the sonoluminescing xenon bubble in sulfuric acid solutions (Kim et al., 2006). Furthermore, the calculated bubble radius-time curve displays alternating pattern of bubble motion which is apparently due to the heat transfer for the sonoluminescing xenon bubble, as observed in experiment (Hopkins et al., 2005). The bubble dynamics model presented in this study has also revealed that the sonoluminescence for an air bubble in water solution occurs due to the increase and subsequent decrease in the bubble wall acceleration which induces pressure non-uniformity for the gas inside the bubble during ns range near the collapse point (Kwak and Na, 1996). The calculated sonoluminescence pulse width from the instantaneous gas temperature for air bubble is in good agreement with the observed value of 150 *ps* (Byun et al., 2005). Due to enormous heat transfer the gas temperature inside the sonoluminescing air bubble at the collapse point is about 20000~40000 K instead of 10^7 K (Moss et al., 1994) which is estimated to be in the adiabatic case. Molecular dynamics (MD) simulation results for the sonoluminescing xenon bubble were compared to the theoretical predictions and observed results (Kim et al., 2007, Kim et al., 2008).

2. TEMPERATURE PROFILE IN THERMAL BOUNDARY LAYER

A sketch of the bubble model employed is given in Fig.1, which shows a spherical bubble in liquid temperature T_∞ and liquid pressure P_∞. Heat transfer is assumed to occur through the thermal boundary layer of thickness δ(t). The temperature profile in this layer is assumed to be quadratic (Theofanous et al., 1969).

$$\frac{T - T_\infty}{T_{bl} - T_\infty} = (1 - \xi)^2 \tag{1}$$

where T_{bl} is the temperature at the bubble wall and T_∞ is the ambient temperature in Eq.(1). The parameter ξ in Eq.(1) is given as $\xi = (r - R_b)/\delta$ and $R_b(t)$ is the instantaneous bubble radius. Such a second order curve satisfies the following boundary conditions:

$$T(R_b,t) = T_{bl}, \quad T(R_b + \delta,\, t) = T_\infty \quad \text{and} \left(\frac{\partial T}{\partial r}\right)_{r=R_b+\delta} = 0 \tag{2}$$

The heat transfer conducted through this thermal boundary layer whose thickness is δ(t) can be obtained by applying the Fourier law at the bubble wall, or

$$\dot{Q}_b = k_l A_b \left(\frac{\partial T}{\partial r} \right)_{r=R_b} = - \frac{8\pi R_b^2 k_l (T_{bl} - T_\infty)}{\delta} \tag{3}$$

where A_b is the surface area of bubble and k_l is the conductivity of liquid. The bubble model including such liquid phase zone has been verified experimentally (Suslick et al., 1986).

3. CONSERVATION EQUATIONS FOR THE GAS INSIDE BUBBLE

The hydrodynamics related to studying the bubble behavior in liquid involves solving the Navier-Stokes equations for the gas inside the bubble and the liquid adjacent the bubble wall. Especially the knowledge of the behavior of gas or vapor inside evolving bubble is a key element to understand the bubble dynamics. Firstly, various conservation laws for the gas are considered to obtain the density, pressure and temperature distributions for the gas inside the bubble.

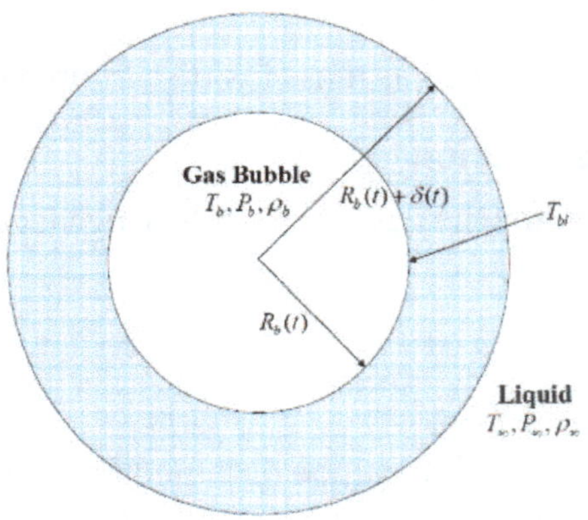

Fig. 1. A physical model with thermal boundary layer for a spherical bubble in liquid.

3.1 Mass conservation

The mass conservation equation for the gas inside a bubble is given by

$$\frac{D\rho_g}{Dt} + \rho_g \nabla \cdot \bar{u}_g = 0 \tag{4}$$

where ρ_g and u_g are gas density and velocity, respectively. With decomposition of the gas density in spherical symmetry as $\rho_0(t) + \rho_r(r,t)$, the continuity equation becomes

$$\left[\frac{d\rho_o}{dt} + \rho_o \nabla \cdot \bar{u}_g \right] + \frac{D\rho_r}{Dt} + \rho_r \nabla \cdot \bar{u}_g = 0 \tag{5}$$

where ρ_0 is the gas density at the bubble center and ρ_r is the radial dependent gas density inside the bubble and the notation of the total derivative used here is D/Dt = ∂/t + u(∂/∂r). The rate of change of the density of a material particle can be represented by the rate of volume expansion of that particle in the limit V→ 0 (Panton, 1996). Or

$$u_g = \frac{\dot{R}_b}{R_b} r \tag{6}$$

With this velocity profile, the density profile can be obtained as

$$\rho_g = \rho_o + \rho_r \tag{7-1}$$

$$\rho_0 R_b^3 = \text{const.} \tag{7-2}$$

$$\rho_r = ar^2 / R_b^5 . \tag{7-3}$$

The constant, a is related to the gas mass inside a bubble (Kwak and Yang, 1995).

3.2 Momentum conservation

The momentum equation for the gas when neglecting viscous forces may be written as

$$\rho_g \frac{D\bar{u}_g}{Dt} = -\nabla P_b \tag{8}$$

The gas pressure P_b inside bubble can be obtained from this equation by using the velocity and density profiles given in Eq. (6) and (7), respectively. Or

$$P_b = P_{b0} - \frac{1}{2}\left(\rho_0 + \frac{1}{2}\rho_r \right) \frac{\ddot{R}_b}{R_b} r^2 \tag{9}$$

Note that the linear velocity profile showing the spatial inhomogeneities inside the bubble is a crucial ansatze for the homologous motion of a spherical object, which is interestingly encountered in another energy focusing mechanism of gravitational collapse (Jun and Kwak, 2000). The quadratic pressure profile given in Eq. (9), was verified recently by comparison with direct numerical simulations (Lin et al., 2002).

3.3 Energy conservation

Assuming that the internal energy for the gas inside a bubble is a function of gas temperature only as de = $C_{v,b} dT_b$, the energy equation for the gas inside the bubble may be written as

$$\rho_g \frac{De}{Dt} = \rho_g C_{v,b} \frac{DT_b}{Dt} = -P_b \nabla \cdot \bar{u}_g - \nabla \cdot \bar{q} \tag{10-1}$$

where Cv,b is the constant-volume specific heat and q is heat flux. The viscous dissipation term in the internal energy equation also vanishes because of the linear velocity profile. Since the solutions given in Eqs. (6), (7) and (9) also satisfy the kinetic energy equation, only the internal energy equation given in Eq. (10-1) needs to be solved. On the other hand, Prosperetti et al.(1988) solved the internal energy equation combined with the mass and momentum equation numerically to consider heat transport inside the bubble using a simple assumption. However, heat transfer through the liquid layer, which is very important in obtaining the temperature at the bubble wall, was not considered in their study.

Using the definition of enthalpy, the internal energy equation for the gas can also be written as

$$\rho_g \frac{Dh}{Dt} = \rho_g C_{p,b} \frac{DT_b}{Dt} = \frac{DP_b}{Dt} - \nabla \cdot \bar{q} \tag{10-2}$$

where $C_{p,b}$ is the constant-pressure specific heat. Eliminating DT_b/Dt from Eqs. (10-1) and (10-2), one can obtain the following heat flow rate equation for the gas pressure inside the bubble (Kwak et al., 1995, Kwak and Yang, 1995)

$$\frac{DP_b}{Dt} = -\gamma P_b \nabla \cdot \bar{u}_g - (\gamma - 1)\nabla \cdot \bar{q} \tag{11}$$

Rewriting Eq. (11), we have

$$(\gamma - 1)\nabla \cdot \bar{q} = -\frac{1}{R_b^{3\gamma}} \frac{D}{Dt}(P_b R_b^{3\gamma}) \tag{12}$$

which implies that the relation $P_b V^3\gamma$= const. holds if ∇·q = 0 inside the bubble. Substituting Eq. (12) into Eq. (10), and rearranging the equation, we have

$$\frac{D}{Dt}\left\{\ln\left(\frac{P_b R_b^3}{T_b}\right)\right\} = \left(\frac{\rho_g R_g}{P_b} - \frac{1}{T_b}\right)\frac{DT_b}{Dt} \tag{13}$$

where R_g is gas constant. If the equation of state for ideal gas, $P_{bo} = \rho_g R_g T_{bo}$ holds at the bubble center, LHS in Eq. (13) vanishes. The result can be written as

$$P_{bo} R_b^3 / T_{bo} = const., \tag{14}$$

which is consistent with Eq.(7-2).

Note that the time rate change of the pressure at the bubble center can be written as with help of Eqs. (3) and (11).

$$\frac{dP_{bo}}{dt} = -\frac{3\gamma P_{bo}}{R_b}\frac{dR_b}{dt} - \frac{6(\gamma-1)k_l(T_{bl}-T_\infty)}{\delta R_b} \tag{15}$$

The time rate change of the temperature at the bubble center can be obtained with help of Eq. (14). That is

$$\frac{dT_{bo}}{dt} = -\frac{3(\gamma-1)T_{bo}}{R_b}\frac{dR_b}{dt} - \frac{6(\gamma-1)k_l(T_{bl}-T_\infty)}{\delta R_b P_{bo}} \tag{16}$$

4. TEMPERATURE PROFILES INSIDE THE BUBBLE

4.1 Uniform pressure profile inside the bubble

A temperature profile can be obtained by solving Eq. (12) with the Fourier law by assuming that the conductivity of gas inside the bubble is constant and the gas pressure inside the bubble is uniform (Kwak et al., 1995). That is

$$T_b = \frac{r^2}{6(\gamma-1)k_g}\left[\frac{dP_b}{dt} + 3\gamma P_b\frac{\dot{R}_b}{R_b}\right] + T_{bo}(t) \tag{17}$$

where kg heat conductivity of gas inside the bubble. The above equation can be written, with help of Eq.(12), as follows:

$$T_b = (T_{bo}-T_\infty)\left[1-\left(\frac{r}{R_b}\right)^2 / (1+\eta)\right] + T_\infty \tag{18}$$

where $\eta = \frac{k_g}{k_l}\frac{\delta}{R_b}$

The temperature at the bubble wall can be obtained easily from the above equation. That is

$$T_{bl} = (T_\infty + \eta T_{bo})/(1+\eta) \tag{19}$$

The above relation shows how the bubble wall temperature is related to the temperature at the bubble center and the ambient temperature. Assigning an arbitrary value on Tbl is not permitted as a boundary condition.

An uniform temperature distribution also occurs when there is no heat flux inside the bubble. This can be achieved when the bubble oscillating period is much longer than the characteristic time of heat diffusion so that the gas distribution function depends only on thermal velocity (thermal equilibrium case). In this limit, we may obtain the gas temperature inside the bubble by taking the value of the gas conductivity as infinity in Eq. (19). That is $T_b=T_{bl}=T_{bo}$, which validates the bubble dynamics formulation with an assumption of uniform vapor temperature inside the bubble (Kwak et al., 1995). The heat transfer through the thermal boundary layer adjacent to the bubble wall determines the heat exchange between the bubble and medium in this case.

However, the temperature gradient inside the bubble should exist, provided that characteristic time of bubble evolution is much shorter than the relaxation time of the vibration motion of the gases inside the bubble, which is of the order on 10-6s for high gas temperature. If the temperature gradient inside the bubble exists inside the bubble, the heat transfer through the bubble wall depends on both the properties of the gas inside the bubble and the liquid in the thermal boundary layer. In this case one may rewrite Eq.(3), with help of Eq.(19). That is

$$\dot{Q}_b = -\frac{2A_b k_g (T_{bo} - T_\infty)}{R_b} / (1+\eta) \tag{20}$$

As long as the value of η is finite, there exists a temperature distribution inside the bubble. For a very small value of η, the heat flow rate from the bubble is solely determined from the temperature gradient of the gas inside the bubble (Prosperetti et al., 1988).

Assume the thermal conductivity for the gas inside the bubble is linearly dependent on the gas temperature such as

$$k_g = AT_b + B \tag{21}$$

For air A=5.528x10^{-5} W/mK^2 and B=1.165x10^{-2}W/mK (Prosperetti et al., 1988) and for xenon, A=1.031×10^{-5} W/mK^2 and B=3.916×10^{-3} W/mK were used. With this approximation and Fourier law, one can obtain the following temperature profile by solving Eq. (12) with uniform pressure approximation, which is quite good until the acceleration and deceleration of the bubble wall is not significant. Thus

$$T_b(r) = \frac{B}{A} \cdot \left[-1 + \sqrt{(1+\frac{A}{B}T_{b0})^2 - 2\frac{A}{B}(T_{bl} - T_\infty)(\frac{r}{R_b})^2 / \eta'} \right] \tag{22}$$

where $\frac{1}{\eta'} = \frac{B}{k_l}\frac{\delta}{R_b}$.

The temperature distribution is given in Eq. (22) is valid until the characteristic time of bubble evolution is of the order of the relaxation time for vibrational motion of the molecules (Vincenti and Kruger, 1965) and/or is much less than the relaxation time of the translational motion of the molecules (Batchelor, 1967). The temperature at the bubble wall T*bl* can also be obtained with the thermal boundary conditions given in Eq.(2). That is

$$T_{bl} = -\frac{B}{A}(1+\frac{1}{\eta'}) + \frac{B}{A}\sqrt{\left(1+\frac{1}{\eta'}\right)^2 + 2\frac{A}{B}\left(T_{bo} + \frac{A}{2B}T_{bo}^2 + \frac{T_\infty}{\eta'}\right)} \tag{23}$$

For constant gas conductivity limit, or A→0 and B→ k_g the temperature distribution inside the bubble, Eq. (22), reduces to Eq. (18).

4.2 Non-uniform pressure profile

If the bubble wall acceleration has significant value, for example, the value exceed 10^{12} m/S$_2$, the second term is comparable to the first term in RHS of Eq.(9). This occurs for the sonoluminescing air bubble during few nanoseconds of collapse phase. Taking into account the bubble wall acceleration, the heat flow rate equation given in Eq. (12) may be rewritten as with help of Eqs. (6), (7) and (9)

$$(\gamma-1)\nabla\cdot\bar{q}=-\left[\frac{dP_b}{dt}+3\gamma P_{bo}\frac{\dot{R}}{R_b}\right]+\frac{1}{2}(\rho_o+\rho_r)\left[(3\gamma-1)\frac{\dot{R}\ddot{R}}{R_b}+\frac{\dddot{R}}{R_b}\right]r^2 \tag{24}$$

Since the temperature rise due to the bubble wall acceleration is transient phenomenon occurred during few nanoseconds around the collapse point of the bubble, the above equation may be decomposed into

$$(\gamma-1)\nabla\cdot\bar{q}_o=-\left[\frac{dP_b}{dt}+3\gamma P_{bo}\frac{\dot{R}}{R_b}\right] \tag{25-1}$$

and

$$(\gamma-1)\nabla\cdot(\bar{q}-\bar{q}_o)=\frac{1}{2}(\rho_o+\rho_r)\left[(3\gamma-1)\frac{\dot{R}\ddot{R}}{R_b}+\frac{\dddot{R}}{R_b}\right]r^2 \tag{25-2}$$

Abrupt temperature rise and subsequent rapid quenching due to the bubble wall acceleration and the increase and decrease in the acceleration may be treated in another time scale (Davidson, 1972), different from the bubble wall motion. A solution of Eq.(25-2) with no temperature gradient at the bubble center is given as

$$T_b^{'}(r)=-\frac{1}{40(\gamma-1)k_g^{'}}\left(\rho_o+\frac{5}{21}\rho_r\right)\left[(3\gamma-2)\frac{\dot{R}_b\ddot{R}_b}{R_b}+\frac{\dddot{R}_b}{R_b}\right]r^4+C(t) \tag{26}$$

The coefficient C may be determined from a boundary condition $k_g dT_b/dr=k_1 dT_1/dr$ at the wall where T_1 is the temperature distribution in the thermal boundary layer with different thickness δ›. Note that the boundary conditions employed for solving Eq. (25-1) and (25-2) are the same at the bubble center. However different properties of the gas was employed at the bubble wall so that the coefficient C(t) given in Eq. (26) is given as

$$C(t)=-\frac{1}{20(\gamma-1)}\left[(3\gamma-2)\dot{R}_b\ddot{R}_bR_b+\dddot{R}_bR_b^2\right]\left[\frac{\delta'}{k_l}\left(\rho_o+\frac{5}{14}\rho_{r=R_b}\right)+\frac{R_b}{2k_g^{'}}\left(\rho_o+\frac{5}{21}\rho_{r=R_b}\right)\right] \tag{27}$$

The temperature distribution from Eq. (22) with low thermal conductivity kg can be regarded as background one because the duration of the thermal spike represented by Eq. (26) is so short less than 500 ps. The gas conductivity at ultra-high temperature kg' may be obtained from collision integrals (Boulos et al., 1994). The value of δ' may be chosen so that proper bouncing motion results after the collapse and is about 0.1 μm. The final solution of the heat transport equation can be represented by the superposition of the temperature distributions caused by the uniform pressure and by the radial pressure variation induced by the rapid change of the bubble wall acceleration, as can be seen in equation (28); that is,

$$T(r) = T_b(r) + T_b'(r) \tag{28}$$

5. NAVIER-STOKES EQUATION FOR THE LIQUID ADJACENT TO THE BUBBLE WALL

5.1 Bubble wall motion

The mass and momentum equation for the liquid adjacent bubble wall provides the well- known equation of motion for the bubble wall (Keller and Miksis, 1980), which is valid until the bubble wall velocity reaches the speed of sound of the liquid. That is

$$R_b\left(1-\frac{U_b}{C_b}\right)\frac{dU_b}{dt}+\frac{3}{2}U_b^2\left(1-\frac{U_b}{3C_b}\right)=\frac{1}{\rho_\infty}\left(1+\frac{U_b}{C_b}+\frac{R_b}{C_b}\frac{d}{dt}\right)\left[P_B-P_s\left(t+\frac{R_b}{C_b}\right)-P_\infty\right] \tag{29}$$

The liquid pressure on the external side of the bubble wall P_B is related to the pressure inside the bubble wall P_b according to:

$$P_B = P_b - \frac{2\sigma}{R_b} - 4\mu\frac{U_b}{R_b} \tag{30}$$

The driving pressure of the sound field Ps may be represented by a sinusoidal function such as

$$P_s = -P_A \sin \omega t \tag{31}$$

where $\omega = 2\pi f_d$.

The Keller-Miksis equation reduces to the well known Rayleigh equation which is valid at the incompressible limit without forcing field (Batchelor, 1967). That is

$$R_b\frac{dU_b}{dt}+\frac{3}{2}U_b^2=\frac{1}{\rho_\infty}(P_B-P_\infty) \tag{32}$$

5.2 Thermal boundary layer thickness

The mass and energy equation for the liquid layer adjacent to the bubble wall with the temperature distribution given in Eq. (1) provides a time dependent first order equation for the thermal boundary layer thickness (Kwak et al., 1995, Kwak and Yang, 1995). It is given by

$$\left[1+\frac{\delta}{R_b}+\frac{3}{10}\left(\frac{\delta}{R_b}\right)^2\right]\frac{d\delta}{dt}=\frac{6\alpha}{\delta}-\left[2\frac{\delta}{R_b}+\frac{1}{2}\left(\frac{\delta}{R_b}\right)^2\right]\frac{dR_b}{dt}$$
$$-\delta\left[1+\frac{1}{2}\frac{\delta}{R_b}+\frac{1}{10}\left(\frac{\delta}{R_b}\right)^2\right]\frac{1}{T_{bl}-T_\infty}\frac{dT_{bl}}{dt} \quad (33)$$

The above equation which was discussed in detail by Kwak and Yang (1995) determines the heat flow rate through the bubble wall. The instantaneous bubble radius, bubble wall velocity and acceleration, and thermal boundary thickness obtained from Eqs. (29) and (33) provide the density, velocity, pressure and temperature profiles for the gas inside the bubble with no further assumptions. The gas temperature and pressure at the bubble center can be obtained from Eqs. (15) and (16), respectively.

The entropy generation rate in this kind of oscillating bubble-liquid system, which induces lost work for bubble motion needs to be calculated by allowing for the rate change of entropy for the gas inside the bubble and the net entropy flow out of the bubble as results of heat exchange (Bejan, 1988). That is

$$\dot{S}_g=\frac{DS_b}{Dt}-\frac{\dot{Q}_b}{T_\infty} \quad (34)$$

6. CALCULATED EXAMPLES

6.1 An evolving bubble formed from the fully evaporated droplet at its superheat limit - Uniform temperature and pressure distributions for the vapor inside the bubble

It is well known that one may heat a liquid held at 1 atm to a temperature far above its boiling point without occurrence of boiling. The maximum temperature limit at which the liquid boils explosively is called the superheat limit of liquid (Blander and Katz, 1975). It has been verified experimentally that, when the temperature of a liquid droplet in an immiscible medium reaches its superheat limit at 1 atm, the droplet vaporizes explosively without volume expansion and the fully evaporated droplet becomes a bubble (Shepherd and Sturtevant, 1982). Since the internal pressure of the fully evaporated droplet is very large (Kwak and Panton, 1985, Kwak and Lee, 1991), the droplet expands spontaneously. At the initial stage of this process, the fully evaporated droplet expands linearly with time. However, its linear growing fashion slows down near the point where the nonlinear growing starts. The pressure inside the bubble may be taken as the vapor pressure given temperature with saturated vapor volume at the start of the nonlinear growing. Since the vapor pressure inside the bubble is still much greater than the ambient pressure, the bubble expands rapidly so that it overshoots the mechanical equilibrium condition and its size oscillates. In this case, since the temperature of the vapor inside the bubble is so low that vibrational motion of the vapor is not excited and the characteristic time of bubble evolution of ms range is much longer than the relaxation time of the translational motion of vapor molecules, uniform temperature and pressure distribution for the vapor molecules inside the bubble are achieved (Kwak et al., 1995).

The calculated pressure wave signal from the evolving butane bubble in ethylene glycol at the ambient pressure of 1 atm and at a temperature of 378 K is shown in Fig. 2, together with the observed data (Shepherd and Sturtevant, 1982). In this case heat transfer occurs through the thermal boundary layer. Thermal damping due to finite heat transfer (Moody, 1984) is barely seen in this Figure. In Fig. 3(a), the time rate change of the vapor temperature during the bubble evolution is shown. The bubble evolution is neither isothermal nor adiabatic. In Fig. 3(b), the entropy generation rate due to finite heat transfer for the evolving butane bubble is shown. As expected, the entropy generation during the bubble oscillation is always positive. More clear thermal damping can be observed from the far-field pressure signal of the evaporating droplet and evolving bubble formed from a cyclohexane droplet at its droplet, 492.0 K (Park et al., 2005) as shown in Fig.4. After first two volume oscillations, the original bubble has begun to disintegrate into a cloud of bubbles so that the far-field pressure signal becomes considerably smaller compared to the calculation results.

The far field pressure signal from the evolving bubble at a distance rd from the bubble center can be written in terms of the volume acceleration of the bubble (Ross, 1976). Or

$$p(t) = \frac{\rho_\infty \ddot{V}_b}{4\pi r_d} = \frac{\rho_\infty}{r_d}(2R_b\dot{R}_b^2 + R_b^2\ddot{R}_b) \tag{35}$$

For the uniform temperature and pressure distribution, the bubble behavior can be calculated from Eqs. (15), (16), (19), (32) and (33) with appropriate initial conditions.

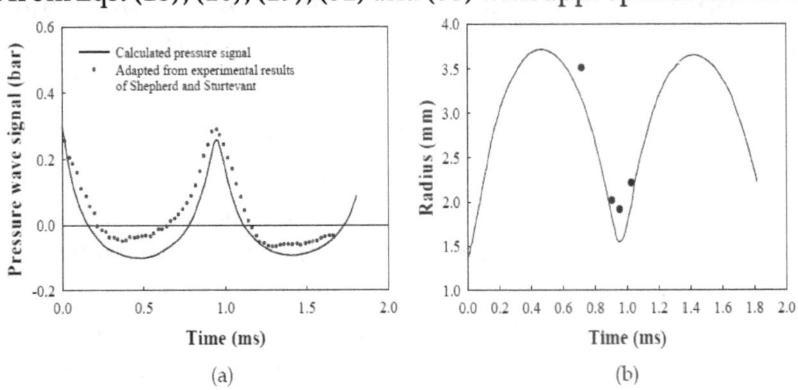

Fig. 2. Pressure wave signal from a oscillating butane bubble in ethylene glycol at 1.013 bar (a) and radius-time curve for the butane bubble (b) with the observed results (full circles).

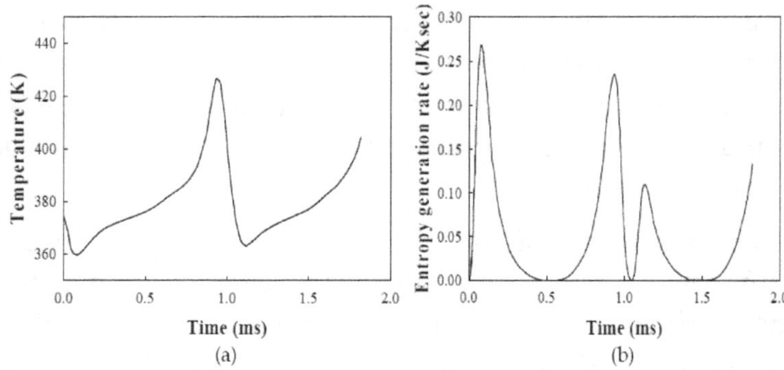

Fig. 3. Time dependence of the temperature inside the bubble (a) and time dependent entropy generation rate for the butane bubble (b) shown in Fig. 2.

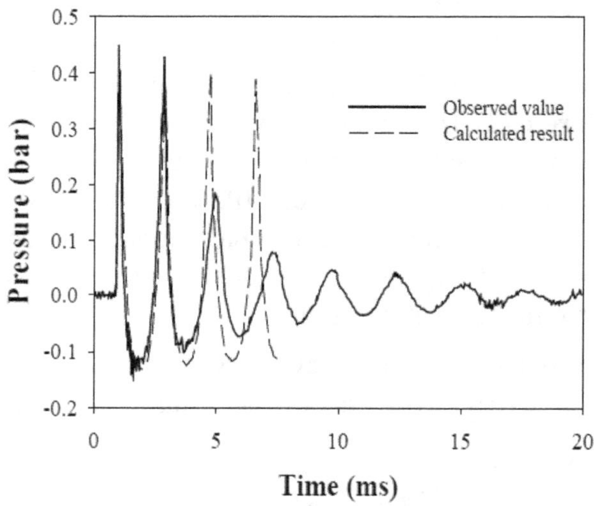

Fig. 4. Far-field pressure signal from an evolving bubble formed from a fully evaporated cyclohexane droplet at its superheat limit, 492.0 K in glycerine.

6.2 An air bubble oscillation under ultrasonic field---Non-uniform temperature and uniform pressure distribution for the gas inside bubble

If the gas temperature is above 2000 K, the vibrational modes of polyatomic molecules are expected to be excited. Since the relaxation time of the vibrational motion is rather long (10-6 s) compared with that of the translational motion, the perfect thermal equilibrium cannot be achieved for the duration in which mechanical equilibrium prevails. Certainly the temperature gradient for the gas inside bubble exists in this situation which is the case of an air bubble of micro size oscillates under ultrasonic field of amplitude below 1.2 atm and frequency of 26.5 kHz (Kwak and Yang, 1995).

The calculated radius–time curves for the bubble with an equilibrium radius of 8.5 μm , driven by the ultrasonic field with a frequency of 26.5 kHz and an amplitude of 1.075 atm which is certainly below the sonoluminescence threshold is shown in Fig.5. As shown in Table 1, the calculated values of the maximum radius and the period for each bouncing motion are in good agreement with the observed one (Loefstedt et al., 1993) can be seen. However, the bubble radius-time curve obtained by the Rayleigh equation with a polytropic relation of $P_b V_b^{1.4}$=const. shows 10 number of bouncing motions rather than 7.

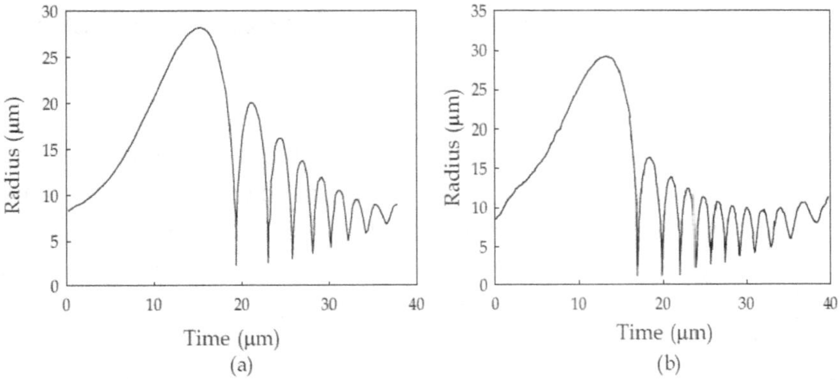

Fig. 5. Theoretical radius-time curves for an air bubble of R_0=8.5 μm at P_A=1.075 atm and f=26.5 kHz in water by our model (a) and by the Rayleigh-Plesset equation with polytropic relation (b).

Also the magnitude of the maximum bubble radius at the first bounce is significantly less than the observed one. The radius, the center temperature and pressure of the bubble at the collapse point are 2.23 μm, 2354 K and 518 atm, respectively (Kwak and Yang, 1995). The minimum bubble wall velocity at the bubble collapse point is about -111.5 m/s. In this case it is better to use Eq. (23) to obtain the bubble wall temperature and Eq. (29) to obtain the instantaneous bubble radius and velocity.

Table. 1. Calculated and measured maximum bubble size and the corresponding period of bouncing motion after the first bubble collapse for air bubble of R_0 = 8.5 μm at P_A = 1.075 atm and f = 26.5 kHz.

Bouncing number	Maximum bubble size (μm)		Corresponding period (μs)	
	Measured value	Calculated value	Measured value	Calculated value
1	19.9	20.0	3.56	3.62
2	15.6	16.2	2.43	2.76
3	13.0	13.6	2.24	2.24
4	11.1	12.1	2.06	2.07

Bouncing number	Maximum bubble size (μm)		Corresponding period (μs)	
	Measured value	Calculated value	Measured value	Calculated value
5	10.4	10.5	1.87	1.90
6	9.7	9.5	1.87	1.82

6.3 Sonoluminescing xenon bubble in sulfuric acid solutions—Non-uniform temperature and almost uniform pressure distribution for the gas inside the bubble

Sonoluminecence (SL) phenomena associated with the catastrophic collapse of a gas bubble oscillation under ultrasonic field (Gaitan et al., 1992) have been studied extensively during last 20 years or so for their exotic energy focusing mechanism (Putterman and Weinger, 2000, Young, 2005). The SL from gas bubble in water is characterized by ten to hundred picoseconds flash (Gompf et al., 1997, Hiller et al., 1998), the bubble wall acceleration exceeding 1012 m/S_2 (Kwak and Na, 1996, Wininger et al., 1997), and submicron bubble radius at the collapse point (Weninger et al., 1997). On the other hand, the SL in sulfuric acid solution revealed rather longer flash widths of ns (Jeon et al., 2008), mild wall acceleration of 1010 m/S_2 (Hopkins et al., 2005) and micron bubble radius at the collapse point under similar conditions of ultrasonic field for the case of sonoluminescence in water. The calculated radius-time curve along with observed results for a xenon bubble with R_o= 15 μm , driven by the ultrasonic field with a frequency of 37.8 kHz and an amplitude of 1.5 atm in aqueous solution of sulfuric acid is shown in Fig. 6. With air data for the thermal conductivity, the calculated radius-time curve which exactly mimics the alternating pattern of for the observed result shows two different states of bubble motion. With xenon data for the thermal conductivity, however, slight different pattern for the bubble motion was obtained. These calculation results imply that the bubble behavior, consequently the sonoluminescence phenomena depends crucially on the heat transfer in the gas medium as well as in the liquid layer and that the xenon bubble may contain a lot of air molecules. On the other hand the Rayleigh-Plesset equation with a polytropic relation, a conventional method used to predict the sonoluminescence phenomena cannot predict the two states of bubble motion as shown in Fig. 6(b). The alternating pattern for the bubble motion may happen due to the entropy generation by the finite heat transfer through the bubble wall, which produces lost work: less entropy generation in one cycle having lower maximum bubble radius provides more expansion work to the bubble next cycle, while larger amplitude motion experiencing more entropy generation provides less expansion work to the subsequent motion (Kim et al., 2006).

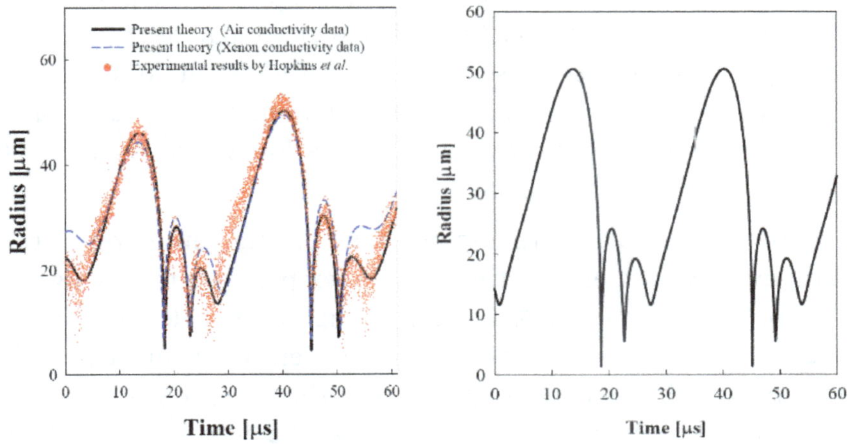

(a) (b)

Fig. 6. Theoretical radius-time curve (a) along with observed one by Hopkins et al. (2005) for xenon bubble of R_0 =15.0 μm at P_A =1.50 atm and f_d =37.8 kHz in sulfuric acid solutions and the curve calculated by polytropic relation (b).

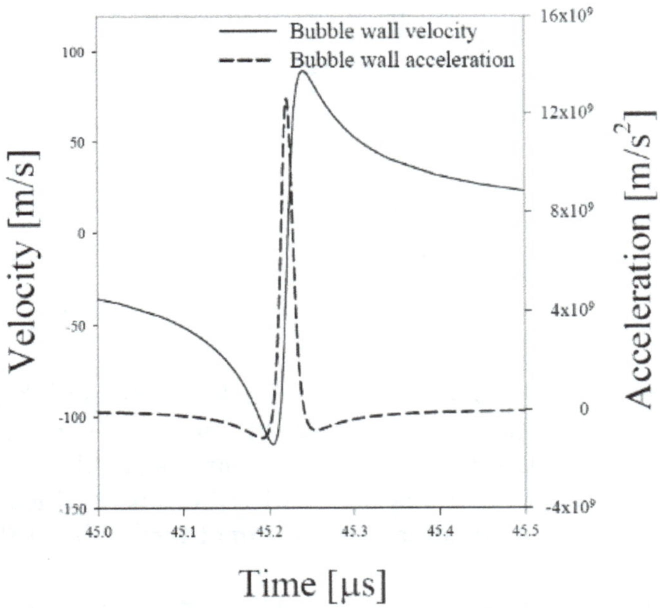

Fig. 7. Calculated bubble wall velocity (line) and acceleration (dash) near the collapse point for the bubble shown in Fig. 6.

Figure 7 shows the time-dependent bubble wall velocity and the variation of the bubble wall acceleration around the collapse point for the bubble shown in Fig. 6. The calculated magnitude of the minimum velocity at the collapse point for the light emitting cycles is about 115 m/s which is close to the observed velocity of 80 m/s. Whereas the maximum bubble wall velocity for non-light-emitting cycle is about 88 m/s, which is also close to the observed results of 60 m/s (Hopkins et al., 2005). However, the magnitude of the minimum velocity calculated by the Rayleigh-Plesset equation with the polytropic relation, which is about 900 m/s, is much higher than the observed value.

Figure 8 shows the calculated time-dependent bubble center temperature and the temporal emissive power with the average temperature for the light-emitting cycle of the bubble shown in Fig.6. The peak temperature calculated at the bubble center is about 8200 K, which is excellent agreement with the observed value of 6000~7000 K. In fact, the average temperature at the collapse point is about 6000 K because considerable temperature drop occurs at the bubble wall as shown in the insert. However, the pressure inside the bubble is almost uniform as expected as shown in Fig. 8 (b).

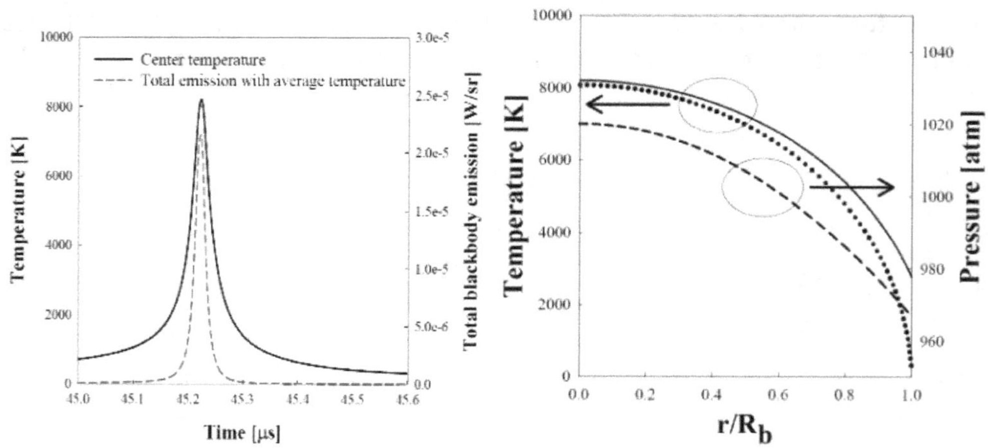

(a) (b)

Fig. 8. Time dependent gas temperature at the bubble center and the corresponding total blackbody emission (a) with the average temperature and the temperature and pressure distributions at the collapse point for the bubble shown in Fig.6. The dotted curve in indicates the temperature distribution obtained by Prosperetti eat al.'s (1988) boundary condition at the bubble wall.

6.4 Sonoluminescing air bubbles in water—Non-uniform temperature and non-uniform pressure distribution for the gas inside the bubble

When the bubble wall acceleration exceeds 10^{12} m/S$_2$, the pressure distribution for the gas inside the bubble is no longer uniform, as clearly confirmed by Eq. (9). In fact, the SL was observed for an air bubble in water or in sulfuric acid solution when the bubble wall acceleration exceeds 10^{12} m/S$_2$ ((Kim et al., 2006).

Figure 9 shows the density, pressure and temperature distributions inside the bubble at 400 ps before the collapse. Certainly, uniform pressure approximation is no longer valid during the collapsing phase for a sonoluminescing air bubble in water solution. Time dependent bubble wall acceleration and the gas temperature at the bubble center are shown in Fig. 10. As can be seen in this Figure, sudden increase and subsequent decrease in acceleration of the bubble wall results in rapid quenching of the gas followed by the substantial temperature rise up to 25000 K, which can be regarded as a thermal spike. Considerable increase in the gas temperature due to the bubble wall acceleration can be seen clearly in this Figure. The maximum bubble wall acceleration achieved near the bubble collapse point is over 1011 g (Kwak and Na, 1996), which is consistent with the observed value (Weininger et al., 1997). With uniform pressure approximation, the maximum temperature achieved is only 5800 K. On the other hand, the temperature of the gas inside the bubble goes to infinity without heat loss to the environment, that is kg goes to zero. One may obtain the maximum temperature up to 107 K if one uses kg= 0.01 W/mK, which indicates that heat transfer is very important also in this case. The heat flux at the collapse point is as much as 10 GW/m2, however, the heat flow rate is about 2 mW.

The intense local heating and high pressure inside bubble and liquid adjacent to the bubble wall from such collapse that it can give arise to unusual effects in chemical reactions (Suslick, 1990), and the sonochemical process has been proven to be useful technique in making specialty nanomaterials (Kim et al., 2009). Note that the estimated temperature and pressure in the liquid zone around the collapsing bubble is about 1300 K and 1000 atm, respectively (Kwak and Yang, 1995).

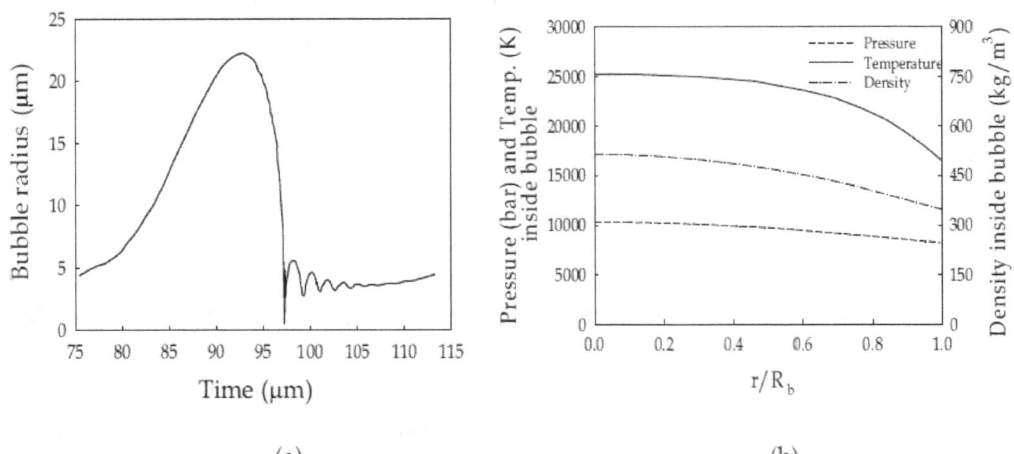

Fig. 9. Theoretical radius-time curve for an air bubble of R_0 =4.5 μm at P_A =1.30 atm and f_d =26.5 kHz in water (a) and density , pressure and temperature distributions inside this bubble at 400 ps prior to the collapse point.

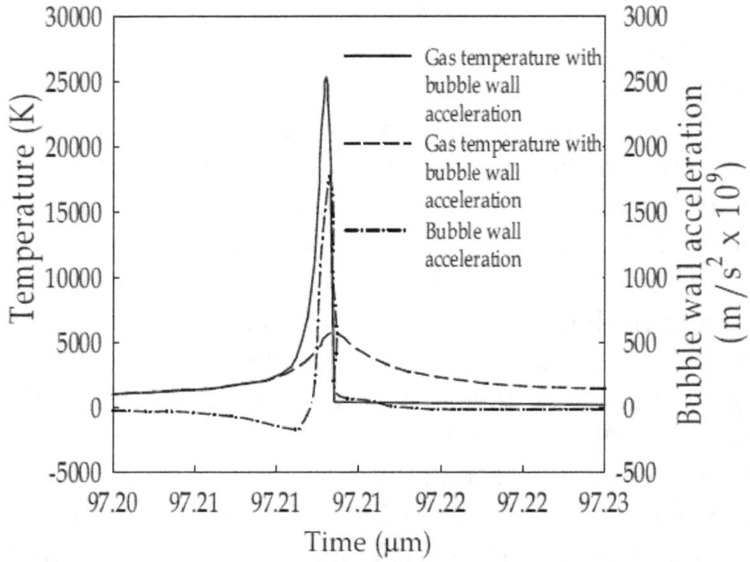

Fig. 10. Time dependent bubble wall acceleration and gas temperature around the collapse point for the case shown in Fig. 9.

7. MOLECULAR DYNAMICS SIMULATION OF SONOLUMINESCENT XENON BUBBLES

The gas temperature within the bubble is significantly influenced by the thermal conductivity of the gas contained in the bubble and the thermal conductivity of the surrounding liquids at the thermal barrier. layer. However, transport property values for the gas and liquid at the extreme condition are not available. As a consequence of this fact, validation of theoretical results was attempted through the MD simulation. The bubble radius was determined by the pressure obtained from MD simulation using Keller-Miksis equation. The mass transfer through the bubble wall, which does not affect the thermal properties at the collapse point very much (Kim and Kwak, 2007) was not considered in this simulation.

7.1 Molecular dynamics simulation

The principles and procedures for molecular dynamics simulation are well documented in standard text books (Haile, 1992, Rapaport, 1995). In this study, numerical integration of one million hard spheres was done in order to count the impulsive collisions between molecules moving by Newton's second law. As is well known, the kinetic events driven by Newton's law can be described by the Vlasov equation or the collisionless Boltzmann equation (Krall and Trivelpiece, 1973), which also yields the conservation equations for mass, momentum, and internal energy. This fact implies that MD simulation of hard sphere molecules might capture the physics related to SL phenomena. Furthermore, the soft parts of the potential are not significant for high energy collisions which occur near the collapse point (Ruuth et al., 2002).

7.2 Collision between molecule and bubble wall

The molecules inside the bubble may hit the molecules at the layer of the gas-liquid interface. In this simulation, the interface was assumed to be a hard wall, so that when a molecule hits the bubble wall, it reflects from the wall, as shown in Fig. 11. When considering the velocity of the bubble wall and assigning the thermal energy of the temperature at the bubble wall T_{bl}, the velocity of the reflected molecules $\vec{\mathbf{v}}_f$ may be obtained by the following equation for the heat bath boundary condition. In this case, the direction of the reflected particle may be assigned randomly or specularly (Kim et al., 2007).

$$\left|\vec{\mathbf{v}}_f - U_b \hat{\mathbf{r}}_i\right|^2 / 2 = (3/2)(k_B T_{bl} / m) \tag{36}$$

where U_b is the instantaneous bubble wall velocity, m is the mass of a molecule and kB is the Boltzmann constant.

For the adiabatic boundary condition, the particles were assumed to be reflected from the wall with speed equal to the pre-collision speed in the local rest frame of the wall. The direction of reflected particles was determined according to the reflection law at the planar interface in this case. Explicitly, the velocity of the reflected particles is given by

$$(\vec{\mathbf{v}}_f - U_b \hat{\mathbf{r}}_i) = (\vec{\mathbf{v}}_i - U_b \hat{\mathbf{r}}_i) - 2\left[(\vec{\mathbf{v}}_i - U_b \hat{\mathbf{r}}_i) \cdot \hat{\mathbf{r}}_i\right] \hat{\mathbf{r}}_i \tag{37}$$

However, it has been found that neither the adiabatic nor the heat bath boundary condition is appropriate for treating the collapsing process of sonoluminescing gas bubbles. The heat bath boundary condition means that heat flow exists at the bubble wall while adiabatic means no heat flow at the boundary. The degree of the adiabatic change during the collapsing process can be described by the following effective accomodation coefficient (Yamamoto et al., 2006):

$$\alpha = (T_{in} - T_{out}) / (T_{in} - T_{bl}) . \tag{38}$$

where T_{in} is the temperature of the gas particle moving toward the bubble wall and Tout is the temperature of particle leaving the wall. Two limiting cases are $\alpha = 0$ for adiabatic boundary condition and $\alpha = 1$ for the heat bath boundary condition.

For the adiabatic boundary case, a molecule does not lost its kinetic energy after a collision made with a molecule at the bubble wall. On the other hand, for the heat bath boundary case, a molecule having kinetic energy of T_i lost its kinetic energy to become a molecule having kinetic energy of T_{bl} after collision. The value of α defined in Eq. (38) determines the instantaneous gas temperature and pressure as well as the bubble radius and wall velocity via the heat transfer through the bubble wall.

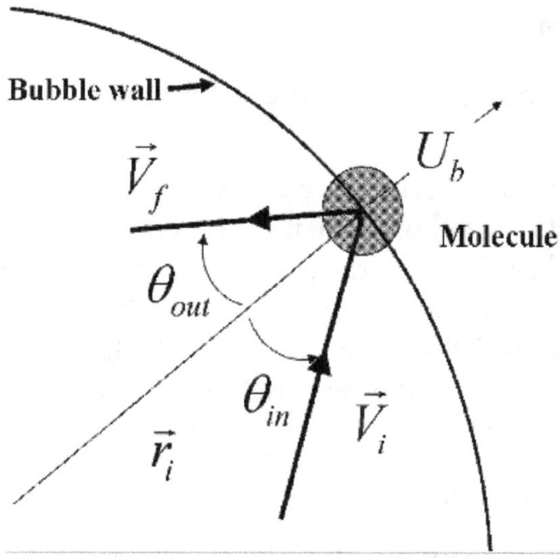

Fig. 11. A collision model between molecules and the bubble wall where v_i and v_f denote the velocity if incident and reflected molecule, respectively.

7.3 Calculation procedures

In this study, a sonoluminescing gas bubble in sulfuric acid solution was considered. The number of molecules in the sonoluminescing bubble may be obtained from the equation of state for an ideal gas.

$$P_\infty\left(\frac{4\pi}{3}R_0^{\ 3}\right) = Nk_BT_\infty \tag{39}$$

where *N* is the number of molecules inside the bubble and P_∞ and T_∞ are the ambient pressure and temperature, respectively. For a sonoluminescing xenon bubble with an equilibrium radius of 0.7 μm in water at T_∞=300 K and P_∞=1 atm, the number of molecules occupying the bubble is about 3.5×10^7, which is difficult to handle with today's computing power.

One may use a scaling transformation for the molecular volume such as $V_gN = const.$ where V_g is the volume of gas molecule and *N* is the number of molecules occupying the bubble and that reduces the number of molecules that are handled in the MD simulation (Metten and Lauterborn, 2000). With this scaling transformation, of course, the mass and the energy of the system should be conserved so that *mN* = *const.* and *TN* = *const.*

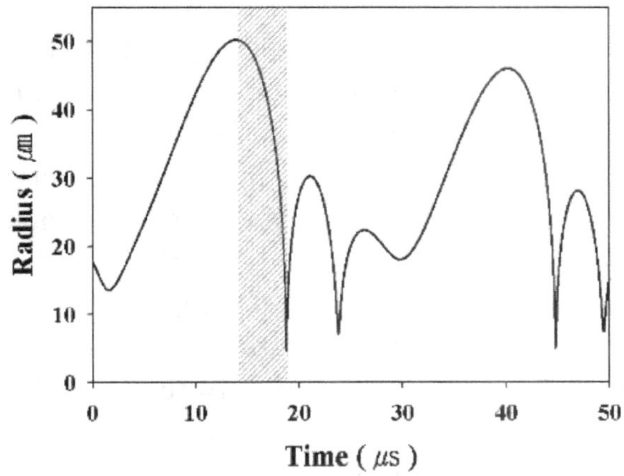

Fig. 12. Theoretical radius-time curves for a xenon bubble of R_0 = 15 µm at P_A = 1.5 atm and f_d = 37.8 kHz in a sulfuric acid solution. Shaded area indicates the time span for MD simulations.

In this study, we chose number of simulated molecules to be 10^6, so the hard sphere diameter of a molecule increases to 71.79 σ for the xenon bubble of 15 µm in radius, where σ is the hard sphere diameter of the xenon molecule, taken to be 0.492 nm . The number 71.79 can be obtained from a relation $\sigma^3 N = \sigma_s^3 N_s$ with the number of simulated molecules of N_s=106 and the total number of xenon molecules inside the bubble of equilibrium radius of 15 µm, N= 3.46x10^{11}. The conversion factors for the bubble radius, pressure and temperature, which were used in this MD simulation are σ, ε/ σ^3 and ε/ kB, respectively.

7.4 Results and discussion

MD simulation for a collapsing xenon bubble with equilibrium radius of 15.0 µm, driven by an ultrasound frequency of 37.8 kHz and amplitude of 1.5 atm in sulfuric acid solution, was started at its maximum radius of 50.3 µm. The xenon bubble was chosen in this study because observed data for the bubble wall velocity and the peak pressure and temperature at the collapse point have been reported (Hopkins et al., 2005), which is same as the case shown in Fig. 6. One to three million particles were used for a scaled-down MD simulation of the bubble that actually had 35 million molecules. The initial gas state with temperature of 297.8 K and pressure of 3.64 × 10^{-2} atm was obtained from theoretical results obtained using a set of solutions of the Navier-Stokes equations for the gas inside the bubble with consideration of heat transfer through the bubble wall (Kwak and Yang, 1995, Kwak and Na, 1996). The cross-hatched region in Fig.12 indicates the time period of MD simulation for the xenon bubble which shows an on/off luminescence pattern in sulfuric acid solution (Hopkins et al., 2005).

In Fig.13, the bubble radius-time curve(a) and the time-dependent bubble wall velocity (b) near the collapse point, calculated by MD simulations with α =0.15 are plotted along with the theoretical results. The time-dependent gas temperature at the bubble wall, used in the MD simulation for employing the heat bath boundary condition, was calculated by theory. A very similar trend between MD simulation and theoretical calculation in the bubble radius-time curve and the time-dependent bubble wall velocity were obtained. It is noted that the theoretical minimum velocity at the collapse point, which is about -115 m/s, is close to the observed value of -80 m/s (Hopkins et al., 2005). The calculated value of the minimum velocity by MD is about -130 m/s. The best agreement between the MD simulation and the theoretical results was obtained with a value of α =0.15 among the trial of α values, 0.1, 0.15, 0.3 and 0.5. The minimum radius of the at the collapse point for the xenon bubble is about 4.56 μm so that the packing fraction at the point is about 0.058.

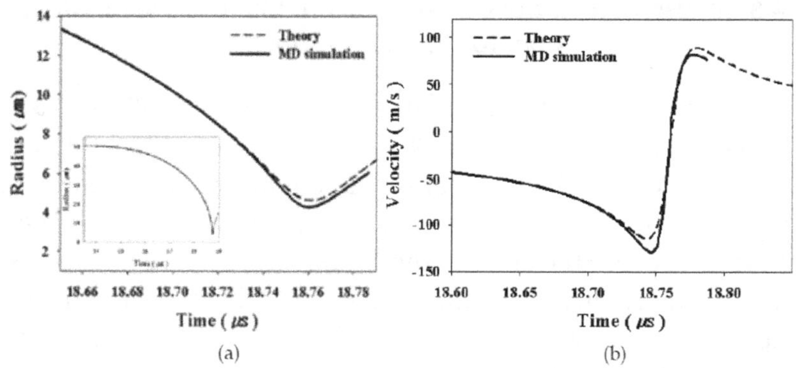

Fig. 13. Calculated radius–time curves (a) and time-dependent bubble wall velocity (b) obtained by MD simulation with α =0.15 and theory for the bubble shown in Fig. 12.

The time-dependent gas temperature and pressure inside the bubble around the collapse point, which were obtained by MD simulation with α =0.15 are shown in Fig.14. The peak temperature and pressure estimated by MD simulation are about 7,500 K and 1,260 atm, respectively, and are very close to the theoretically predicted values of 8,200 K and 1,010 atm, respectively. The MD simulation results shown in Figs.13 and 14 indicate that the collapsing process of a sonoluminescing micron bubble undergoes an almost adiabatic change although a large amount of heat transfer through the bubble wall occurs. At the moment of collapse, the heat flux at the bubble wall is much as 0.6 GW/m^2. A quite different collapsing process was obtained from the MD simulation with the heat bath boundary condition (α =1.0) (Kim et al., 2007): the compression was started very slowly at first and the final stage occurs very rapidly so that the estimated full width at half maximum (FWHM) of the luminescence pulse is about 5 ns, which is quite less than the value of the theoretical estimate of 20 ns (Kim et al., 2006).

Once the time dependent temperature for the gas inside the bubble around the collapse point, the FWHM of the light pulse can be estimated from the temperature profile with assumption of mechanism of the light emission (Kwak and Na, 1996, 1997). In Fig.15, the

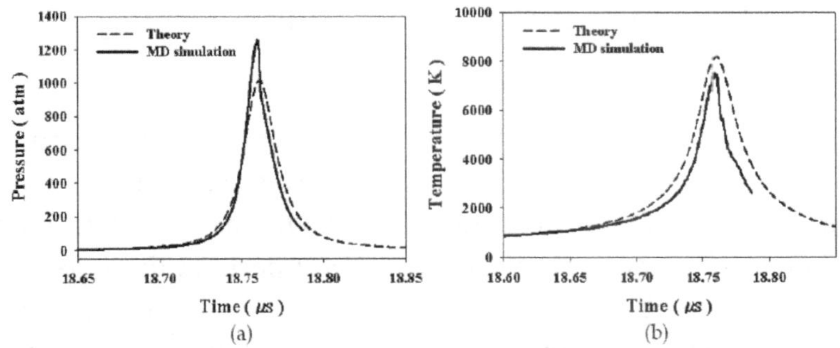

Fig. 14. Time-dependent center pressure (a) and temperature (b) obtained by MD simulation with α =0.15 and theory near the collapse point for the bubble shown in Fig. 12.

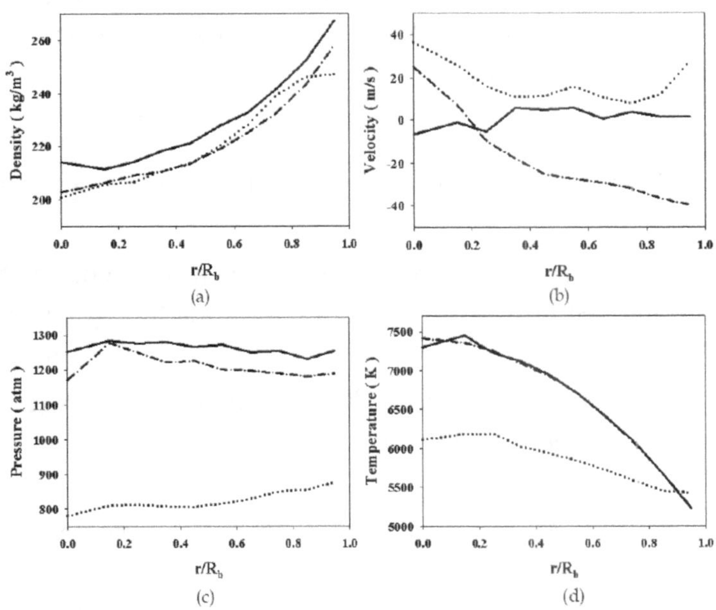

Fig. 15. Gas density(a), velocity(b), pressure(c), and temperature(d) distributions near the bubble collapse point obtained by MD simulation for the bubble shown in Fig.2 before 2.5 ns (— • — •), at (—) , and after 2.5 ns (••••) the collapse.

MD simulation results with α =0.15 for distribution of gas density(a), velocity(b), pressure(c), and temperature(d) are plotted before and at the collapse point for the bubble shown in Fig.12. The gas density increases radially during both the compression and expansion phases, as theorized. The gas velocity magnitude grows radially, reaching its peak at the bubble wall and exhibiting a uniform null value at the collapse point, as expected by the theoretically derived linear velocity profile. The pressure exhibits an almost uniform distribution prior to collapse. The temperature exhibits a quadratic distribution, with its minimum occurring near the bubble wall due to heat transfer, a phenomenon also supported by theoretical predictions. (Kwak and Yang, 1995, Kwak and Na, 1996, Kim et al., 2006).

8. CONCISE SUMMARY

A bubble dynamics model incorporating heat transmission within the bubble and the thermal boundary layer of the neighboring liquid was examined. Thermal damping resulting from finite heat transmission is a crucial factor for the developing bubble generated from a fully evaporated droplet at its superheat limit, as well as for the sonoluminescing xenon bubble exhibiting two states of bubble motion in sulfuric acid solutions. Molecular dynamics simulations have been conducted to validate our theoretical model of bubble dynamics. In conclusion, the nonlinear dynamics of an ultrasonically driven bubble and the sonoluminescence properties of the bubble in sulfuric acid solutions have been accurately predicted by a series of solutions to the Navier-Stokes equations for the gas within the bubble, accounting for heat transfer through the bubble wall.

REFERENCES

Batchelor, G.K. (1967). *An Introduction to Fluid Dynamics*, Cambridge University Press, London.

Barber, B.P. & Putterman, S.J., Light scattering measurements of the repetitive supersonic implosion of a sonoluminescing bubble, *Phys. Rev. Lett.* Vol. 69, 3839, (1992).

Bejan, A. (1988). *Advanced Engineering Thermodynamics*, John Wiley & Sons, New York. Blander , M. & Katz, J. L. (1975). Bubble nucleation in liquids, *AIChE J* Vol. 21, No. 5, pp.833-848.

Boulos, M.I., Fauchais, P. & Pfender, E. (1994). *Thermal Plasma Vol.1* , Plenum Press, New York.

Byun,K., Kim, K.Y. & Kwak, H. (2005). Sonoluminescence characteristics from micron and submicron bubble, *J. Korean Phys. Soc.* Vol. 47, No. 6, pp.1010-1022.

Davidson, R.C. (1972). *Methods in Nonlinear Plasma Theory*, Academic Press, New York.

Gaitan, K.F., Crum, L.A., Church, C.C. & Roy, R.A.(1992). Sonoluminescence and bubble dynamics for a single, stable, cavitation bubble, *J. Acoust. Soc. Am.*, Vol.91, pp.3166- 3183.

Gompf, B., Guenter, R., Nick, G., Pecha, R. & Eisenmenger, W. (1997). Resolving sonoluminescence pulse width with time-correlated single photon counting, *Phys. Rev. Lett.*, Vol. 79, pp. 1405-1407.

Haile , M. (1992). *Molecular Dynamics Simulation* ,John Wiley & Sons.

Hiller, R., Putterman, S.J. & Wininger, K.R. (1998). Time resolving spectra of sonoluminescence, *Phys. Rev. Lett.* Vol.80, pp. 1090-1093.

Hopkins, S.D., Putterman, S.J., Kappus, B.A., Suslick, K.S. & Camara, C.G. (2005). Dynamics of a sonoluminescing bubble in sulfuric acid, *Phys. Rev. Lett.* Vol.95, paper # 254301.

Jeon,J.,Lim,C.&Kwak,H.(2008).Measurement of pulse width from sonoluminescing gas bubble in sulfuric acid solution, *J. Phys. Soc. Jpn*, Vol. 77, paper # 033703.

Jun, J. & Kwak, H. (2000). Gravitational collapse of Newtonian stars, *Int. J. Mod. Phys. D*, Vol.9, pp.35-42.

Keller, J.B. & Miksis, M. (1980). Bubble oscillations of large amplitude, *J. Acoust. Soc. Am.*, Vol.68, pp.628-633.

Kim, H., Kang K. & Kwak, H. (2009). Preparation of supported Ni catalysts with a core/shell structure and their catalytic tests of partical oxidation of methane, *Int. J. Hydrogen Energy*, Vol. 34, pp. 3351-3359.

Kim, K., Byun, B. & Kwak, H. (2006). Characteristics of sonoluminescing bubbles in aqueous solutions of sulfuric acid, *J. Phys. Soc. Jpn.*, Vol. 75, paper # 114705.

Kim, K. & Kwak, H. (2007). Predictions of bubble behavior in sulfuric acid solutions by a set of solutions of Navier-Stokes equations, Chem Eng. Sci., Vol. 62, pp. 2880-2889.

Kim, K., Kwak, H. & Kim, J.H. (2007). Molecular dynamics simulation of collapsing phase for a sonoluminescing gas bubble in sulfuric acid solutions : A comparative study with theoretical results, *J. Phys. Soc. Jpn.* Vol.76, 024301.

Kim, K.Y., Lim, C., Kwak, H. & Kim J.H. (2008). Validation of molecular dynamic simulation for a collapsing process of sonoluminescing gas bubble, *Mol. Phys.*, Vol. 106, No. 8, pp. 967-975.

Krall, N.A. &Trivelpiece, A.W.(1973). *Principles of Plasma Physics* ,McGraw-Hill, New York.

Kwak, H. & Lee, S. (1991). Homogeneous bubble nucleation predicted by molecular interaction model, *ASME J. Heat Trans.* Vol. 113, pp. 714-721.

Kwak, H. & Na, J. (1996). Hydrodynamic solutions for a sonoluminescing gas bubble, *Phys. Rev. Lett.*, Vol.77, pp.4454-4457.

Kwak, H. & Na, J. (1997). Physical processes for single bubble sonoluminescence, *J. Phys. Soc, Jap.* Vol.66, pp.3074-3083.

Kwak, H., Oh, S. & Park, C. (1995). Bubble dynamics on the evolving bubble formed from the droplet at the superheat limit, *Int. J. Heat Mass Transfer*, Vol.38, pp.1707-1718.

Kwak, H. & Panton, R.L. (1985). Tensile strength of simple liquids predicted by a model of molecular interactions, *J. Phys. D: Appl. Phys.*, Vol. 18, pp.647-659.

Kwak, H. & Yang,H. (1995). An aspect of sonoluminescence from hydrodynamic theory, J. Phys. Soc. Jpn. Vol. 64, 1980-1992.

Lauterborn, W. (1976). Numerical investigation of nonlinear oscillation of gas bubbles in liquids, *J. Acoust. Soc. Am.*, Vol. 99, pp. 283-293.

Lin, H., Storey, B.D. & Szeri, A.J.(2002). Inertially driven inhomogenieties in violently collapsing bubbles: the validity of the Rayleigh-Pessset equation, *J. Fluid Mech.*, Vol.452, pp.145-162.

Loefstedt, R., Barber, B.P. & Putterman, S.J. (1993). Toward a hydrodynamic theory of sonoluminescence, *Phys. Fluids*, Vol. A5, pp. 2911-2928.

Metten, B. & Lauterborn, W. (2000). In Nonlinear Acoustics at the Turn of the Millennium: ISNA 15, 15th International Symposium on Nonlinear Acoustics edited by W. Lauterborn and T. Kurz, AIP Conf. Proc. No. 524, pp. 429-432.

Moody, F.J. (1984). Second law thinking-example application in reactor and containment technology, In *Second Law Aspects of Thermal Design* (Edited by A. Bejan and R.C. Reid) Vol. HTD-33, pp. 1-9. ASME, New York.

Moss,W.C., Clarke, D.B., White,J.W. & Young, D.A.(1994). Hydrodynamic simulations of bubble collapse and picosecond sonoluminescence, *Phys. Fluids*, Vol.6, No. 9, pp. 2979-2985.

Na,J., Byun, G. & Kwak, H. (2003) Diffusive stability for sonoluminescing gas bubble. *J. Kor. Phys. Soc.* Vol. 42,pp.143-152.

Panton, R.L. (1996). *Incompressible Flow*, John Wiley & Sons, 2nd Edition, New York.

Park, H., Byun, K. & Kwak, H. (2005). Explosive boiling of liquid droplets at their superheat limits, *Chem. Eng. Sci.*, Vol. 60, pp. 1809-1821.

Prosperetti, A., Crum, L. A. & Commander, K. W. (1988). Nonlinear bubble dynamics, *J. Acoust. Soc. Am.*, Vol.88, pp.1061-1077.

Putterman, S.J. & Weininger, K.R. (2000). Sonoluminescence: How bubbles turn sound into light. Annual Review of Fluid Mech., Vol. 32, pp. 445-476.

Rapaport,D.C.(1995). *The Art of Molecular Dynamics Simulation*, Cambridge University Press, London.

Ross, D. (1976). *Mechanics of Underwater Noise*, Pergamon Press, Oxford.

Ruuth, S.J., Putterman, S. & Merriman, B. (2002). Molecular dynamics simulation of the response of a gas to a spherical piston: Implications for sonoluminescence, *Phys. Rev. E*, Vol. 66, 036310.

Shepherd, J.E. & Sturtevant, B. (1982). Rapid evaporation at the superheat limit, *J. Fluid Mech.*, Vol. 121, pp. 379-402.

Suslick, K.S. (1990). Sonochemistry, Science, Vol. 247, pp. 1439-1445.

Suslick, K.S., Hammerton, D.A. & Cline, R.E. (1986). The sonochemical hot spot, *Am. Chem. Soc.*, Vol. 108, pp. 5641-5642.

Theofanous, T., Biasi, L. & Isbin, H.S.(1969). A theoretical study on bubble growth in constant and time-dependant pressure fields, *Chem. Eng. Sci.*, Vol.24, pp.885-897.

Vincenti, W.G. & Kruger, C.H. (1965). *Introduction to Physical Gas Dynamics*, New York.

Weninger, K.R., Barber, B.P. & Putterman, S.J. (1997). Pulsed Mie scattering measurements of the collapse of a sonoluminescing bubble, Phys. Rev. Lett., Vol. 78, pp. 1799-1802.

Yamamoto,K., Takeuchi,H. & Hyakutake, T. (2006). Characteristics of reflected gas molecules at a solid surface, *Phys. of Fuids*, Vol.18,046103.

Young, F.R.(2005). *Sonoluminescence* , CRC Press, Boca Raton.

PART 2
BOILING, FREEZING AND CONDENSATION HEAT TRANSFER

8 Chapter

BOILING NUCLEATE POOL IN MICROGRAVITY

1. INTRODUCTION: AN OVERVIEW

Every day, nucleate pools boil, efficiently conveying large amounts of heat flux. However, due to the interplay of multiple factors and effects, including the nucleate process, bubble growth, interaction between liquid and vapor on the heater's surface, evaporation at the liquid-vapor interface, and transport of hot liquid and vapour away from the heater's surface, it is a very complex and elusive process. Gravity is one of the many sub-processes that might contribute to the boiling phenomenon and potentially obscure the true mechanism that underlies it. Many careful studies conducted in a normal gravity setting on the ground, where gravity is the primary factor, have contributed to our current understanding of the nucleate pool boiling phenomenon. By generating forces in the systems that propel motions, define boundaries, and compress fluids, gravity has a significant impact on the boiling phenomenon. Moreover, ever-present but relatively tiny effects can be obscured by the presence of gravity. Because of gravity's masking influence, knowledge of the boiling phenomenon has not progressed very far. A unique chance to investigate the intricate relationships without the influence of outside factors like buoyancy—which can alter the dynamics of bubbles and the associated heat transfer—is provided by microgravity studies. Moreover, they can offer a way to investigate the real impact of gravity on boiling. On the other hand, pool boiling in microgravity has grown in importance as a research topic because of its numerous possible uses in space and among nearby planets because of its great heat transmission efficiency. As a result, studies on microgravity will help identify the mechanism causing the occurrence and subsequently create better mechanistic models for applications that are relevant to both Earth and space. With the development of diverse microgravity facilities and expanded experimental opportunities, particularly in the last two decades, research on boiling heat transfer in microgravity has advanced during the course of its more than 50-year existence, with a brief hiatus in the 1970s. There are currently a number of in-depth reviews and monographs published on the developments in this sector. A few of the many people that summarized the global experimental and theoretical efforts at the outset of our research were Straub (2001), Di Marco (2003), Kim (2003), and Ohta (2003a, b). Our group at the National Microgravity Laboratory/CAS has worked on two research projects on nucleate pool boiling in microgravity over the last ten years, both on Chinese recoverable spacecraft. Additionally, studies in the Beijing drop tower have been conducted on the ground in both normal gravity and brief microgravity. The main conclusions are outlined in this chapter, and Zhao (2010) has just offered a concise overview of the space experiment outcomes.

2. POOL BOILING ON A CABLE IN MICROGRAVITY

A TCPB (Temperature-Controlled Pool Boiling) device was developed to study heat transfer of pool boiling on thin wires both on the ground and aboard the 22nd Chinese recoverable satellite (RS-22) (Wan et al., 2003). A platinum wire of 60 μm in diameter and 30 mm in length was simultaneously used as a resistance heater and a resistance thermometer to measure the temperature of the heater surface. The heater resistance, and thus the heater temperature, was kept constant by a feedback circuit, which was similar to that used in constant-temperature hot-wire anemometry. Each step of the heater temperature lasted about 30 seconds in order to obtain steady pool boiling according to Straub (2001). The boiling chamber was filled with degassed R113 and was pressurized in an airproof container. A bellows connected with the chamber and the surrounding housing allowed the pressure in the chamber to be practically constant.

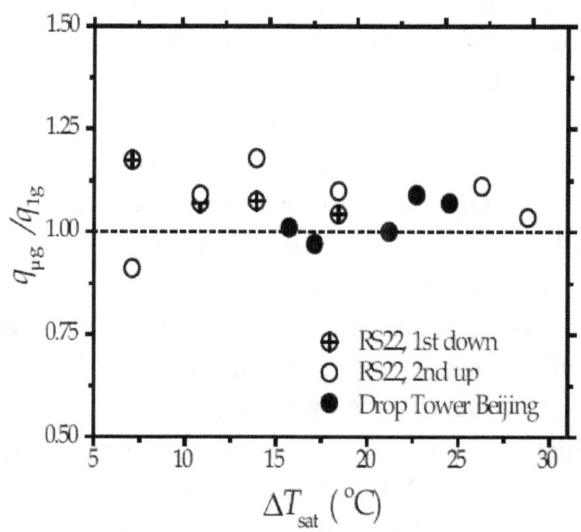

Fig. 1. Microgravity efficiency on heat transfer of nucleate boiling in microgravity (Zhao et al., 2009d).

Several preliminary experimental runs at subcooling condition were conducted in short- term microgravity utilizing the drop tower Beijing, which provides a course of about 3.6 s for microgravity experiments (Zhao et al., 2004). The space experiment was carried out aboard the 22nd Chinese recoverable satellite (RS-22) in September 2005 (Liu, 2006). The level of residual gravity was estimated in the range of 10^{-3}~10^{-5} g_0. Before and after the space flight, ground control experiments using the same facility were also conducted. Comparing with those in normal gravity, the heat transfer of nucleate boiling was slightly enhanced in short- and long-term microgravity (Fig. 1), while about 20% and 40% decrease of heat flux was observed for two-mode transition boiling in short- and long-term microgravity, respectively.

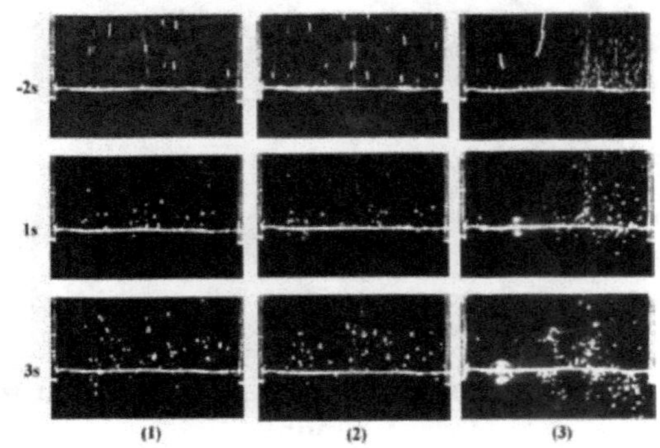

Fig. 2. Bubble behaviors on thin wire in different gravity conditions (Zhao et al., 2004).

In the drop tower tests, bubble behaviors were dramatically altered by the variation of the acceleration (Fig. 2). It was difficult to observe the lateral oscillation of bubbles along the wire in nucleate boiling regime in normal gravity, but this kind of motion was always able to observe in both short- and long-term microgravity. It could lead to the lateral coalescence between adjacent bubbles, and then detached the coalesced bubble from the wire. Sometimes, the coalesced bubble could enclose the wire and a bright spot appeared there. It couldn't, however, last long period and the boiling continued as nucleate boiling. In the two-mode transition boiling regime, the Taylor instability disappeared in microgravity, and then the surface tension reformed the shape of the wavy film appeared in normal gravity to a large spheroid bubble encircling the wire. Then the film part receded after releasing the drop capsule, while the part of nucleate boiling expanded along the wire. The centre of the large spheroid bubble wiggled along the wire and its size increased slowly. Sometimes, the wire near the centre of the large spheroid bubble brightened up, but no real burn-out was observed in the short-term microgravity experiments.

In the space experiment in long-term microgravity, special bubble behaviors were observed firstly (Zhao et al., 2007). There existed three critical bubble diameters in the discrete vapor bubble regime in microgravity, which divided the observed vapor bubbles into four regions (Fig. 3): Tiny bubbles were continually forming and growing on the surface before departing slowly from the wire when their sizes exceeded the first critical value. The bigger bubbles, however, were found staying on the surface again when their diameters were larger than the second critical value. If they grew further larger than the third critical value, departure would be observed once again. Furthermore, the first critical value exhibited no obvious difference between in normal gravity and in microgravity.

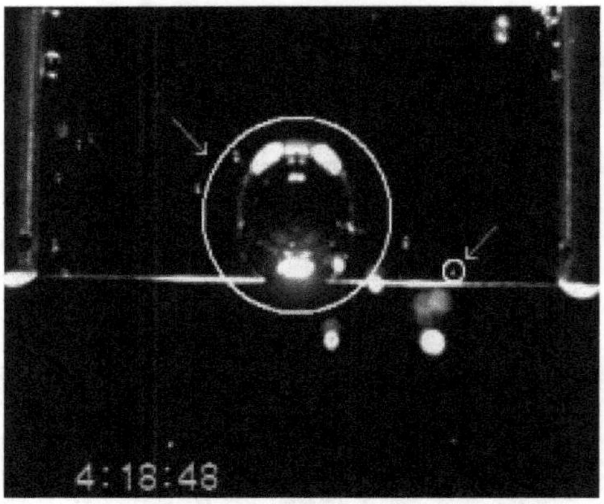

Fig. 3. Special bubble behaviors on thin wire in long-term microgravity (Zhao et al., 2007).

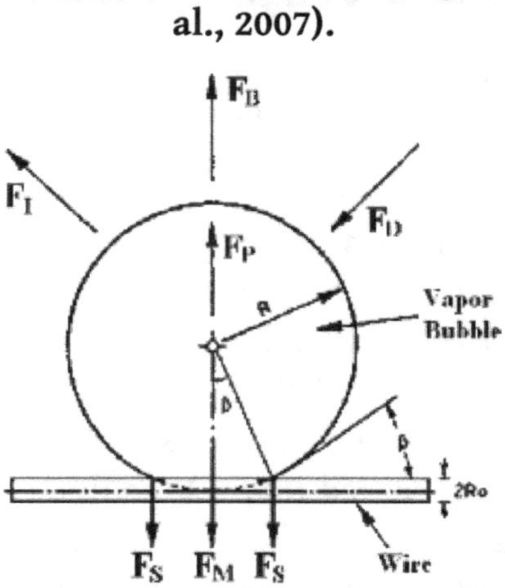

Fig. 4. Forces acted upon a vapour bubble growing on thin wire (Zhao et al., 2008).

Among the commonly used models for bubble departure, no one can predict the whole observation. A qualitative model was proposed by Zhao et al. (2008), in which the Marangoni effect was taken into account (Fig. 4)

$$f(y) = C_4 y^4 + C_3 y^3 + C_1 y + C_0 \tag{1}$$

where,

$$y = \tau^{1/2} \tag{2}$$

$$C_4 = \frac{4}{3}\pi E^3 (\rho_L - \rho_V) g \tag{3}$$

$$C_3 = -2K\pi|\sigma_T| E^2 \nabla T \tag{4}$$

$$C_1 = 4\sigma R_0 \sin^2 \beta + \frac{\pi}{3} \rho_L E^4 \tag{5}$$

$$C_0 = R_0 E^3 \rho_L \sin^2 \beta \left(\frac{1}{3} - \frac{3}{8} C_d \right) \tag{6}$$

$$E = \frac{1}{2\sqrt{\pi}} Ja \sqrt{\alpha_L} \tag{7}$$

where τ, σ_T, σ, ρ, β, α, R_0, C_d and *Ja* denote the growing time of bubble, surface tension and its temperature coefficient, density, contact angle, heat diffusivity coefficient, wire radius, drag coefficient and the Jacob number, respectively. *K* is an empirical parameter to count the departure from the linear theory for the case of finite Reynolds and Marangoni numbers. The subscripts L and V denote liquid and vapour phases, respectively.

According to Eq. (1), the following conclusion can be obtained: If $f(y)<0$, the departure force is larger than the resistant force, so the bubble will stay on the heater's surface; if $f(y)>0$, the departure force is smaller than the resistant force, so the bubble will depart from the heater's surface. Fig. 5 also shows the predictions of Eq. (1) in microgravity. In normal gravity, the function for the total forces acting on the growing bubble, *f(y)*, has only one zero-value point, indicting only one critical diameter for bubble departure. When the residual gravity decreases to no more than $1.36\times10^{-4}\ g_0$, the second and third zero-value points will be predicted by the new model. Comparing the prediction at $g=10^{-4}\ g_0$ **with the observation, the agreement is quite evident.**

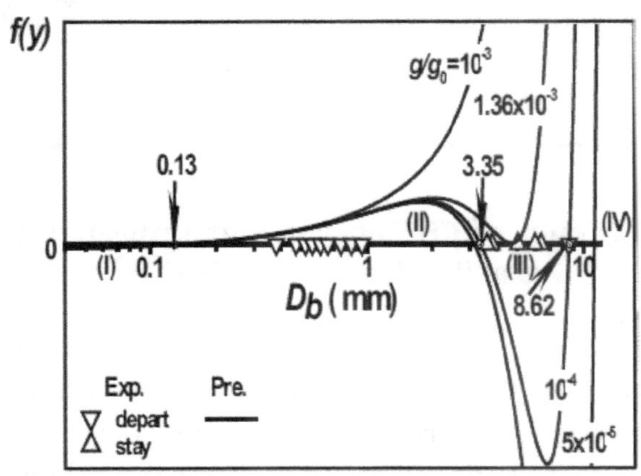

Fig. 5. Bubble departure in the discrete vapor bubble regime in microgravity. (Zhao et al., 2008).

The scaling of CHF with the gravity based on the data obtained both in the present study and in other researches reported in the literature was shown in Fig. 6. It was found that the Lienhard-Dhir-Zuber model (Lienhard & Dhir, 1973), established on the mechanism of hydrodynamic instability, can provide a relative good prediction on the trend of CHF in different gravity conditions, though the value of dimensionless radius $R' = R\sqrt{(\rho_L - \rho_G)g/\sigma}$ was far beyond the initial application range of the model. This observation was consistent with Straub (2001).

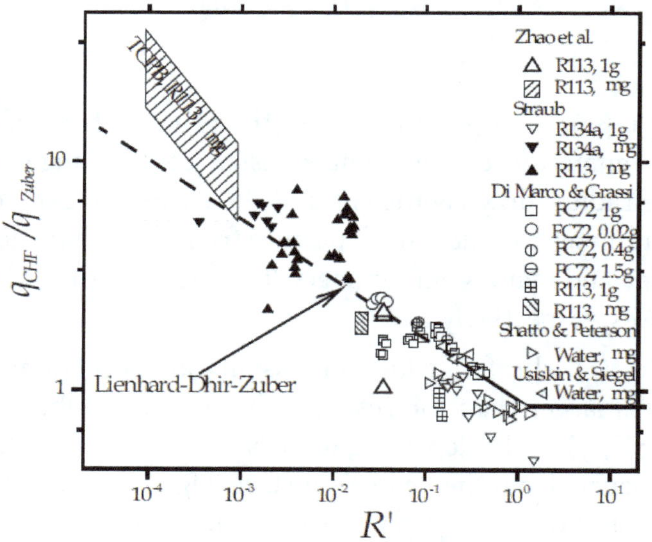

Fig. 6. Scaling of CHF with gravity (Zhao et al., 2009d).

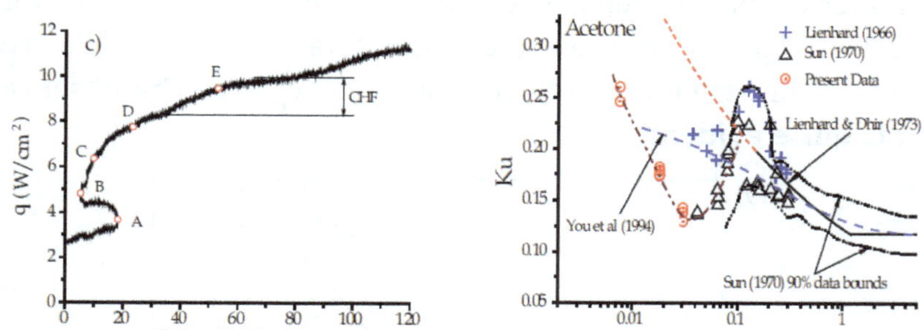

Fig. 7. Scaling behaviours of CHF on wires at saturated condition in normal gravity (Zhao et al., 2009b, c).

However, comparing the trend of CHF in Fig. 6 with the common viewpoint on the scaling of CHF, which was built upon a large amount of experimental data with variable heater diameter on the ground, it was inferred, as pointed out by Di Marco & Grassi (1999), that the dimensionless radius *R'*, or equivalently the Bond number, may not be able to scale adequately the effects and to separate groups containing gravity due to the competition of different mechanisms for small cylinder heaters. Furthermore, Zhao et al. (2009b, c) revisited the scaling behaviours of CHF with respect to *R'* at small value of the Bond number in normal gravity conditions. It has been found that interactions between the influences of the subcooling and size on CHF will be important for the small Bond number, and that there may exist some other parameters, which may be material-dependant, in addition to the Bond number that play important roles in the CHF phenomenon with small Bond number (Fig. 7)

A parameter, named as the limited nucleate size d_{LN}, and a non-dimensional coefficient $\Gamma = d_{LN}/d_{wire}$ were introduced to interpret this phenomenon (Zhao et al., 2009b). It was assumed that the limited nucleate size is not dependent with gravity but with the other parameters of the boiling system, such as the material parameters of the working fluid and the heater, the heater surface condition, an so on. If Γ is small enough, the initial vapour bubbles will be much smaller than the heater surface and then the occurrence of the CHF will be caused by the mechanism of hydrodynamic instability. On the contrary, it will be caused by the mechanism of local dryout if Γ is so large that the initial bubble larger than the wire diameter d_{wire} may easily encircle the heater. Further researches, however, are needed for the delimitation of the two mechanisms.

3. POOL BOILING ON A SURFACE IN MICROGRAVITY

A QSPB (quasi-steady pool boiling) device was developed to study heat transfer of pool boiling on plane plate both in normal and in microgravity, which was flown aboard the Chinese recoverable satellite SJ-8 in September 2006 (Yan, 2007).

To avoid large scatterance of data points measured in steady state boiling experiments and to obtain continuous boiling curves in the limited microgravity duration, a transient heating method was adopted, in which the heating voltage was controlled as an exponential function with time, namely

$$U = U_0 \exp(\tau/\tau_0) \tag{8}$$

where τ denotes the heating time, and the period τ_0 determines the heating rate. In the space experiment aboard SJ-8 and the ground control experiments before the space flight, the period was set for τ_0 = 80 s in order to make the heating process as a quasi-steady state, which was verified in the preliminary experiments on the ground. Furthermore, the period used in the present study was about 3~4 order of magnitude larger than those in Johnson (1971), which guaranteed the fulfillment of quasi-steady condition, though different structure of the heater and working fluid employed here.

The heater used in the study had an Al^2o_3 ceramic substrate with a size of 28×20×1 mm^3 embedded in a PTFE base with a thickness of 25 mm. An epoxy-bonded composite layer of mica sheets and asbestos was set between the ceramic substrate and the PTFE base to reduce the heat loss. The effective heating area with an area of 15×15 mm^2 was covered by a serpentine strip of multi-layer alloy film with a width of 300 μm and a thickness about 10 μm. The space between the adjacent parallel strips is about 70 μm. In addition, the multi- layer alloy film also served simultaneously as a resistance thermometer. The averaged temperature of the heater surface in the experiments was calculated using the correlation between the temperature and the resistance of the multi-layer alloy film, which was calibrated prior to the space flight. In the data reduction, the data of the averaged temperature of the heater surface were filtered to remove noise effects. The total heat flux was transported into both the liquid and the Al^2o_3 ceramic substrate, while the heat loss to the PTFE base and the surrounding was neglected. The filtered temperature data was used to compute the increase of the inner energy of the Al^2o_3 ceramic substrate using appropriate numerical computations. Subtracting the increase of the inner energy of the Al^2o_3 ceramic substrate from the total heat flux input provided the heat flux to the liquid and the transient mean heat transfer coefficient.

Degassed FC-72 was used as the working fluid. The pressure was controlled by a passive control method similar with that used in the TCPB device. Venting air from the container to the module of the satellite decreased the pressure inside the boiling chamber from its initial value of about 100 kPa to the same as that in the module of the satellite, i.e. 40 ~ 60 kPa. An auxiliary heater was used for adjusting the temperature of the bulk liquid from the ambient temperature to about the middle between the ambient and saturation temperature at the corresponding pressure. Except the first run without pre-heating phase, each of the following runs consists of pre-heating, stabilizing and boiling phases, and lasts about one hour. The corresponding experimental conditions are listed in Table 1, in which the estimated values of the critical heat flux (CHF) and the corresponding superheats are also listed. Figs. 8 and 9 show some typical processes of bubble growth, heating history, and the corresponding boiling curves in the space experiments.

Table 1. Space experimental conditions and the estimated CHF values (Zhao et al., 2009a).

Run#	Pressure p (kPa)	Subcooling ΔT_{sub} (K)	CHF q_{CHF} (W/cm^2)	Superheat ΔT_{sat} (K)
I-1	90.8	36.9	8.3 ~ 10.0	28 ~ 66
I-2	97.3	25.8	6.6 ~ 9.1	34 ~ 76
I-3	102.3	21.8	7.0 ~ 7.6	40 ~ 56
I-4	105.7	19.5	7.7 ~ 8.2	20 ~ 29
I-5	111.7	18.4	8.6 ~ 8.9	11 ~ 17
II-1	57.2	24.5	5.7 ~ 6.9	24 ~ 42
II-2	91.1	18.8	7.4 ~ 9.5	26 ~ 55
III-1	65.5	27.5	6.3 ~ 6.6	30 ~ 35

Because of the residual gravity which was estimated in the range of 10^{-3}~10^{-5} $\mathbf{g}_0$, there could exist a week single-phase natural convection before the incipience of boiling. Due to the experimental schedule, different behaviours of the incipience of boiling were observed in the space experiment. In the first run I-1, a great amount of vapor appeared abruptly and explosively at the incipience of boiling. Surface tension then compelled the vapor to form several segregate bubbles. An obvious drop of the heater temperature was observed in the curve of the heating history, correspondingly. This drop caused an additional heat flux from the ceramic substrate to the liquid, and may result in a local maximum of the heat flux to the liquid in the transition region from the incipience to quasi-steady nucleate boiling despite of the monotonous increasing of the heating rate. On the contrary, a gradual growth of the first bubble was observed in the following runs. The process of bubble growth even appeared an obvious standstill after its first appearance. Correspondingly, no over-shooting or drop of the heater temperature can be observed in the curves of the heating history in the following runs. The first appearance of bubbles in the first five runs was observed at 21.89 s, 8.68 s, 8.12 s, 4.54 s, and 4.84 s, respectively. Comparing with the first run, the nucleate boiling occurred significantly earlier in the following runs. Considering the experimental procedure, it may indicate that there could be residual micro-bubbles in cavities after the preceding runs. These micro-bubbles would make the cavities easier to be activated, and the boiling would thus be initiated at a lower wall superheat. Furthermore, bubbles attached on the surface seemed to be able to suppress the activation of the cavities in the neighborhoods according to the detailed analyses of the video images.

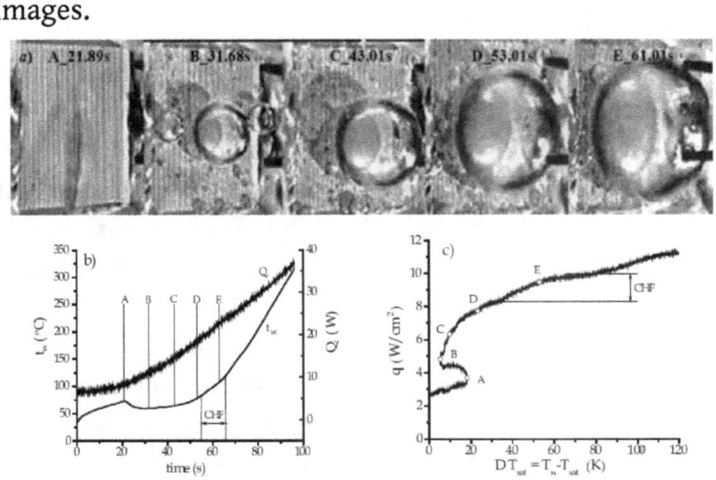

Fig. 8. Bubble dynamics (a), heating history (b), and boiling curve (c) in the run I-1 aboard SJ-8 (Zhao et al., 2009a).

It was observed that primary bubbles generated consistently, slid on the surface, and coalesced with each other to form a larger coalesced bubble. Although the video images were taken only from the sole direction of 45° with respect to the heater surface, it was able to be observed that some primary bubbles generated under the coalesced bubble. The coalesced bubble also engulfed small bubbles around it. It can be inferred that, as pointed out by Ohta et al. (1999), a macro-layer may exist underneath the coalesced bubble, where primary bubbles are forming.

For the cases of higher subcooling, the coalesced bubble with a relative smooth surface was observed oscillating near the center of the heater surface. Higher was the subcooling, smaller and smoother at the same heating time. The coalesced bubble shrank to an elliptical sphere under the action of surface tension. Its size increased with the increase of the surface temperature, but it was very difficult to cover the whole surface. Thus, the bottom of the coalesced bubble may dry out partly at high heat flux, while the other places, particularly in the corners of the heater surface were still in the region of nucleate boiling. Unfortunately, dry spot was not able to be observed directly in the present study. The fact, however, that there existed a much smooth increase of the averaged temperature of the heater surface and no turning point corresponding to CHF in boiling curves indicated a gradual transition to film boiling along with the developing of the area of local dry area, as described by Oka et al. (1995). In this case, it was difficult to determine the accurate value of CHF. However, the trend of the increasing heater temperature with the heating time provided some information of CHF. Supposing the rapid increase of heater temperature corresponds to the beginning of the transitional boiling while a constant slope of the temperature curve to the complete transition to film boiling, the range of CHF and the corresponding superheat were estimated, which were also marked in Fig. 8.

The bubble behaviours and the characteristics of the boiling curves at lower subcooling were different from those at higher subcooling. In these runs, *e.g.* the run I-4 shown in Fig. 9, the size of the coalesced bubble increased quickly, and a strong oscillation appeared on its surface. Higher was the pressure, stronger the surface oscillation. Furthermore, before the abrupt transition to film boiling, the heat flux remained increasing though the surface temperature rose slowly or even fell down along with the heating time. The above observations can be interpreted as follows. Because of the decrease of surface tension with the increase of the saturation temperature and the corresponding pressure, local dry spots underneath the coalesced bubble with a strong surface oscillation cannot develop steadily. They may be re-wetted by the surrounding liquid, and nucleate boiling will remain on the heater surface. Furthermore, even more nucleate sites could be activated under the action of the strong oscillation of the coalesced bubble. Thus, heat transfer was enhanced.

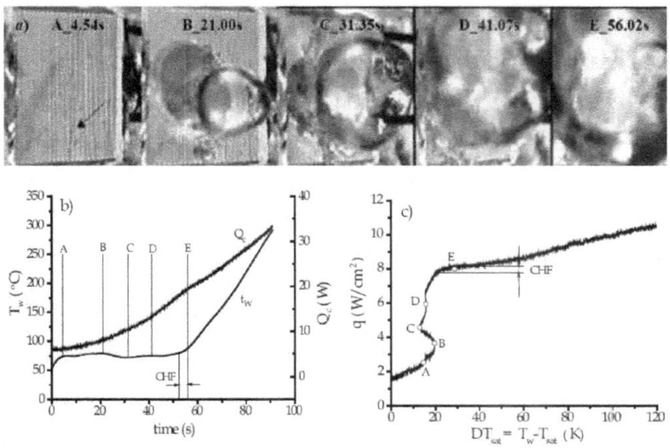

Fig. 9. Bubble dynamics (a), heating history (b), and boiling curve (c) in the run I-4 aboard SJ-8 (Zhao et al., 2009a).

Comparative analyses of boiling curves in microgravity indicated that heat transmission degraded with reduced subcooling at constant pressure, whereas it improved with increased pressure at constant subcooling. The calculated values of CHF in microgravity rose with increased subcooling at constant pressure and likewise elevated with pressure at fixed subcooling. These tendencies resemble those observed under normal gravity conditions. The value of CHF in microgravity, however, was merely one third of that observed under comparable pressure and subcooling in terrestrial conditions. Regrettably, the pressure and temperature of the liquid cannot be entirely segregated due to the passive regulation of pressure within the boiling chamber employed in this context. Consequently, there were certain interdependencies between pressure and subcooling on critical heat flux (CHF).

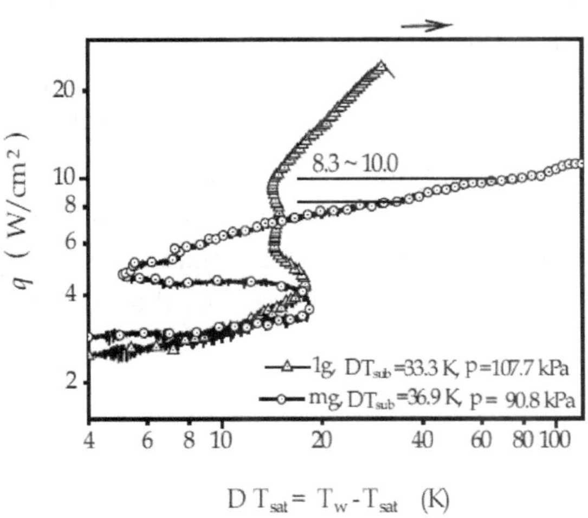

Fig. 10. Comparison of boiling curves in different gravity (Zhao et al., 2009a).

In Fig. 10, boiling curves in different gravity were compared with each other at the similar pressure and subcooling conditions. Generally, boiling heat transfer in microgravity was deteriorated comparing with that in normal gravity, particularly at high superheats or heat fluxes. Much obvious enhancement, however, could be observed just beyond the incipience, which was consistent with those in steady state pool boiling experiments, such as reported by Lee et al. (1997). It was also observed that the incipience of boiling occurred in microgravity at the same superheat as that in normal gravity, which was in agreement with Straub (2001).

Recently, a new serial of experiments of pool boiling of FC-72 with non-condensable gas on smooth surface (denoted as chip S) in short-term microgravity have been conducted utilizing the drop tower Beijing (Xue et al., 2010). The boiling vessel was filled with about 3 L of FC-72 as the working liquid. The test chip was a P doped N-type silicon chip with the dimensions of 10×10×0.5 mm^3, which was set horizontally upward. The chip was Joule heated by a direct current. Two 0.25-mm diameter copper wires were soldered by a low temperature solder to the chip side surfaces at the opposite end for power supply. A programmable DC power supply was used to provide constant heating electric current for the chip. A nearly atmospheric pressure is maintained by attaching a rubber bag to the test vessel. A K-type thermocouple was used for measuring the local temperature of the test liquid at the chip level, and directly connected to a temperature display monitor for visual observation through a CCD camera. Besides, the local wall temperature at the center of the chip and the local temperature of the test liquid at about 40 mm from the edge of the chip were measured by two 0.13 mm-diameter T-type thermocouples which were connect with a data acquisition system (DI710-UHS). A high speed video camera (VITcam CTC) imaging 250 frames per second at a resolution of 1024×640 pixels with a shutter speed of 1/2000s was used along with a computar lens (MLM-3XMP) to obtain images of the boiling process. The high-speed camera was positioned in front of the test vessel at a 30° angle relative to the heating surface. Given the brief period of microgravity (about 3.6 seconds), boiling commenced prior to the deployment of the drop capsule and was maintained at a stable state for a sufficient duration. Subsequently, the drop capsule was deployed, and the experiment was conducted in a microgravity environment. The experimental run concluded following the retrieval of the drop capsule.

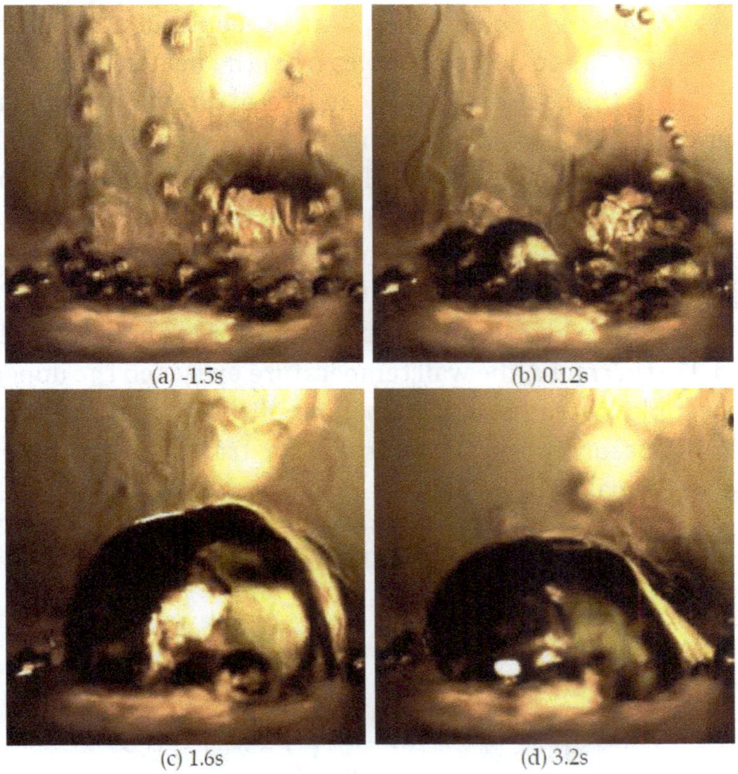

Fig. 11. Bubble behaviors on chip S (Wei et al., 2010).

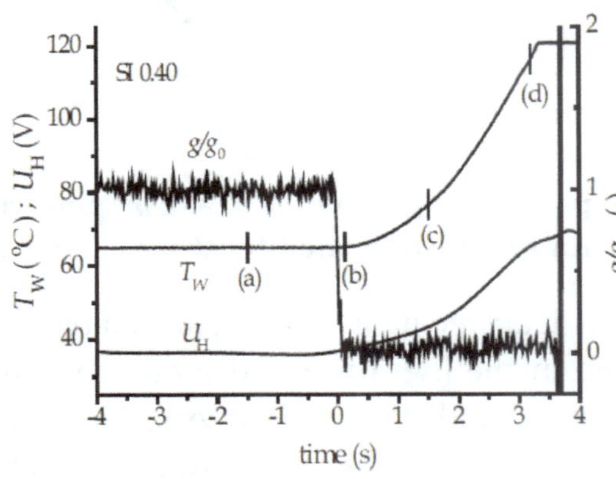

Fig. 12. Variations of surface temperature, heating voltage, and gravity for chip S (Wei et al., 2010).

Steady- or quasi-steady nucleate pool boiling was observed in the experiments for low and intermediate heat flux in the short-term microgravity conditions. At low heat fluxes in microgravity condition, the vapor bubbles increase in size but little coalescence occurs among bubbles due to large space between adjacent bubbles on the heater surface, thus the steady nucleate pool boiling can be obtained. As the heat flux increases, the vapor bubbles number as well as their size significantly increase in microgravity. Coalescence occurs continuously among adjacent bubbles. Departure of the coalesced bubbles from the heater surface caused by the surface oscillation of the coalesced bubble in lateral direction results in a constant heater temperature and heat flux in microgravity compared to that in normal gravity. The steady-state pool boiling still can be maintained. At high heat fluxes, a large coalesced bubble forms quickly and covers the heater surface completely in microgravity, followed by shrinking to an oblate in shape and smooth in contour due to the highly subcooled condensation (Fig. 11). An obvious increase of the heater temperature (Fig. 12), which indicates deterioration of boiling heat transfer, is then observed. Furthermore, the wall temperature exceeded the upper limit cutoff of the thermocouple instrument near the end of the short-term microgravity. It is possible for the occurrence of local dry-out or transition to film boiling at the bottom of the large coalesced bubble.

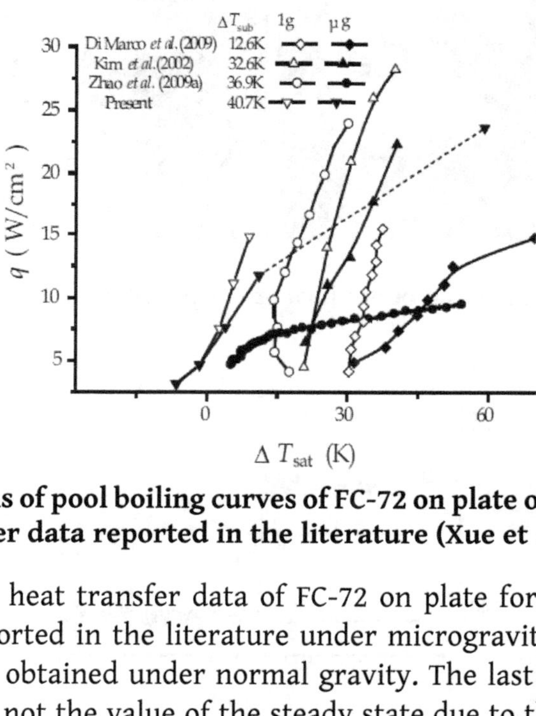

Fig. 13. Comparisons of pool boiling curves of FC-72 on plate of the present results with other data reported in the literature (Xue et al., 2010).

Figure 13 plots the heat transfer data of FC-72 on plate for the present results and other data reported in the literature under microgravity together with the pool boiling curves obtained under normal gravity. The last data point with the highest heat flux is not the value of the steady state due to the short duration of microgravity, and then a dashed line was used to connect it with the other data points. It can be clearly see that the influence of subcooling on nucleate pool boiling heat transfer in microgravity appears similar to that in earth gravity, *i.e.* the heat transfer increase distinctly as the subcooling increases. This agrees with the results of studies made by Lee et al. (1997). But, since the gravity level also greatly influences the average heater surface temperature, especially in the high heat flux region, the tendency of the influence of subcooling on nucleate pool boiling in microgravity is still required to further investigate. In particular, the results of

Zhao et al. (2009a) in microgravity changes greatly from that in normal gravity and the value of CHF is only about one third of that at the similar pressure and subcooling in terrestrial condition. Besides, the trend of the present data result is consistent with that the Di Marco & Grass (2009), and the slopes of the heat transfer curves decrease in the high heat flux region in microgravity. However, the result of Kim et al. (2002) in microgravity changes slightly from that in normal gravity, which is different from the present and other studies. Thus, it can be inferred that the heat transfer of nucleate pool boiling in microgravity is related with the liquid subcooling, the size of the heater, the heating method, content of non-condensable gas, and so on.

4. POOL BOILING ON MICRO-PIN-FINNED SURFACE IN MICROGRAVITY

Very recently, a serial of experiments on boiling enhancement in microgravity by use of micro-pin-fins which were fabricated by dry etching have been performed in the drop tower Beijing (Wei et al., 2010). This project was motivated by the following observations in space experiments of boiling. Vapour bubbles cannot depart easily from the heater surface in microgravity, and then can grow attaching to the surface and coalesced with each other. As the increase of their sizes, the coalesced bubbles can cover the heater surface and prevent the fresh liquid from moving to the heater surface, thus local dryout may occur, resulting in deterioration of heat transfer. On the contrary, if plenty of fresh liquid can be supplied to the superheated wall for vaporization, the efficient nucleate pool boiling can be maintained and then no deterioration of heat transfer can occur. Following the enhanced boiling heat transfer mechanisms for the micro-pin-finned surfaces (Wei et al., 2009), it is supposed that although the bubbles staying on the top of the micro-pin-fins cannot be detached soon in microgravity, the fresh bulk liquid may still access to the heater surface through interconnect tunnels formed by the micro-pin-fins due to the capillary forces, which is independent of the gravity level.

The experimental facility and schedual are similar with those used in Xue et al. (2010). The test chip was a P doped N-type silicon chip with the dimensions of $10\times10\times0.5$ mm^3. Micro-pin-fins were fabricated on the chip surface for enhancing boiling heat transfer. The fin thickness is 50 **μm** and fin height is 60 **μm** (denoted as chip PF50-60). The test chip was heated by setting a constant electric current for the desired heat flux to initiate boiling on the heater surface. After the heat transfer reached a steady state in about two minute, the free falling of drop capsule started which could provide approximately 3.6 s effective microgravity environment. The high-speed video camera could work for a duration time of 8 s, which was divided into two half sections by an external trigger signal. The bubble behaviours in normal gravity before the release of the drop capsule was recorded in the first half section, while those in microgravity after the release was recorded in the other one. Moreover, the data measurement and the video recording were operated simultaneously.

The transition of vapor bubble behaviors and the mean surface temperature of the micro- pin-finned chip responding to the variation of gravity level for the heating current of 0.42 A (corresponding to a heat flux of 19.4 W/cm^2), similar to that shown in Figs. 11 and 12, are shown in Figs. 14 and 15, respectively. The positions of Fig. 14 a-d are marked on the curves of the mean temperature of Chip PF50-60 shown in Fig. 15. The liquid subcooling keeps at about 41K, also as the same as that shown in Figs. 11 and 12.

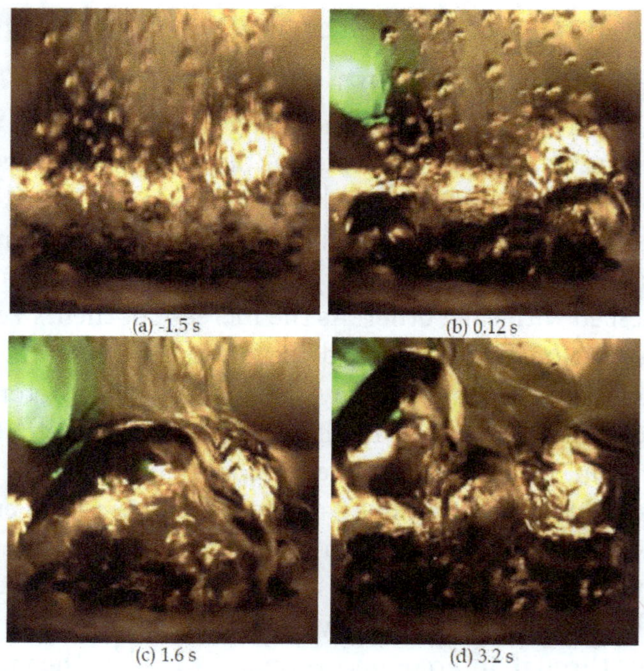

Fig. 14. Bubble behaviors on chip PF50-60 (Wei et al., 2010).

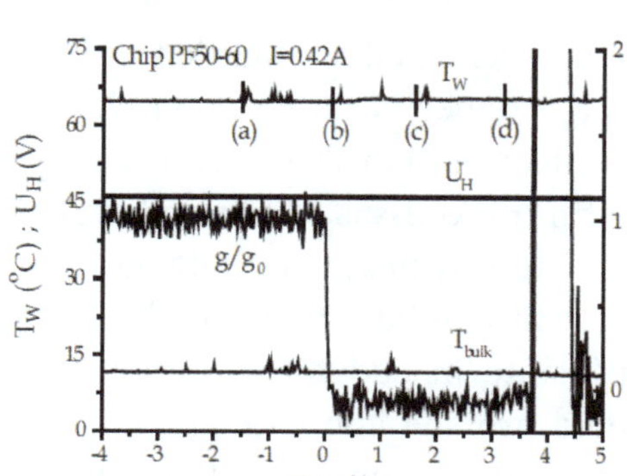

Fig. 15. Variations of surface temperature, heating voltage, and gravity for chip PF50-60 (Wei et al., 2010).

Similar to the smooth chip scenario, bubbles constantly form and detach from the heating surface due to buoyancy forces under normal gravity prior to the discharge of the drop capsule (Fig. 14a). Nevertheless, the bubble count is significantly more than that of the smooth chip, suggesting that the micro-pin-finned surface offers a greater number of nucleation sites to improve boiling heat transfer efficiency. Approximately 0.12 seconds after entering the microgravity environment, the vapor bubbles commence coalescing to create many big bubbles adhering to the chip surface (Fig. 14b). Small bubbles are present in the departure state upon entering the microgravity condition, allowing for their further detachment from the heater surface at this moment. Over time, the bubbles merge to create a big spherical bubble (Fig. 14c). Nonetheless, the substantial bubble coverage on the heater surface does not result in a significant elevation of wall temperature (Fig. 15).

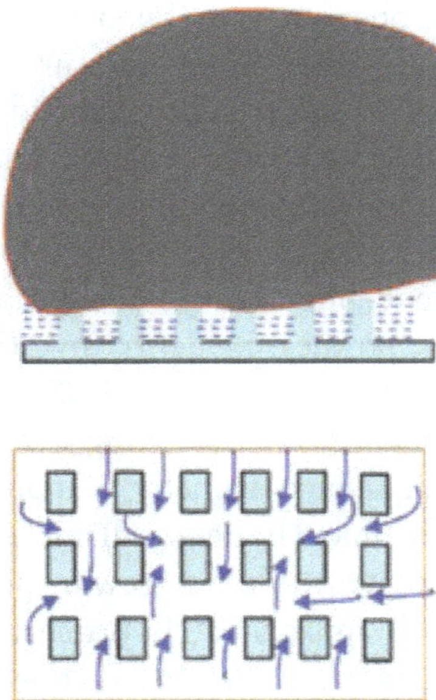

Fig. 16. Bulk liquid supply and micro-convection caused by capillary force (Wei et al., 2009).

The capillary force generated by the interface between the large bubble and the liquid of the micro-layer beneath the bubble drives plenty of fresh liquid to contact with the superheated wall for vaporization through the regular interconnected structures formed by the micro- pin-fins, as well as improves the micro-convection heat transfer by the motion of liquid around the micro-pin-fins, as shown schematically in Fig. 16. The sufficient supply of bulk liquid to the heater surface guarantees the continuous growth of the large bubble. Therefore, contrary to boiling on chip S, there is no deterioration of boiling heat transfer performance for the micro-pin-finned surface in microgravity, and the heater surface temperature can keep almost constant in both gravity and microgravity conditions.

In summary, the micro-pin-fined surface structure generates significant capillary force and minimal flow resistance, facilitating substantial bulk liquid access to the heater surface for evaporation in high heat flux regions, hence enhancing boiling heat transfer considerably. The capillary force is independent of gravitational levels, making the micro-pin-fined surface a promising enhanced surface for efficient cooling of electronic components in both normal and microgravity conditions, thereby aiding in the reduction of cooling system weight in space and on planetary bodies.

5. FUTURE INVESTIGATIONS ON BOILING IN MICROGRAVITY

A forthcoming project, DEPA-SJ10, is scheduled for deployment on the Chinese recoverable satellite SJ-10 (Wan & Zhao, 2008). The project will utilize boiling in a singular artificial cavity as a model to investigate subsystems in nucleate pool boiling of pure substances. Transient phenomena of bubble formation, development, and detachment will be examined, while the temperature distribution at the active nucleation site will be quantified under subcooled and saturated circumstances. The primary objective is to elucidate bubble dynamics and convection surrounding the expanding vapor bubble in microgravity, to comprehend microscale heat transfer mechanisms, and to uncover the physical phenomena that regulate nucleate boiling. Numerical modeling of single bubble boiling has been developed, utilizing this phenomena as a physical model to investigate the thermodynamic behaviors of bubbles, heat transport, and the associated gravitational effects in nucleate pool boiling. (Zhao et al., 2010). According to some preliminary results, it was indicated that the growing bubble diameter is approximately proportional to the 0.4-th power of the growing time. The detach diameter of bubble is proportional to the -1/3-th power of the gravity, while the growing period to the -4/5-th power of the gravity. The heat flux is approximately proportional to the 1.5-th power of wall superheat with a fixed number density of active nucleation sites in all the studied gravity levels. The heat transport in the micro-wedge region significantly influences the overall efficacy of boiling. Additional experimental investigations on the performance of micro-pin-finned surfaces are scheduled to be conducted at the drop tower in Beijing. The objective is to examine the behavior at elevated heat flux levels surrounding the critical heat flux phenomenon and to ascertain the optimal configuration of the micro-pin-fins. These projects will enhance the comprehension of such phenomena and contribute to the advancement of space systems that incorporate boiling phenomena.

6. CONCISE SUMMARY

Nucleate pool boiling is a common occurrence that efficiently transfers high heat flow. Nonetheless, it is a highly intricate and elusive process. Gravity can significantly influence many sub-processes in the boiling phenomenon, perhaps obscuring the true mechanisms involved. Microgravity experiments provide a distinctive chance to investigate complex interactions devoid of external forces, such as buoyancy, which can influence bubble dynamics and associated heat transfer. Moreover, they can offer a method to examine the true impact of gravity on boiling. Conversely, the numerous possible applications in space and on neighboring planets, attributed to its great heat transmission efficiency, have rendered pool boiling in microgravity an increasingly prominent area of research. Over the past decade, two research initiatives on nucleate pool boiling in microgravity have been executed onboard Chinese recoverable satellites. Ground-based tests have been conducted in both standard gravity and brief microgravity conditions at the Beijing drop tower, with numerical models. The principal findings are encapsulated in this chapter. The continuous boiling of R113 on slender platinum wires was examined using a temperature-regulated heating technique, whereas the quasi-steady boiling of FC-72 on a flat plate was analyzed with an exponentially escalating heating voltage. The bubble dynamics in microgravity differs significantly from that in standard gravity, and the heat transfer characteristics are contingent upon the bubble dynamics. Bubbles exhibited lateral movements on the heaters prior to their departure in microgravity. The surface oscillation of the coalesced bubbles, resulting from lateral merging with neighboring bubbles, caused their detachment from the heaters. The impact of Marangoni effects has elucidated the distinct properties of bubble behavior in microgravity. A novel bubble departure model has been suggested, capable of predicting the entire observation in both microgravity and normal gravity conditions. A minor improvement in heat transmission on wires is noted in microgravity, whereas a reduction is shown for high heat flux in the plate scenario. The variations in characteristics may be attributed to the differing liquid supply beneath the developing bubbles in the two distinct scenarios. It is proposed that optimal heat transfer performance in nucleate pool boiling under microgravity conditions will be achieved if an adequate supply of liquid is delivered to the base of the expanding bubbles. A series of studies on pool boiling on a micro-pin-finned surface have been conducted using the drop tower in Beijing. While bubbles cannot separate in microgravity and remain atop the micro-pin-fins, the new liquid can still reach the heater surface via interconnecting tunnels generated between the micro-pin-fins due to capillary forces, which are unaffected by gravitational conditions. Consequently, no degradation of heat transfer in microgravity is noted, even at much elevated heat flux levels approaching the critical heat flux reported under normal gravity conditions. The value of CHF in microgravity is less than that in normal gravity; yet, it may still be accurately predicted using the correlation established by Lienhard & Dhir (1973), despite the dimensionless radius in this instance exceeding its original applicability range. The scaling of CHF with gravity differs significantly from conventional perspectives, and a potential mechanism is proposed based on experimental results.

REFERENCES

Di Marco, P., 2003. Review of reduced gravity boiling heat transfer: European research. J. Jpn. Soc. Microgravity Appl., 20(4), 252–263.

Di Marco, P., Grassi, W., 1999. About the scaling of critical heat flux with gravity acceleration in pool boiling. In: Proc. XVII UIT Nat. Heat Transfer Conf., Ferrara, pp.139-149.

Di Marco, P., Grassi, W., 2009. Effect of force fields on pool boiling flow patterns in normal and reduced gravity. Heat Mass Transfer, 45: 959-966.

Johnson, H.A., 1971. Transient boiling heat transfer to water. Int. J. Heat Mass Transfer, 14, 67–82.

Kim, J., 2003. Review of reduced gravity boiling heat transfer: US research. J. Jpn. Soc. Microgravity Appl., 20(4), 264–271.

Kim, J., Benton, J., Wisniewski, D., 2002. Pool boiling heat transfer on small heaters: effect of gravity and subcooling. Int. J. Heat and Mass Transfer, 45: 3919-3932.

Lee, H.S., Merte, H., Jr., Chiaramonte, F., 1997. Pool boiling curve in microgravity. J. Thermophy. Heat Transfer, 11(2), 216–222.

Lienhard, J.H., Dhir, V.K., 1973. Hydrodynamic prediction of peak pool boiling heat fluxes from finite bodies. J. Heat Transfer, 95, 152–158.

Liu, G., 2006. Study of subcooled pool boiling heat transfer on thin platinum wires in different gravity conditions. M. Sc. Thesis, Institute of Mechanics, Chinese Academy of Sciences, Beijing, China.

Oka, T., Abe, Y., Mori, Y.H., Nagashima, A., 1995. Pool boiling of n-pentane, CFC-113 and water under reduced gravity: parabolic flight experiments with a transparent heater. J. Heat Transfer Trans. ASME, 117: 408-417.

Ohta, H., Kawasaki, K., Azuma, H., Yoda, S., and Nakamura, T., 1999. On the heat transfer mechanisms in microgravity nucleate boiling. Adv. Space Res., 24(10): 1325-1330.

Ohta, H., 2003a. Review of reduced gravity boiling heat transfer: Japanese research. J. Jpn. Soc. Microgravity Appl., 20(4), 272–285.

Ohta, H., 2003b. Microgravity heat transfer in flow boiling. Adv. Heat Transfer, 37, 1–76.

Ohta, H., Kawasaki, K., Azuma, H., Yoda, S., Nakamura, T., 1999. On the heat transfer mechanisms in microgravity nucleate boiling. Adv. Space Res., 24(10), 1325–1330.

Straub, J., 2001. Boiling heat transfer and bubble dynamics in microgravity. Adv. Heat Transfer, 35, 57–172.

Sun, K.H., Lienhard, J.H., 1970. The Peak Pool Boiling Heat Flux on Horizontal Cylinders, Int. J. Heat Mass Transfer, 13: 1425-1439.

Wan, S.X., Zhao, J.F., 2008. Pool boiling in microgravity: recent results and perspectives for the project DEPA-SJ10. Microgravity Sci. Tech., 20(3-4), 219-224.

Wan, S.X., Zhao, J.F., Liu, G., Li, B., Hu, W.R., 2003. TCPB device: description and preliminary ground experimental results. In: 54th Int. Astronautical Cong., Sep. 29– Oct 3, Bremen, Germany.

Xue, Y.F., Zhao, J.F., Wei, J.J., Li, J., Guo, G., Wan, S.X., 2010. Experimental Study of FC-72 Pool Boiling on Smooth Silicon Chip in Short-Term Microgravity. Submitted to J. Heat Transfer Trans. ASME.

Yan, N., 2007. Experimental study on pool boiling heat transfer in microgravity. M. Sc. Thesis, Institute of Mechanics, Chinese Academy of Sciences, Beijing, China.

You, S.M., Hong, Y.S., O'Connor, J.P., 1994. The Onset of Film Boiling on Small Cylinders: Local Dryout and Hydrodynamic Critical Heat Flux Mechanisms, Int. J. Heat Mass Transfer, 37: 2561-2569.

Wei, J.J., Zhao, J.F., Yuan, M.Z., Xue, Y.F., 2009. Boiling heat transfer enhancement by using micro-pin-finned surface for electronics cooling. Microgravity Sci. Tech., 21(Suppl. 1): S159 – S173.

Wei, J.J., Xue, Y.F., Zhao, J.F., Li, J., 2010. High efficiency of heat transfer of nucleate pool boiling on micro-pin-finned surface in microgravity. submitted to Chin. Phys. Lett.

Zhao, J.F., Wan, S.X., Liu, G., Hu, W.R., 2004. Subcooled pool boiling in microgravity: results of drop tower testing. In: 7th Drop Tower Days, Sep. 12–25, Bremen, Germany.

Zhao, J.F., Liu, G., Li, Z.D., Wan, S.X., 2007. Bubble behaviors in nucleate pool boiling on thin wires in microgravity. In: 6th Int. Conf. Multiphase Flow, July 9–13, Leipzig, Germany.

Zhao, J.F., Liu, G., Wan, S.X., Yan, N., 2008. Bubble dynamics in nucleate pool boiling on thin wires in microgravity. Microgravity Sci. Tech., 20(2), 81-89.

Zhao, J.F., Li, J., Yan, N., Wang, S.F., 2009a. Bubble behavior and heat transfer in quasi- steady pool boiling in microgravity. Microgravity Sci. Tech., 21(Suppl. 1): S175 – S183.

Zhao, J.F., Lu, Y.H., Li, J., 2009b. CHF of pool boiling on microwires. ASME 2009 2nd Micro/Nanoscale Heat Mass Transfer Int. Conf., December 18-21, 2009, Shanghai, China.

Zhao, J.F., Lu, Y.H., Li, J.,2009c. CHF on cylinders - revisit of influences of subcooling and cylinder diameter. ECI Int. Conf. on Boiling Heat Transfer, May 3-7, 2009, Florianópolis, Brazil.

Zhao, J.F., Wan, S.X., Liu, G., Yan, N., Hu, W.R., 2009d. Subcooled pool boiling on thin wire in microgravity. Acta Astronautica, 64(2-3): 188 – 194.

Zhao, J.F., 2010. Two-phase flow and pool boiling heat transfer in microgravity. Int. J. Multiphase flow, 36(2): 135-143.

Zhao, J.F, Li, Z.D, Li, J. 2010. Numerical simulation of single bubble boiling in different gravity conditions. In: 8th Japan-China-Korea Workshop on Microgravity Sciences for Asian Microgravity Pre-Symposium, September 22 – 24, 2010, Akiu, Sendai, Japan.

9 Chapter THERMAL TRANSFER IN FILM BOILING OF FLOWING WATER

1. INTRODUCTION: AN OVERVIEW

Film boiling is a post-critical heat flux (CHF) regime characterized by elevated surface temperatures, preventing wall contact with the liquid, resulting in vapor coverage and consequently diminished heat transfer efficiency due to the vapor's low thermal conductivity. Film boiling occurs in different applications, including as metallurgy, refrigeration, chemical engineering, and power engineering. During a hypothesized loss of coolant event in nuclear reactors, the exposed core would undergo this condition, with the peak fuel temperature predominantly influenced by the heat transfer associated with film boiling. The film boiling has undergone intensive experimental and theoretical investigations due to its considerable significance to applications. It was one of three subjects in a coordinated research program on Thermal-hydraulic relationships for advanced water- cooled reactors, which was organized by the International Atomic Energy Agency (1994 – 1999). A comprehensive review on these investigations has been presented in the technical document (IAEA-TECDOC-1203, 2001)

In film boiling the heat is transferred from the wall to the vapor, then from the vapor to the liquid, characterized by non-equilibrium. The interaction between two phases dominates the vapor generation rate and the superheat, associated with extremely complicated characteristics. This presents a major challenge for the estimation of heat transfer because of less knowledge on the interfacial processes. In particular, due to the peculiar feature of the boiling curve it is difficult to establish the film boiling regime at stable condition in a heat flux controlled system by using a conventional experimental technique. As shown in Fig.1, the stable film boiling regime can only be maintained at a heat flux beyond the CHF, which associates with an excessively high surface temperature for water. But for a heat flux, q, below the CHF, the regime can not be maintained stably at the post-CHF region (F or T), but at the pre-CHF region (N).

The experimental data on film boiling were mostly obtained with refrigerant or cryogenic fluids, and the data of water were generally obtained in a temperature-controlled system or at transient condition with less accuracy. Since a so-called hot patch technique was developed for establishment of the stable film boiling regime (Groeneveld, 1974, Plummer, 1974, Groeneveld & Gardiner, 1978), a large number of experimental data have been obtained (Stawart & Groeneveld, 1981, Swinnerton et al., 1988, Mossad, 1988). Based on the data base various physical models have been proposed (Groeneveld & Snoek, 1984, Groeneveld, 1988, Mossad & Johannsen, 1989), and the tabular prediction methods have been developed for fully-developed film boiling heat transfer coefficients (Leung et al., 1997, Kirillov et al., 1996).

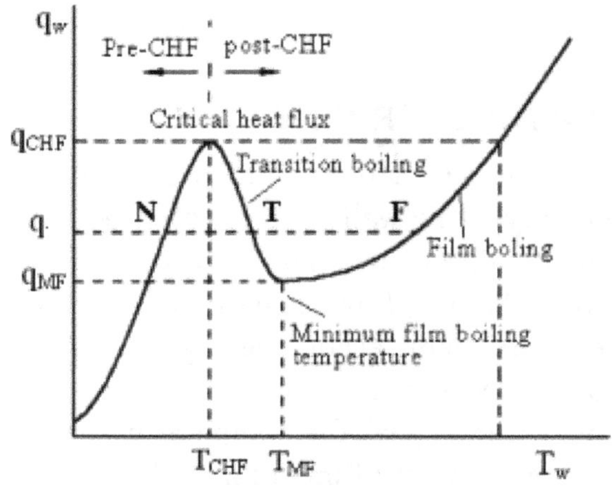

Fig. 1. Typical boiling curve

In 1984 a directly heated hot patch technique was applied by the authors to reach higher heat flux, enabling the steady-state experiment to cover extended range of conditions (Chen & Li, 1984). The results fill the gaps of data base, especially in the region of lower flow, where thermal non-equilibrium is significant, associated with much complicated parametric trends and strongly history-dependent features of the heat transfer coefficient (Chen, 1987, Chen et al., 1989, Chen & Chen, 1994). With these unique data the film boiling has been studied systematically and the prediction methods have been suggested, as will be shown in the following paragraphs.

2. STEADY-STATE EXPERIMENTAL TECHNIQUE

The hot patch technique is to supply separate power to a short section just ahead of the test section to reach CHF, preventing the rewetting front from moving forward. It was first used in freon and nitrogen experiments (Groeneveld, 1974, Plummer, 1974). To increase the power of hot patch for the experiment of water, it was improved by Groeneveld & Gardiner (1978), using a big copper cylinder equipped with a number of cartridge heaters.

To reach further high heat flux, a directly heated hot patch technique was applied by the authors (Chen & Li, 1984). As shown schematically in Fig.2, the test section included two portions, AB and BC, with each heated by a separate supply. The length of section AB was 10 – 25 mm. Near the end (B) the wall thickness was reduced locally, so that a heat flux peak can be created there by electric supply due to higher electric resistance.

During experiment, at first the inlet valve of the test section was closed, and the water circulation was established in a bypass at desired pressure, flow rate and temperature. The test section was then heated by switching on two supplies with it in empty of water. When the wall temperature reached above 500 °C, the flow was switched from the bypass to the test section. As the rewetting front moved upward the power to the upstream section was increased to reach CHF at the end (B), where the rewetting front was arrested without an excessive increase in the wall temperature as a result of axial heat conduction. In the same way, another rewetting front was arrested at the end of section BC by the upper hot patch. Therefore, the stable film boiling regime was maintained on the section BC with heat flux below the CHF. Shown in Fig.3 are the pictures of stable film boiling in an annulus for different water temperatures with the hot patch on and a reflooding transient with the hot patch off.

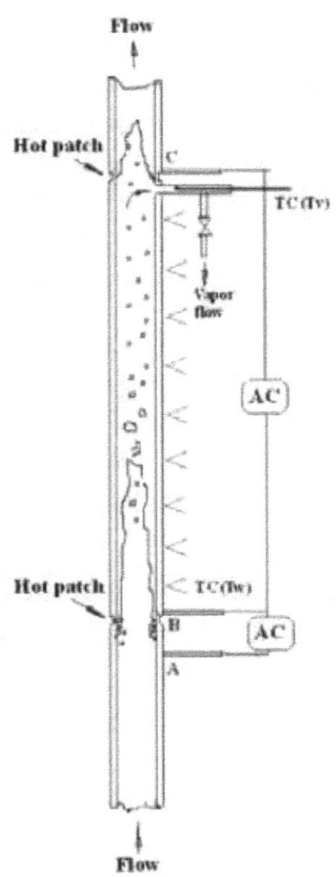

Fig. 2. Schematic of the test section with measurements of both the wall and vapor temperatures

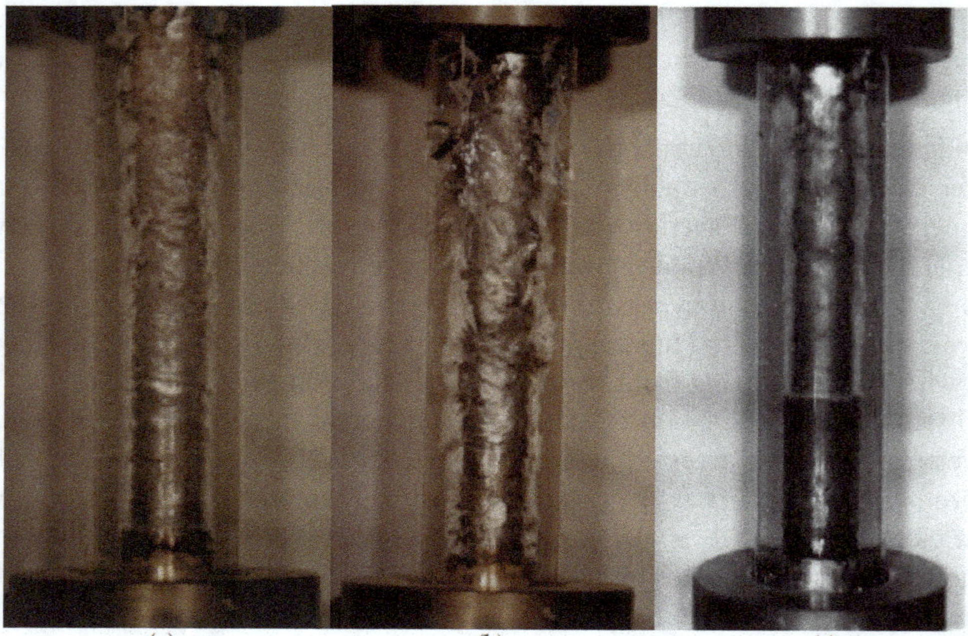

(a) (b) (c)

Fig. 3. Inverted annular film boiling in an annulus with water flowing upward (a) and (b): Stable regime (with the hot patch on), $T_{l,a}<T_{l,b}$, (c): Reflooding transient (with the hot patch off)

The steady-state film boiling experiments have been performed with water flowing upward in tubes of 6.7 – 20 mm in diameter and 0.15 – 2.6 m in length, covering the ranges of pressure of 0.1 – 6 MPa, mass flux of 23 – 1462 kg/m^{2}s and inlet quality of -0.15 – 1.0.

3. ATTRIBUTES OF HEAT TRANSFER DURING FILM BOILING

The term "film boiling" was originally used for a post-CHF regime in a pool characterized by the wall separated from the stagnant liquid by a continuous vapor film. It was then used in forced flow, though the flow pattern varied with the enthalpy in the channel. It includes two major regimes: 1) the inverted annular film boiling (IAFB), which occurs at subcooled or low quality condition, and 2) the dispersed flow film boiling (DFFB), which occurs at saturated condition with the void fraction larger than around 0.8. In IAFB the vapor film separates the wall from the continuous liquid core, in which some bubbles might be entrained for saturated condition. The DFFB is characterized by liquid droplets entrained in the continuous vapor flow. It can be resulted from break-up of the IAFB or from dryout of the liquid film in an annular flow. Fig.4 shows the film boiling regimes in a bottom reflooding transient at different flooding rates.

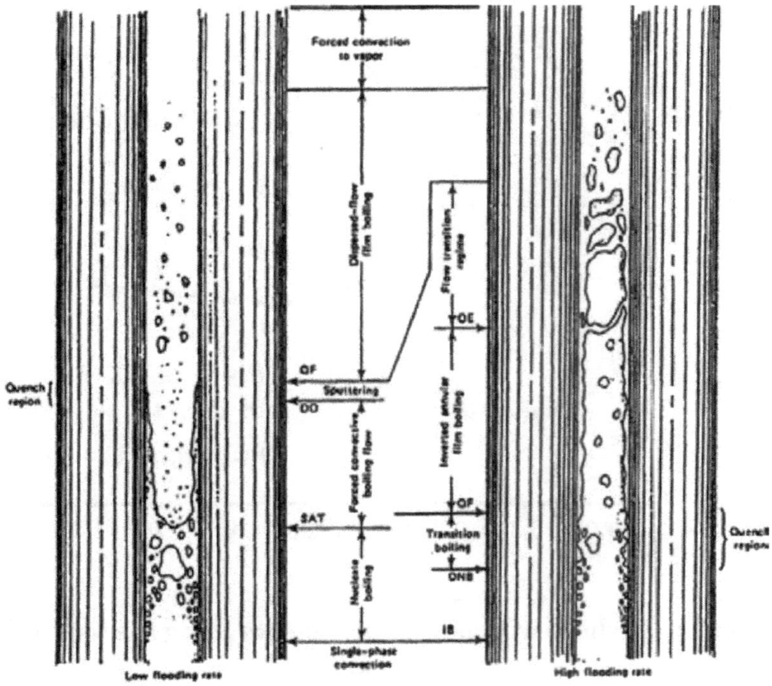

(a) lower inject rate (b) higher inject rate

Fig. 4. Film boiling regimes during reflooding with different flooding rates (Arrieta & Yadigaroglu, 1978)

Typical experimental results are exemplified in Fig.5, where the heat transfer coefficient distributions in a tube for different inlet qualities are displayed by h (= $q_w/(T_w-T_s)$) versus x_E. For subcooled (run no.1) and low quality (run no. 2) inlet condition the post-CHF region initiates with IAFB followed by DFFB. While for relatively high inlet quality (run no. 3 and 4) the DFFB covers the whole post-CHF region. As seen, lower heat transfer coefficients are attained in the transition region.

3.1 Inverted annular film boiling

In IAFB the heat is transferred by convection and radiation from the wall to the vapor, subsequently from the vapor to the interface with liquid. For subcooled condition it is then partially transferred to the liquid core. At the interface the vaporization takes place and the vapor generation rate is determined by the heat flux to the interface minus that to the liquid core. As the increase of vapor generation the vapor flow in the film may transit from laminar to turbulent. Furthermore, the interaction between two phases could result in interface oscillation, having enhancement effect on the heat exchange in both the vapor film and the liquid core.

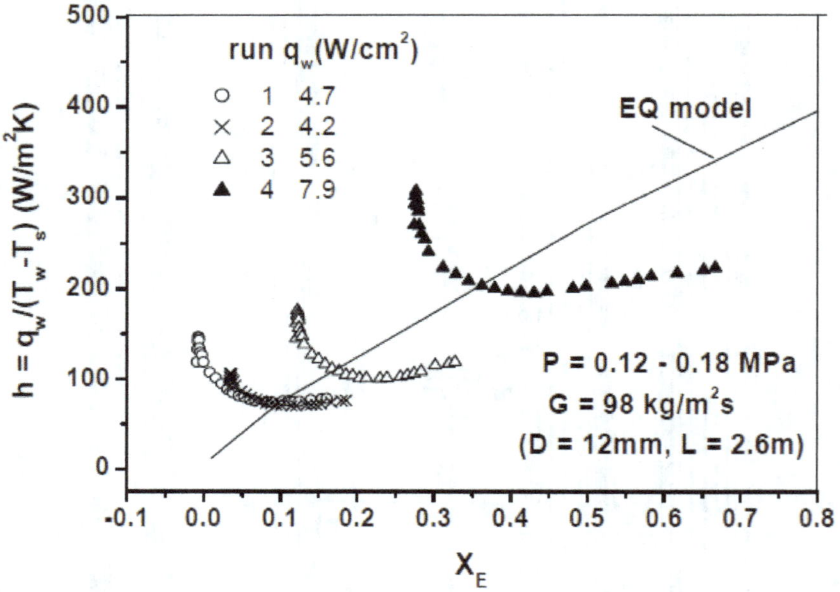

Fig. 5. Distributions of the heat transfer coefficients along the length for different inlet qualities

3.1.1 Effects of the pressure, mass flux and subcooling

Fig.6 shows the distributions of heat transfer coefficients ($h = q_w/(T_w-T_s)$) under different conditions. For lower flow with higher subcooling the h decreases rapidly with distance, while as subcooling decreasing the h decreases, and the trend becomes mild (Fig.6(a)). For higher flow with higher subcooling a maximum h is attained at a few centimeters from the dryout point (Fig.6(b,c)). In this case the thickness of vapor film is very small, so the interface oscillation could lead to dry-collision between liquid and wall, resulting in a substantial increase in the h. For low inlet subcooling the variation of h along the length is not substantial (Fig.6(f)) This suggests that as the distance increases the negative effect of the increase in thickness and the positive effect of disturbance in the vapor film are comparable on the heat transfer.

At higher pressure the heat transfer coefficients are generally higher than those at lower pressure for low subcooling or saturation condition (Fig. 6(e, f). An opposite effect is observed for higher flow and higher subcooling (Fig. 6(d)). This can be explained in terms of the thickness of vapor film and the interface oscillation. Higher pressure corresponds to smaller volumetric vapor generation and thus smaller thickness of the film.

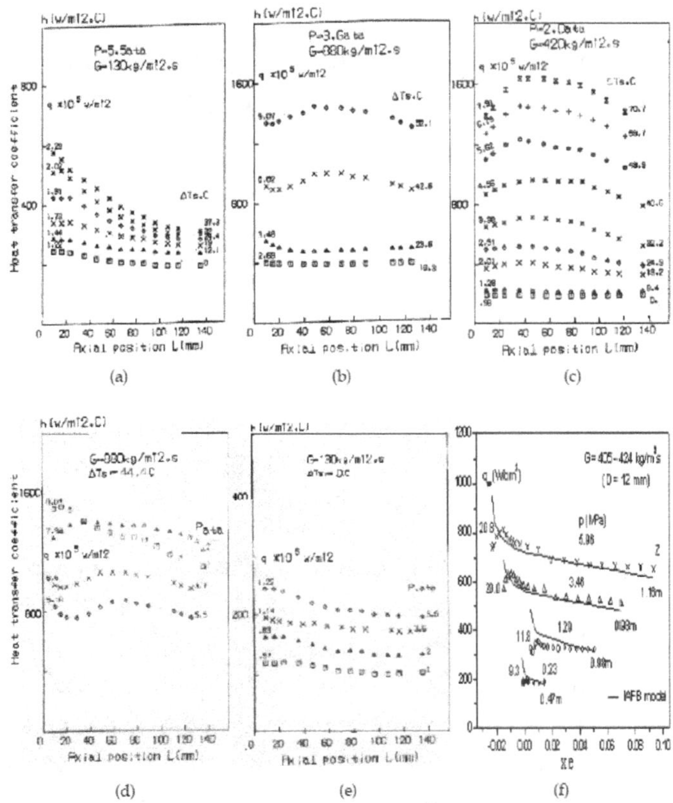

Fig. 6. Variation of the heat transfer coefficient in IAFB under different conditions (Chen, 1987)

It results in higher h for low subcooling or saturated condition. Nevertheless at higher flow and higher subcooling the film is very thin, and for lower pressure there could exist stronger interface oscillation, even dry-collision of liquid to wall, which has predominant effect on the heat exchange in both the vapor film and the liquid core. While for higher pressure this effect is less important due to less interface oscillation.

3.1.2 Effect of the preceding heating

To clarify the effect of preceding heating, an additional power supply was provided to a section of L = 225 mm immediately ahead of AB (with heat flux q_0). When the q_0 exceeded a value for the onset of boiling a substantial fall in the T_w was attained over the first about 100 mm for fixed p, G and ΔTs at the dryout point, as shown in Fig.7 (Chen, 1987). In this case a bubble layer was produced upstream, which was determinant for the vapor flow rate and the interfacial oscillation over a certain length near the dryout point. For high subcooling the vapor film was very thing, and this effect could be more substantial. Nevertheless, at a q0 without boiling the Tw near the dryout point was increased slightly. In this case a temperature profile was developed in the subcooled liquid core, which would result in lower heat transfer coefficient from the interface to liquid core, compared to that with uniform core temperature for the same average temperature.

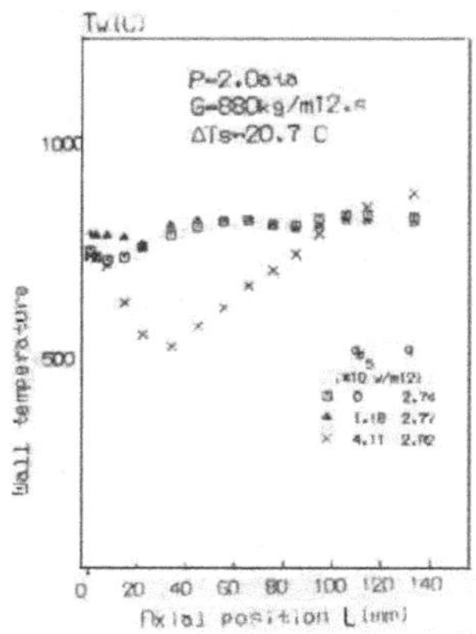

Fig. 7. Effect of the preceding heating power on the wall temperature (Chen, 1987)

3.2 Dispersed flow film boiling

In DFFB the heat is transferred from the wall to the vapor, then from the vapor to the liquid droplets entrained in the continuous vapor flow. The wall temperature is mainly dominated by the vapor convection heat transfer and the vapor temperature. The liquid droplets would induce some disturbance for the vapor convection, and the vapor-droplet interfacial heat transfer determines the vapor temperature. This effect is closely relative with the flow conditions, associated with complicated parametric trends of the wall temperature.

3.2.1 Effects of the pressure, mass flux and inlet quality

Typical distributions of the h (= $q_w/(T_w-T_s)$) along the length are shown in Fig.8. In general, as distance increases from the dryout point, at first the h decreases rapidly. For lower flow it decreases monotonously over the whole length, though the trend becomes milder downstream. For higher flow the h turns to increase after a certain distance. This behavior varies distinctly with pressure. At $p < 0.2$ MPa, for instance, the increase trend in the h is observed at mass flux below 300 kg/m^2s, while for higher pressure it is attained at higher mass flux.

In addition to the local parameters, p, G and x_e, the inlet quality (at the dryout point) has a significant effect on the h. As seen, for the same pressure and mass flux with different inlet quality, different h may be attained at a fixed local x_e, and higher h corresponds to higher inlet quality, exhibiting a strongly history-dependent feature. This is understandable due to the fact that to reach a same x_e the flow with higher inlet quality subjects to less heat transfer and thus less superheat of vapor. At low flow this effect is so significant, that the fully-developed condition cannot be reached even at L > 2 m or L/D > 200.

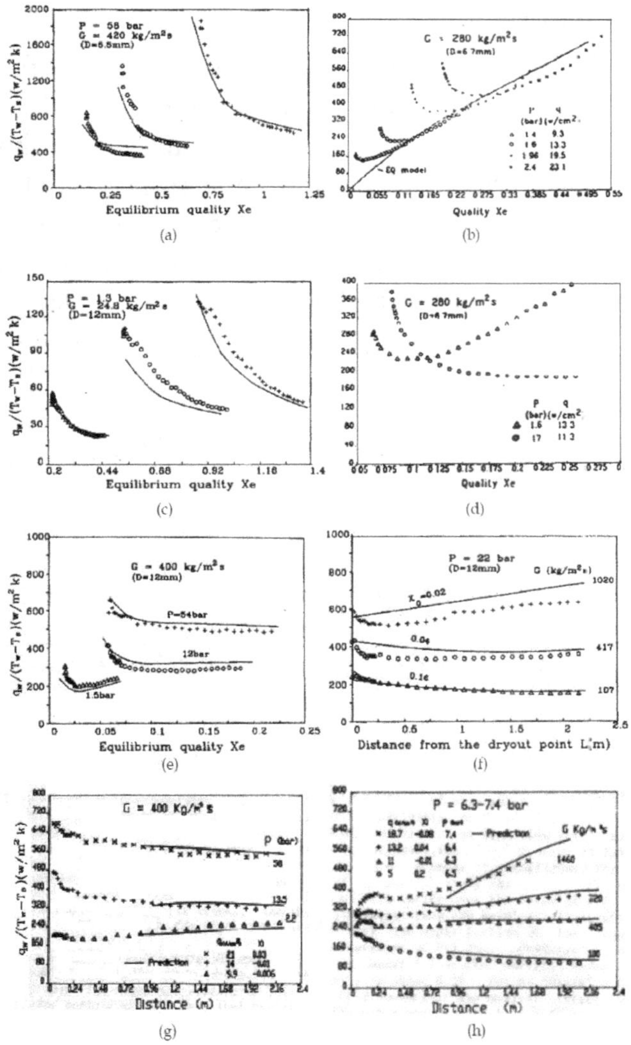

Fig. 8. Variations of heat transfer coefficient for different conditions in DFFB (— mechanistic model), (a-f): DFFB covering the whole post-CHF region; (g,h): DFFB preceded by IAFB (Chen et al., 1991, 1992, 1994b)

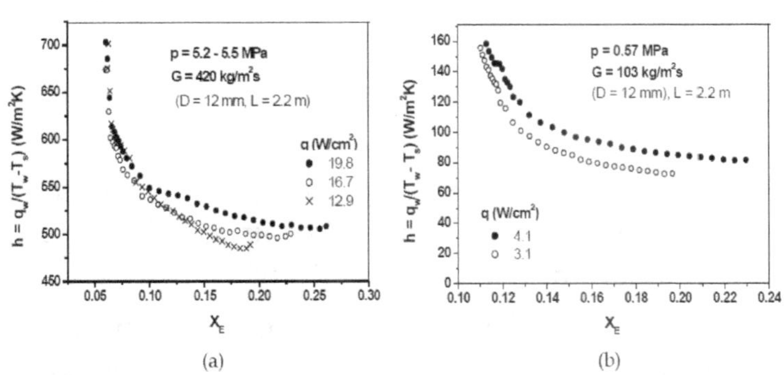

Fig. 9. Effect of heat flux on the heat transfer coefficients in DFFB

3.2.2 Effects of other factors

The effect of heat flux on the h is shown in Fig.9. Higher heat flux corresponds to higher h. This is mainly attributed to the increase in the radiation heat transfer due to higher wall temperature at higher heat flux. Fig.10 shows the effect of diameter on the h. In general, smaller diameter corresponds to lower heat transfer coefficients over the downstream. It is expectable that for same heat flux and mass flux smaller diameter corresponds to greater increase rate of the enthalpy along the length, leading to stronger thermal non-equilibrium and thus lower h.

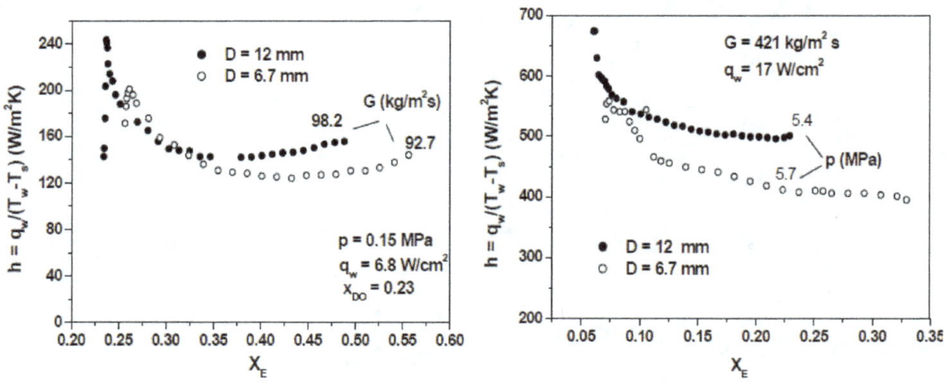

(a) (b)

Fig. 10. Effect of diameter on the heat transfer coefficients in DFFB

3.2.3 Thermal non-equilibrium

The complicated parametric trends of the heat transfer in DFFB are closely related to the thermal non-equilibrium, which is determined by the fraction of total heat to the vapor for superheating. The following thermal non-equilibrium parameter was defined by Plummer et al. (1977),

$$K = \frac{x - x_0}{x_e - x_0} \tag{1}$$

with

$$x = x_e \left(1 + \frac{C_{pg}(T_v - T_s)}{h_{fg}}\right)^{-1}$$

where the T_v and T_s are the vapor temperature and saturation temperature, respectively, the x_0 is the quality at the dryout point, and the x and x_e are the local actual quality and equilibrium quality, respectively.

The case of K = 1 represents the thermal equilibrium, in which all the heat from the wall goes to liquid for evaporation and the vapor temperature keeps at constant

(T_s), so the h increases along the length as the vapor flow rate increasing. The case of K = 0 represents that all the heat goes to the vapor for superheating without vapor generation, so the T_w increases as the T_v increasing and thus the h decreases monotonously. Fig.11 illustrates the substantial effect of the K on both the values and the trends of the heat transfer coefficient.

Using a technique to prevent the probe from striking by the liquid droplets and from the effect of radiation, the data of vapor superheat were successfully obtained in steady-state film boiling experiments near the exit of test section (Chen, 1992, Chen & Chen, 1994a). The values of K and ratio of $(T_v-T_s)/(T_w-T_s)$ were then evaluated from the vapor superheats measured at 2m from the dryout point, as shown in Fig.12. For low X_0, the K decreases as X_0 increasing. At certain increased X_0 the trend becomes milder. It varies distinctly with pressure, and higher K is attained at lower pressure. The ratio $(T_v-T_s)/(T_w-T_s)$ decreases with mass flux. For G < 100 kg/m²s, the $(T_v-T_s)/(T_w-T_s)$ is larger than 0.5, suggesting a major contribution of the vapor superheat to the wall superheat. For G < 50 kg/m²s the thermal non-equilibrium is much significant, so that the T_v and T_w increase significantly along the length, and the h (= $q_w/(T_w-T_s)$) exhibits sharp decrease trend. The effects of various parameters on the thermal non-equilibrium can be explained in terms of droplet size and concentration, the vapor-droplet relative velocity and heat transfer coefficient, the properties, etc.. This is made clear in the analysis with the two-fluid mechanistic model.

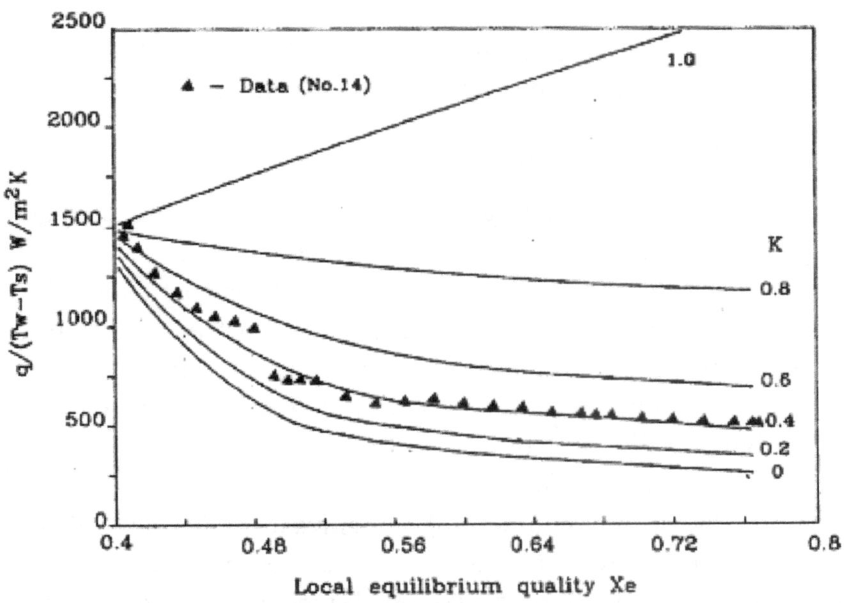

Fig. 11. Variations of the heat transfer coefficient along the length for different K (p = 5.8 MPa, G = 417 kg/m²s, xDO=0.383, D=6.8mm) (Chen, et al., 1992)

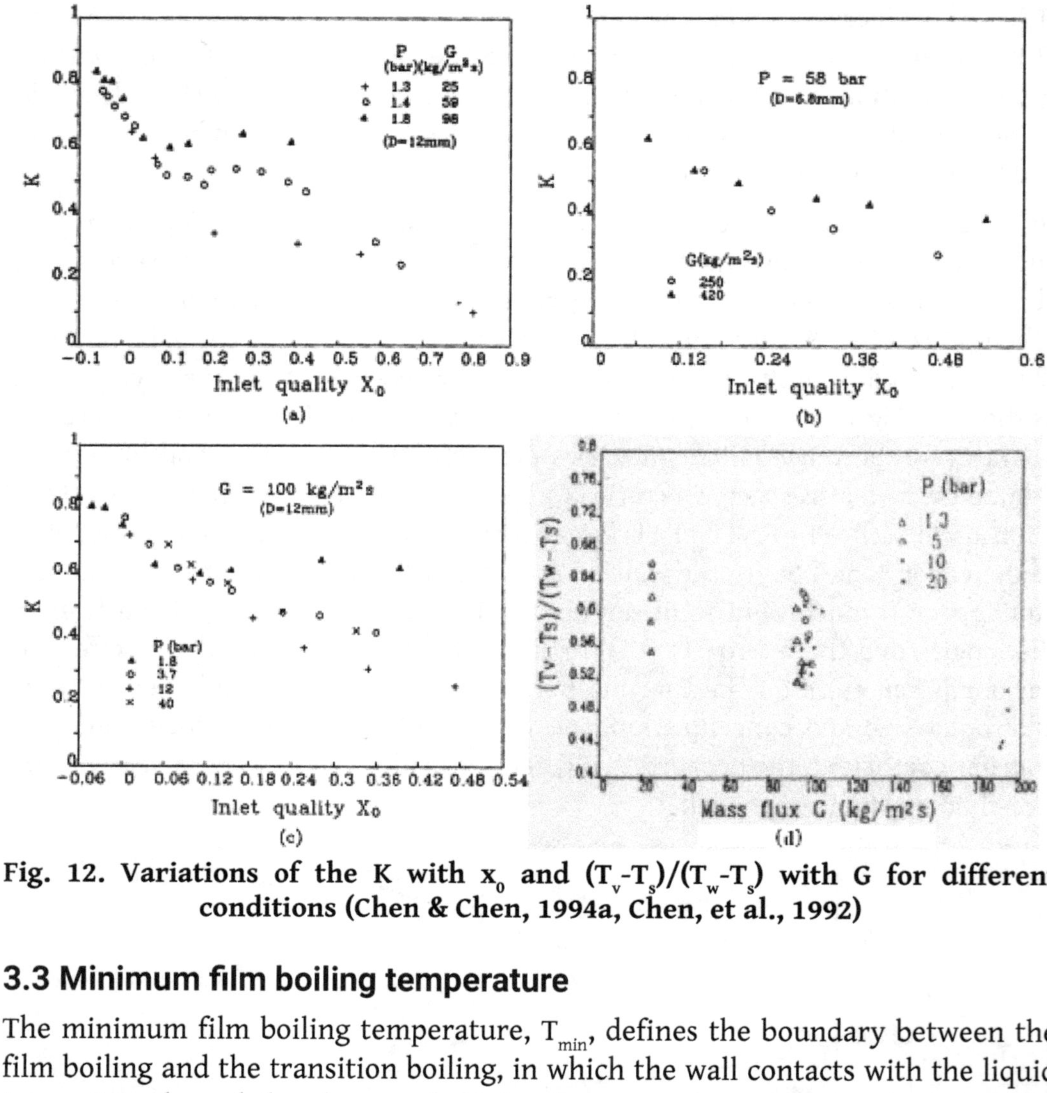

Fig. 12. Variations of the K with x_0 and $(T_v-T_s)/(T_w-T_s)$ with G for different conditions (Chen & Chen, 1994a, Chen, et al., 1992)

3.3 Minimum film boiling temperature

The minimum film boiling temperature, T_{min}, defines the boundary between the film boiling and the transition boiling, in which the wall contacts with the liquid intermittently and thus has much higher heat transfer coefficient than the film boiling. The collapse of film boiling could be resulted from the thermodynamic limit or the hydrodynamic instability. During a fast transient it could be thermodynamically controlled, while for low flow and low pressure it is likely to be hydraudynamically controlled. Six types of the film boiling termination mechanisms have been identified: (1) collapse of vapor film, (2) top flooding, (3) bottom flooding, (4) droplet cooling, (5) Leidenfrost boiling and (6) pool boiling. Significant discrepancies were found among the existing correlations of the T_{min}, and were attributed to the different types of the mechanism and scarcity of reliable data (Groeneveld & Snoek, 1984). With the hot patch technique the minimum film boiling temperatures were measured in steady-state film boiling experiments by decreasing the power to the test section slowly with small steps until the collapse of film boiling occurred. The following empiric correlation was formulated from an experiment over the ranges of p = 115 – 6050 kPa, G = 53 – 1209 kg/m^{2}s, x = -0.055 – 0.08 and ΔT_s = -35 – 25.1 K (Chen, 1989),

$$T_{min} = 363,6 + 38.37 \ln p + 0.02844p - 3.86 \times 10^{-6} p^2 + a\Delta Ts \quad (2)$$

with

$$a = 17.1 / (3.3 + 0.0013p) \qquad \text{for } \Delta Ts > 0$$

and

$$a = 0 \qquad \text{for } \Delta Ts \leq 0$$

where the p is in kPa and the T_{min} **and ΔT_s in K.**

This correlation is in reasonable agreement with that derived from a similar experiment by Groeneveld and Steward (1982). It can be recommended for type (1-4) of film boiling termination.

4. FORECASTS FOR THE THERMAL CONDUCTIVITY COEFFICIENTS

The film boiling is defined by a lack of equilibrium in both velocity and temperature between phases, linked to highly complex parametric trends. The steady-state experimental results acquired in the tube with flowing water were juxtaposed with established correlations, revealing substantial disparities, as illustrated in Fig. 13. This result revealed the suspect of the correlations, and it was attributed to the lack of reliable data base and the difficulty in accounting for various physical mechanisms in a simple correlation (Stewart and Groeneveld, 1982, Groeneveld & Snoek, 1984). With steady- state technique the accuracy of the experimental data was improved substantially. As shown in Fig.14, the present steady-state data are in well agreement with those obtained by Swinnerton et al (1988) using indirectly heated hot patch technique for similar conditions.

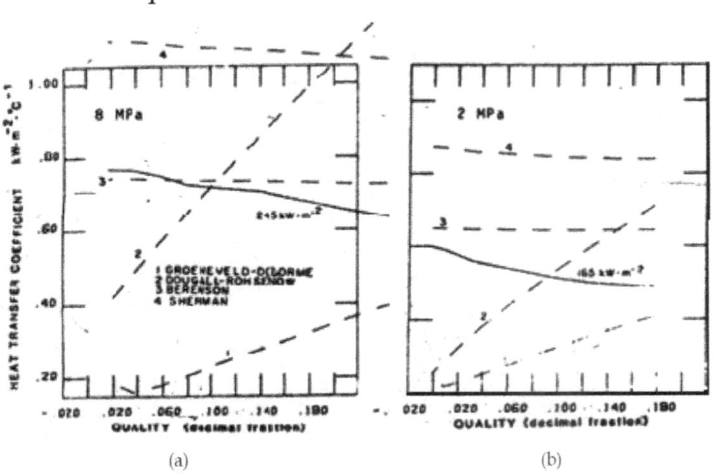

Fig. 13. Comparison of the steady-state experimental data of water with existing correlations (Stewart and Groeneveld, 1982)

To predict the non-equilibrium characteristics in film boiling the two-fluid models are favorable, and have been proposed by many investigators (Groeneveld, 1988, 1992, Mossad & Johannsen, 1989). The major challenge for these models is to simulate the interfacial heat and momentum exchanges. Due to less knowledge on these processes they were generally accounted by empiric or semi-empiric correlations. Therefore, the suitability of this kind of models is heavily determined by the ranges and the accuracy of data base. The following two-fluid models are developed based on the present experimental data.

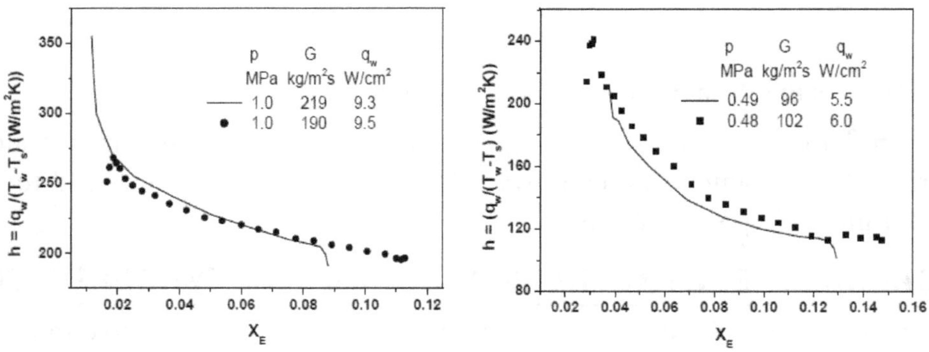

(a) (b)

Fig. 14. Comparison of the results obtained in different steady-state experiments with water flowing in tube (— Swinnerton, et al. D=9.75mm, - Present experiment, D=12mm)

4.1 IAFB model

The schematic of flow structure in IAFB is shown in Fig.15, in which the vapor film is divided into two regions, A and B, bounded with the locus of the maximum velocity line.

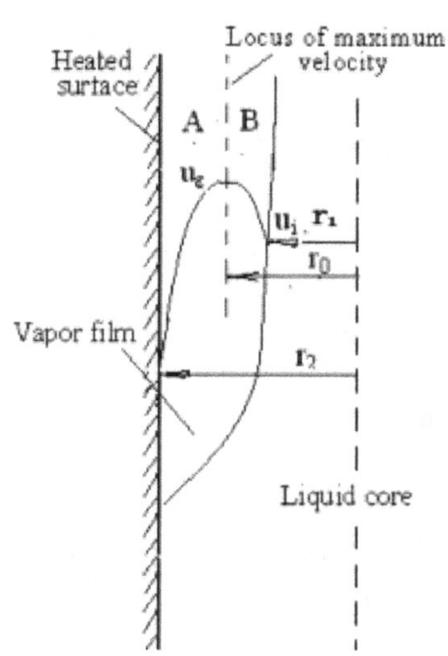

Fig. 15. Inverted annular film boiling mode

The assumptions are as follows:

- The pressure over a cross section is uniform.
- The kinetic energy and viscous dissipation and pressure loss due to acceleration are negligible.
- The properties of vapor are evaluated at $(T_w + T_s)/2$.
- The vapor-liquid interface is treated as smooth, and both the vapor and the liquid at the interface are at saturation.
- The velocity profile in the liquid core is uniform and equal to the vapor velocity at the interface.

In vapor film the force balance gives

$$\mu_v(1+\frac{\varepsilon_v}{\nu_v})\frac{du}{dr}+\frac{r^2-r_0^2}{2r}(\rho_v g+\frac{dp}{dz})=0 \quad (3)$$

with boundary conditions as

$$u=0 \text{ at } r=r_2$$

and

$$u=u_i=\overline{u_l} \text{ at } r=r_i$$

where $\overline{u_l}$ is the average velocity in the core.

Neglecting the weight of vapor, the integration of Eq.(3) gives, for region A

$$u_g=\frac{1}{2\mu_v(1+\varepsilon_v/\nu_v)}\frac{dp}{dz}\left[r_0^2\ln\frac{r}{r_2}+\frac{1}{2}(r_2^2-r^2)\right] \quad (4)$$

and for region B

$$u_g=\frac{1}{2\mu_v(1+\varepsilon_v/\nu_v)}\frac{dp}{dz}\left[r_0^2\ln\frac{r}{r_1}-\frac{1}{2}(r^2-r_1^2)\right]+u_i \quad (5)$$

with

$$\frac{dp}{dz}=\frac{r_1^2\rho_c g+(r_0^2-r_1^2)\rho_v g}{r_0^2} \quad (6)$$

where ε_v is the eddy diffusivity in the vapor film, and the ρc is the average density of the core. Some vapor may be entrained in the core, and ρ_c is evaluated by

$$\rho_c=\alpha_c\rho_v+(1-\alpha_c)\rho_l$$

where α_c is the void fraction of the core. The following expression is attempted with adjustable factor c=40,

$$\alpha_c = \frac{1}{1 + c\rho_v / \rho_l / x}$$

Assuming the momentum eddy diffusivity ε_v independent of r, by integrating Eq. (4) and (5) the vapor flow rate in the film is as

$$\Gamma = \frac{\pi \rho_v}{\mu_v (1 + \varepsilon_v / \nu_v)} \frac{dp}{dz} \left[\frac{r_0^4}{2} \ln \frac{r_2}{r_1} + \frac{1}{8}(r_2^4 - r_1^4) - \frac{1}{2} r_0^2 (r_2^2 - r_1^2) \right] + \pi u_i \rho_v (r_0^2 - r_1^2) \tag{7}$$

The wall heat flux is expressed as

$$q = q_{w-i,c} + q_{w-i,r} + q_v \tag{8}$$

where q_v is the heat for vapor superheating, $q_{w-i,c}$ and $q_{w-i,r}$ are the heat flux to the interface by convection and by radiation, respectively, evaluated by,

$$q_{w-i,c} = h_c (T_w - T_s) \tag{9}$$

with

$$h_c = \frac{k_v}{\delta} \left(1 + \frac{\varepsilon_v}{\nu_v} \frac{\Pr_v}{\Pr_t}\right) \tag{10}$$

and

$$q_{w-i,r} = 0.75 \frac{\sigma_{SB}}{(1/\varepsilon_l + 1/\varepsilon_w - 1)} (T_w^4 - T_s^4) \tag{11}$$

where δ **is the thickness of film, and** $\Pr_t$ is the turbulent Prandtl number. The energy balance equation at the interface can be written as,

$$q_{w-i,c} + q_{w-i,r} = q_g + q_l$$

where q_l is the heat flux from the interface to the core, and q_g is the heat flux for evaporation, as

$$q_g = \frac{d\Gamma}{dz} h_{fg} / 2\pi r_1 \tag{12}$$

The mass equation is written as

$$\Gamma + \pi r_1^2 \overline{u} \rho_c = \pi r_2^2 G \tag{13}$$

From the authors experiments with saturated and low subcooling condition, the following empiric expression for ε_v is proposed with $\Pr_t$ =1.0.

$$\varepsilon_v / \nu_v = 0.011 \mathrm{Re}_v^{0.6} \mathrm{Re}_l^{0.3} \tag{14}$$

with

$$\mathrm{Re}_v = 2Gx(r_2 - r_1) / \mu_v$$

and

$$Re_l = G(1-x)D_1 / \mu_l$$

For the heat transfer from interface to subcooled liquid core the following correlation is available (Alalytis & Yadigaroglu 1987)

$$q_l = 0.06(k_l / D)Re_l^{0.6} Pr_l^{0.3} [\mu_v \rho_v / (\mu_l \rho_l)]^{-0.15} Pr_v^{0.6} (Ts - T_f) \quad (15)$$

It should be noted that for subcooled condition the h near the dryout point is related to the condition at the dryout point, which is determined by the preceding heating, as described above. It is not simulated in the model, therefore, the calculation result could have appreciable uncertainty over a short length, especially for high subcooling.

For low subcooling and saturated condition, this model gives satisfactory calculations for the experimental data of p = 0.1 – 6 MPa, and G = 90 – 1462 kg/m^{2}s, as exemplified in Fig. 6(f).

4.2 DFFB model

The following two-fluid mechanistic model is based on the motion and energy equations, involving various equations for the wall-vapor-droplet heat transfer (Chen & Chen, 1994b). The heat from the wall, qw, is transferred to the vapor by convection, q_c, and radiation, q_r, as

$$q_w = q_c + q_r \quad (16)$$

with

$$q_c = h_c(T_w - T_v) \quad (17)$$

and

$$q_r = \varepsilon \sigma_{SB}(T_w^4 - T_v^4) \quad (18)$$

where h_c is the convective heat transfer coefficient, σ_{SB} is the Boltzman's constant, and ε the emissivity.

The vapor temperature is evaluated by

$$T_v = T_s + (\frac{x_e}{x} - 1)\frac{h_{fg}}{C_{pv}} \quad (19)$$

where x and x_e are the actual quality and equilibrium quality, respectively. From the heat balance equation, we have

$$\frac{dx}{dz} = \frac{6(1-x)q_d''}{u_l h_{fg} \rho_l \delta} \quad (20)$$

where u_l and δ are the droplet velocity and diameter, respectively. The heat flux to the droplet, q_d, includes the convective and radiative components, i.e.,

$$q_d'' = q_{d,c} + q_{d,r} \tag{21}$$

with

$$q_{d,r} = E\sigma_{SB}(T_w^4 - T_s^4) \tag{22}$$

and

$$E = \frac{2}{3}\varepsilon\frac{\delta}{(1-\alpha)D}$$

The vapor to droplet convective heat transfer is evaluated by Frossling correlation, as

$$q_{d,c}'' = \frac{k_v}{\delta}(2 + 0.552(\frac{\rho_v(u_v - u_l)\delta}{\mu_v})^{0.5}\,\mathrm{Pr}_v^{0.33})(T_v - T_s) \tag{23}$$

The motion equation for the droplet gives

$$\rho_l u_l \frac{du_l}{dz} = \frac{3}{2}\frac{C_d}{\delta}\frac{\rho_v(u_v - u_l)^2}{2} - (\rho_l - \rho_v)g \tag{24}$$

where the drag coefficient, C_d, is evaluated by Ingebo's correlation, as

$$C_d = \frac{24(1 + 0.1\mathrm{Re}_d^{0.75})}{\mathrm{Re}_d} \quad \text{for} \quad \mathrm{Re}_d < 1000$$

with

$$\mathrm{Re}_d = \frac{\rho_v(u_v - u_l)\delta}{\mu_v}$$

and

$$C_d = 0.45 \text{ for } \quad \mathrm{Re}_d \geq 1000$$

The wall to vapor convective heat transfer is predicted by the correlation derived from an author's convection heat transfer experiment in pure steam, multiplying an enhancement factor due to the disturbance induced by the droplets, as

$$h_c = h_0 F \tag{25}$$

with

$$h_0 = 0.0175 \frac{k_{v,f}}{D} \left(\frac{\rho_{v,f} u_v D}{\mu_{v,f}} \right)^{0.812} \mathrm{Pr}_{v,f}^{0.33}$$

The factor *F* would be related with pressure and quality. To fit the calculations with the experimental results, the following expression for *F* is proposed with p in MPa, as

$$F = 2.32(1 + 0.1p)e^{-12x} + 1$$

For the DFFB occurring from the dryout of liquid film of annular flow, the initial droplet diameter, δ0 , is evaluated by a modification of Nujiyama-Tanazawa equation, as

$$\delta_0 = 798 \frac{\rho_v}{Gx_0} \left(\frac{\sigma}{\rho_l} \right)^{0.5} \left(\frac{\mu_v}{Gx_0 D} \right)^a \quad (26)$$

where

$$a = 0.63 - 0.1p \quad \text{for} \quad p < 1.3MPa$$

and

$$a = 0.5 \quad \text{for} \quad p \geq 1.3MPa$$

in which p is in MPa, δ and *D* in m, *G* in kg/m²s, *μ* in kg/ms, σ in kg/m, ρ_l and ρ_v in kg/m³.

The droplet has a maximum diameter, determined by the critical Weber number, We_c, as

$$\delta_m = \frac{We_c \sigma}{\rho_v (u_v - u_l)^2} \quad (27)$$

with $We_c = 5.0$.

For the DFFB occurring from the break-up of the IAFB, the initial diameter δ_0 **is taken** as δ_m. The onset of DFFB is defined by the criterion of droplet carryover (Yanomoto et al., 1987), as

$$J_{gc} = \frac{3.57}{\sqrt{\rho_g C_d}} (\sigma g(\rho_l - \rho_g))^{1/4} N_{\mu g}^{1/6} \quad (28)$$

with

$$N_{\mu g} = \frac{\mu_g}{\left[\rho_g \sigma \sqrt{\frac{\sigma}{g(\rho_l - \rho_g)}} \right]^{1/2}}$$

where J_{gc} is critical volumetric flux and $N_{\mu g}$ is viscosity number.

Assuming that the droplet break-up does not occur, so it varies with the actual quality, as

$$\delta = \delta_0 (\frac{1-x}{1-x_0})^{1/3}$$

The present model gives satisfactory predictions for the DFFB data over the range of pressure of 0.1 – 6 MPa, mass flux of 23 – 1020 kg/m^{2}s, as exemplified in Fig.8.

4.3 Tabular method

The film boiling models and phenomenological equations are labor-intensive due to the extensive number of constitutive equations governing heat and momentum exchanges between phases, as well as the assessment of numerous parameters. Moreover, they are solely relevant to the particular testing conditions of each individual investigator. In recent years, the tabular technique has gained widespread acceptance owing to its excellent accuracy, broad applicability, and ease of use and updating.

4.3.1 fully-developed film boiling heat transfer coefficients

Based on over 15000 film boiling data points, Leung et al. (1997) have developed a look-up table for the fully-developed film boiling heat transfer coefficients in tubes with vertical upward flow. It contains a tabulation of normalized heat transfer coefficients at discrete local parameters of pressure (0.1 – 20 MPa in 14 steps), mass flux (0 – 7000 kg/m^{2}s in 12 steps), quality (-0.2 – 1.2 in 11 steps) and heat flux (0.05 – 3 MW/m2 in 9 steps). The agreements of the calculations of this table for surface temperatures with the data are 6.7% in r.m.s.

The range of flow conditions of the data base is presented in Fig.16. It shows significant gaps in the range of low flows and medium pressures, where the thermal non-equilibrium could be significant, and the heat transfer could be strongly history-dependent.

4.3.2 Thermal non-equilibrium parameter

For estimation of the heat transfer coefficient in the region of lower flow, a table for thermal non-equilibrium parameter has been proposed (Chen & Chen 1998). It formulates the non- equilibrium parameter, K_0, for tube of D = 8 mm at concrete values of the pressure p, mass flux G and quality at the dryout point x_0 **(Tab.1). The values of K_0 were produced from the calculations of DFFB model, which was validated by the data of vapor temperature measured at the steady-state film boiling experiments.**

The K is assumed to be related with the diameter, local quality and heat flux, as expressed by

$$K = K_0 F_d F_x F_q$$

where, F_d, F_x and F_q are the correction factors for the effects of diameter, local quality and heat flux, respectively.

For the present experimental results, the F_d is evaluated by

$$F_d = (\frac{D}{0.008})^{0.26}$$

The calculation indicates that from the dryout point the K increases rapidly, while after a short distance it does not vary appreciably. So it is simply accounted by a proper K_0 **with** $F_x = 1$. At present the expression for the F_q is not available and is also set to be 1.0.

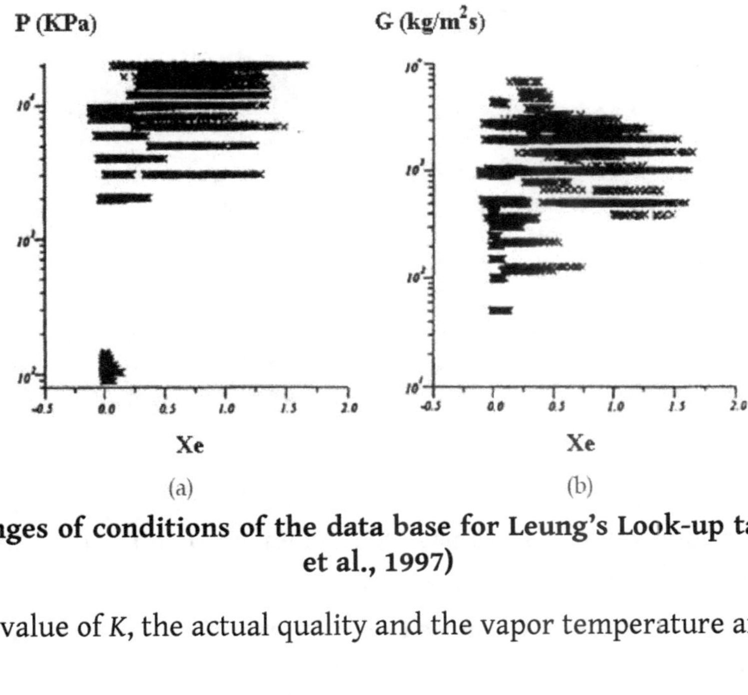

(a) (b)

Fig. 16. Ranges of conditions of the data base for Leung's Look-up table (Leung, et al., 1997)

Having the value of K, the actual quality and the vapor temperature are evaluated by

$$K = \frac{x - x_0}{x_e - x_0}$$

and

$$T_v = T_s + (\frac{x_e}{x} - 1)\frac{h_{fg}}{C_{pg}}$$

Then, the wall temperature is calculated by

$$T_w = T_v + (q_w - q_r) / h_c$$

Table 1. Table of K_0 with parameters p, G and x_0

P (MPa)	G (kg/m²s)	X0: 0.0	0.05	0.1	0.2	0.4	0.6	0.8	1.0
0.1	25	0.68	0.58	0.44	0.31	0.28	0.26	0.18	0.0
	50	0.69	0.59	0.45	0.42	0.48	0.36	0.25	0.0
	100	0.74	0.63	0.60	0.64	0.68	0.56	0.44	0.0
	200	0.78	0.73	0.75	0.78	0.77	0.74	0.60	0.0
	400	0.81	0.81	0.86	0.89	0.90	0.89	0.81	0.0
	600	0.82	0.88	0.92	0.94	0.96	0.95	0.90	0.0
	1000	0.83	0.92	0.96	0.98	0.98	0.98	0.97	0.0
	1500	0.84	0.96	0.98	0.99	0.99	0.99	0.98	0.0
0.3	25	0.72	0.60	0.45	0.40	0.31	0.17	0.10	0.0
	50	0.74	0.61	0.46	0.43	0.36	0.30	0.16	0.0
	100	0.76	0.64	0.54	0.57	0.52	0.47	0.33	0.0
	200	0.78	0.71	0.66	0.70	0.68	0.63	0.47	0.0
	400	0.80	0.78	0.79	0.81	0.83	0.75	0.63	0.0
	600	0.82	0.80	0.84	0.88	0.90	0.86	0.75	0.0
	1000	0.83	0.86	0.89	0.92	0.93	0.92	0.87	0.0
	1500	0.83	0.90	0.95	0.96	0.98	0.97	0.95	0.0
0.5	25	0.74	0.56	0.44	0.35	0.29	0.23	0.12	0.0
	50	0.76	0.58	0.45	0.38	0.33	0.22	0.14	0.0
	100	0.78	0.60	0.50	0.44	0.40	0.36	0.26	0.0
	200	0.79	0.62	0.63	0.64	0.61	0.56	0.40	0.0
	400	0.80	0.67	0.65	0.77	0.74	0.70	0.54	0.0
	600	0.81	0.80	0.81	0.83	0.83	0.80	0.67	0.0
	1000	0.82	0.84	0.86	0.89	0.90	0.88	0.78	0.0
	1500	0.82	0.87	0.92	0.94	0.95	0.94	0.89	0.0
1.0	25	0.76	0.62	0.53	0.42	0.25	0.10	0.08	0.0
	50	0.77	0.63	0.56	0.46	0.28	0.14	0.09	0.0
	100	0.79	0.66	0.58	0.53	0.33	0.22	0.13	0.0
	200	0.80	0.66	0.60	0.50	0.45	0.38	0.25	0.0
	400	0.80	0.68	0.66	0.65	0.62	0.56	0.40	0.0
	600	0.80	0.77	0.74	0.73	0.71	0.63	0.45	0.0
	1000	0.81	0.80	0.79	0.82	0.80	0.73	0.57	0.0
	1500	0.82	0.84	0.85	0.87	0.84	0.82	0.70	0.0
2.0	25	-	-	-	-	-	-	-	-
	50	0.81	0.68	0.65	0.54	-	-	-	-
	100	0.81	0.71	0.68	0.62	0.32	0.16	0.08	0.0
	200	0.81	0.69	0.65	0.53	0.35	0.28	0.16	0.0
	400	0.81	0.70	0.68	0.56	0.52	0.44	0.29	0.0
	600	0.82	0.73	0.71	0.64	0.61	0.53	0.36	0.0
	1000	0.83	0.74	0.74	0.74	0.71	0.65	0.47	0.0
	1500	0.83	0.81	0.80	0.80	0.80	0.71	0.55	0.0
4.0	25	-	-	-	-	-	-	-	-
	50	-	-	-	-	-	-	-	-
	100	0.81	0.72	0.62	0.45	0.30	0.21	0.08	0.0
	200	0.81	0.72	0.63	0.50	0.34	0.26	0.14	0.0
	400	0.82	0.72	0.63	0.55	0.49	0.41	0.26	0.0
	600	0.83	0.77	0.73	0.61	0.56	0.48	0.33	0.0
	1000	0.83	0.78	0.76	0.71	0.70	0.60	0.42	0.0
	1500	0.84	0.79	0.78	0.78	0.77	0.69	0.53	0.0
6.0	25	-	-	-	-	-	-	-	-
	50	-	-	-	-	-	-	-	-
	100	0.81	0.72	0.63	0.45	0.35	0.20	0.11	0.0
	200	0.82	0.73	0.64	0.52	0.42	0.25	0.14	0.0
	400	0.82	0.76	0.67	0.55	0.51	0.39	0.26	0.0
	600	0.83	0.77	0.70	0.60	0.58	0.48	0.31	0.0
	1000	0.84	0.78	0.74	0.69	0.68	0.61	0.43	0.0
	1500	0.84	0.80	0.77	0.76	0.74	0.68	0.53	0.0

where h_c is the convective heat transfer coefficient, and q_r is the radiative heat flux. They are evaluated by Eq.(25) and (18), respectively.

The present tabular method was based on 2192 experimental data in tubes of D = 6.7 and 12 mm for pressure of 0.1 –5.8 MPa, mass flux of 23 – 1462 kg/m²s and local equilibrium quality of 0 – 1.36. The conditions of the data base are shown in Fig. 17. The agreements of the calculations of wall temperatures with the data are 7.2% in r.m.s., as exemplified in Fig 18.

5. CONCISE SUMMARY

Using the directly heated hot patch technique the film boiling experiments have been performed at stead-state in tubes of 6.7 – 20 mm with water flowing upward, covering the ranges of pressure of 0.1 – 6 MPa, mass flux of 23 – 1462 kg/m2s and inlet quality of -0.15 – 1.0. The characteristics of film boiling have been investigated systematically, including the inverted annular film boiling, the dispersed flow film boiling and the minimum film boiling temperature. A great number of data of heat transfer coefficients, the minimum film boiling temperatures and the vapor superheats have been obtained. They fill the gaps of the data base and understanding in the regions of lower flow and medium pressure.

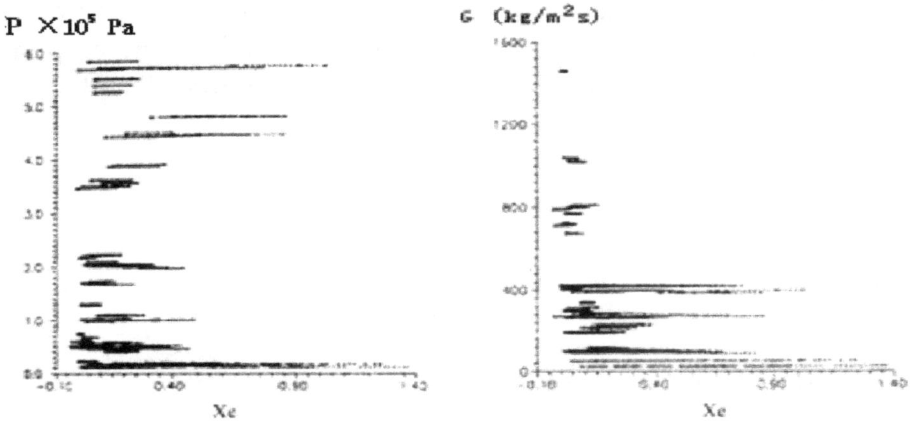

(a) (b)

Fig. 17. Ranges of conditions of the present film boiling data for x_e > 0 (Chen & Chen 1998)

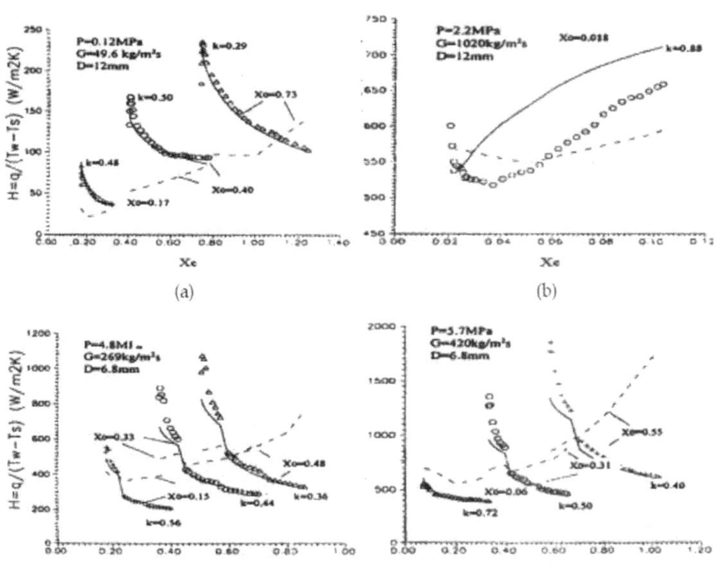

(a) (b) (c) (d)

Fig. 18. Comparison of the calculation by tabular methods with experimental data (Chen & Chen, 1998) — Table of Non-equilibrium parameter, --- Leung's Table ...

Note: in (c,d) two portions of boiling length were heated at different heat fluxes

The film boiling is characterized by non-equilibrium between phases in both the velocity and temperature, associated with extremely complicated parametric trends and strongly history-dependent feature of the heat transfer coefficients. Two-fluid mechanistic models and tabular method have been proposed to predict the heat transfer coefficients satisfactorily for the inverted annular film boiling with saturated or low subcooling conditions and the dispersed flow film boiling.

The results show that the heat transfer coefficient near the dryout point is closely relative to the preceding heating. Therefore the IAFB model without encountering for this effect could have appreciable uncertainties over a short length for higher subcooling condition.

6. NOMENCLATURE

C_d drag coefficient

C_p specific heat

D diameter

F enhancement or correction factor G mass flux

h heat transfer coefficient

h_{fg} **latent heat**

K heat conductivity

L distance from the dryout point

p pressure

q heat flux

r radius

T temperature

u velocity

x quality

J_{gc} critical volumetric flux

$N_{\mu g}$ viscosity number

Nu Nusselt number

Pr Prandtl number

Pr_t turbulent Prandtl numbe

Re Reynolds number

We Weber number;

α void fraction

δ droplet diameter, thickness of film

μ kinetic viscosity

ν dynamic viscosity

ρ density

ε eddy diffusivity, emissivity

σ surface tension

σ_{SB} Stefan-Boltzman constant

Subscript

c critical, convective, core

d droplet

e equilibrium

m maximum

min minimum

r radiation

i interface

l liquid

s saturation

v, g vapor

w wall

0, DO inlet (at dryout point)

7. REFERENCES

Analytis, G. Th. & Yadigaroglu, G. (1987), Analytical Modelling of Inverted Annular Film Boiling, *Nuclear Engineering and Design*, Vol. 99, PP 201 - 212

Arrieta, L. & Yadigaroglu, G. (1978), Analytical Model for Bottom Reflooding Heat Transfer in Light Water Reactor, *EPPI-NP 756*

Chen Y. & Li, J. (1984), Subcooled Flow Film Boiling of Water at Atmospheric Pressure, in *Two-phase Flow and Heat Transfer*, X. J. Chen and T. N. Veziroglu, PP 141 – 150, Hemisphere Pub. Co., ISBN 0-89116-432-4

Chen,Y. (1987), Experimental Study of Inverted Annular Flow Film Boiling Heat Transfer of Water in *Heat Transfer Science and Technology*, Bu-Xuan Wang, PP 627 – 634, Hemisphere Pub. Co., ISBN 0-89116-571-1,

Chen, Y.; Wang, J.; Yang, M. & Fu, X. (1989a), Experimental Measurement of the Minimum Film Boiling Temperature for Flowing Water, in *Multiphase Flow and*

Heat Transfer, Xue-Jun Chen, T. N. Veziroglu and C. L. Tien, Vol. 1, PP 393 - 400 17. Hemisphere Pub. Co., ISBN 1-56032-050-8

Chen, Y.; Cheng, P.; Wang, J. & Yang, M. (1989b), Experimental Studies of Subcooled and Low Quality of Film Boiling Heat Transfer of Water in Vertical Tubes at Moderate Pressure, *Proc. 4th Int. Topical Meeting on Nuclear Reactor Thermal-Hydraulics,* Vol. 2, PP 1106 - 1110

Chen Y., (1991), An Investigation of Dispersed Flow Film Boiling Heat Transfer of Water - Experiment and numerical analysis, *2d World Conf. on experimental fluid, heat transfer and thermodynamics, Yugoslavia*

Chen, Y.; Chen, H.& Zhu, Z. (1992), Post-Dryout Droplet Flow Heat Transfer-Measurements of both wall and vapor superheat at stable condition, in *Transport Phenomena Science and Technology,* Wang B. X. PP 319 – 324, Higher Education Press, ISBN 7-04-004122-7/TH324

Chen, Y & Chen, H. (1994a), An Experimental Investigation of Thermal Non-equilibrium in Dispersed Flow Film Boiling of Water, *Proc. Int. Conf. on New Trends in Nuclear System Thermal-Hydraulics,* Pisa Italy Vol.1 PP 31 - 37

Chen, Y. & Chen, H. (1994b), A Model of Dispersed Flow Film Boiling Heat Transfer of Water, *Proc. 10th Int. Heat Transfer Conf.* Brighton UK, Vol. 7, PP 419 – 424

Chen, Y & Chen, H. (1998), A Tabular Method for Prediction of the Heat Transfer during Saturated Film Boiling of Water in a Vertical Tube, , *Proc. 11th Int. Heat Transfer Conf.,* Kyongju, Korea, Vol. 2, PP 163 – 168

Groeneveld, D. C. (1974), Effect of a Heat Flux Spike on the Downstream Dryout Behavior, *J. of Heat Transfer* PP 121 – 125

Groeneveld, D. C. and Gardiner, R. M. (1978), A Method of Obtaining Flow Film Boiling Data for Subcooled Water, *Int. J. Heat Mass Transfer,* Vol. 21 PP 664 – 665

Groeneveld, D.C. & Steward, J. C. (1982), The Minimum Film Boiling Temperature for Water during Film Boiling Collapse, *Proc. 6th Int. Heat Transfer conf.* PP 393 – 398

Groeneveld, D.C. & Snoek, C. W. (1984), A Comprehansive Examination of Heat Transfer Correlations Suitable for Reactor Safety Analysis, *Multiphase Science and Technology,* Vol.2

Groeneveld, D.C. (1988), Recent Developments in Thermalhydraulic Prediction Methods, *Proc. Int. Topical Meeting on Thermalhydraulics of Nuclear Reactors,* Seoul

Groeneveld, D. C. (1992), A Review of Inverted Annular and Low Quality Film Boiling, in *Post-dryout Heat Transfer,* Hewitt, Delhaye, Zuber, CRC Press, PP 327 – 366

Kirillov, P. L., et al. (1996), The Look-up Table for Heat Transfer Coefficient in Post-Dryout Region for Water Flowing in Tubes, *FEI-2525*

Leung, L. K. H.; Hammouda, N. & Groeneveld, D. C. (1997), A look-up Table for Film Boiling Heat Transfer Coefficients in Tubes with Vertical Upward Flow, *Proc. 8th Int. Topical Meeting on Nuclear Reactor Thermal-Hydraulics*, Kyoto, Japan, Vol. 2, pp 671 – 678

Mossad, M. (1988), Sobcooled Film Boiling Heat Transfer to Flowing Water in Vertical Tube Dr-Ing. *Thesis*, Techn. Univ. Berlin, Berlin,

Mossad, M. & Johannsen, K. (1989), A new Correlation for Subcooled and Low Quality Film Boiling Heat Transfer of Water at Pressures from 0.1 to 8 MPa, *Proc. 4th Int. Topical Meeting on Nuclear Reactor Thermal-Hydraulics*, Vol. 2, PP 1111 - 1117

Plummer, D, N.; Iloeje, O. C.; Rohsenow, W. M.; Griffith, P. & Ganic, E. (1974), Post Critical Heat Transfer to Flowing Liquid in a Vertical Tube, MIT Dpt. Of Mech. Eng. Report 72718-91

Plummer, D. N.; Griffith, P. & Rohsenow, W. M. (1977), Post-Critical Heat Transfer to Flowing Liquid in a Vertical Tube, Trans. *ASME*, Vol. 4, N.3, PP 151 – 158

Stewart, J. C. & Groeneveld, D. C. (1982), Low Quality and Subcooled Film Boiling at Elevated Pressures, *Nucl. Eng. Design*. Vol. 67, PP 259 - 272

Swinnerton, D.; Hood, M. L. & Pearson, K.G. (1988), Steady State Post-Dryout at Low Quality and Medium Pressure –Data Report, *UKAEA AEEW-R 2267*

Thermohudraulic relationships for advanced water cooled reactors, (2001), *IAEA-TECDOC- 1203*

Yonomoto, T.; Koizumi, Y. & Tasaka, K, (1987) Onset Criterion for Liquid Entrainment in Reflooding Phase of LOCA, *Nucl. Science and Technology*, Vol. 24, No. 10, pp 798-810

10 Chapter

TWO-PHASE FLOW BOILING HEAT TRANSFER FOR EVAPORATIVE REFRIGERANTS IN DIVERSE CIRCULAR MINICHANNELS

1. INTRODUCTION: AN OVERVIEW

In recent decades, global awareness of climate change has increased; numerous governments have suggested reducing carbon emissions and regulating refrigerants in accordance with the Montreal Protocol. The loss of the ozone layer has become a significant concern for numerous researchers. Refrigerant has been utilized for an extended period and continues to serve as the primary substance in refrigeration systems. The demand for innovative refrigeration and air conditioning solutions has fostered the development of more effective and efficient refrigeration systems. Compact refrigeration systems may serve as a potential option to mitigate the indirect depletion of the ozone layer. Numerous studies and experimental findings have been published regarding flow patterns in horizontal tubes, two-phase flow boiling heat transfer, and pressure drop utilizing refrigerant as the examined fluid. The pressure drop associated with two-phase flow boiling heat transfer of refrigerants in minichannels has been studied for several decades. Limited research in the literature address the two-phase flow heat transfer and pressure drop of refrigerants in minichannels. In contrast to pure refrigerants in conventional channels, the flow boiling of refrigerants in minichannels exhibits distinct characteristics attributable to the physical and chemical properties of the refrigerants and the dimensions of the minichannels. The primary benefits of minichannels include elevated heat transfer coefficients, substantial reductions in the dimensions of compact heat exchangers, and diminished fluid mass requirements. The enhanced heat transmission in minichannels results from significant ratios of heat transfer surface area to fluid flow volume and its characteristics. The reduction in size enables heat exchangers to attain considerable weight reductions, diminished fluid inventories, reduced capital and installation expenses, and energy savings. Notwithstanding those improvements, the pressure loss in minichannels exceeds that of ordinary tubes due to heightened wall friction.Chisholm (1967) proposed a theoretical basis for the Lockhart–Martinelli correlation for two- phase flow. The Friedel (1979) correlation was obtained by optimizing an equation for the two-phase frictional multiplier using a large measurement database. Many studies have developed pressure drop correlations on the basis of the Chisholm (1967) and Friedel (1979) correlations. Mishima and Hibiki (1996), Yu et al.(2002), and Kawahara et al. (2002) developed pressure drop correlations on the basis of the Chisholm (1967) correlation.

Chang et al. (2000), Chen et al. (2001), and Zhang and Webb (2001) developed pressure drop correlations on the basis of the Friedel (1979) correlation. Tran et al. (2000) measured the two-phase flow pressure drop with refrigerants R-134a, R-12, and R-113 in small round and rectangular channels. They modified Chisholm's (1983) correlation and proposed a new correlation.

This chapter reports on a study whose goals were to present pressure drop experimental data for the refrigerant R-22, and its alternatives, R-134a, R-410A, R-290, R-717 (NH3) and R- 744 (CO_2) which were measured in horizontal and local heat transfers during evaporation in smooth minichannels to establish a new correlation for heat exchangers with minichannel designs. The present experimental data for pressure drop were obtained by the previous experiments compared with existing two-phase pressure drop prediction methods, namely Beattie and Whalley (1982), Cicchitti et al (1960), McAdams (1954), Chang et al. (2000), Dukler et al. (1964), Friedel (1979), Chisholm (1983), Tran et al. (2000), Zhang and Webb (2001), Mishima and Hibiki (1996), Lockhart and Martinelli (1949), Shah (1988), Tran et al. (1996), Jung at al. (1989), Gungor and Winterton (1987), Takamatsu et al (1993), Kandlikar and Stainke (2003), Wattelet et al (1994), Chen (1966), Zhang et al. (2004), Chang and Ro (1996), Yu et al. (2002), Friedel (1979), Kawahara et al. (2002), Mishima (1983), Chisholm et al. (2000), Tran et al. (2000), Chen (2001), and Yoon et al. (2004). A new correlation for two- phase frictional pressure drop was developed on the basis of the Lockhart–Martinelli method using the present experimental data. In the present paper, heat fluxes to make the flow boiling heat transfer coefficients were electrically heated. The experimental results were compared with the predictions of seven existing heat transfer correlations, namely those reported by Mishima and Hibiki (1996), Friedel (1979), Chang at al. (2000), Lockhart and Martinelli (1949), Chisholm (1983), Zhang and Webb (2001), Chen (1966), Chen et al. (2001), Kawahara et al. (2002), Tran et al. (1996), Tran et al. (2000), Wattelet et al. (1993), Wattelet et al. (1994), Gungore-Winterton (1986), Gungore-Winterton (1989), Zhang et al (2004),

Kandlikare-Steinke (1996), Kandlikare-Steinke (2003), Jung et al.(1989), Jung et al. (2004), Shah (1988), Gungore-Winterton (1987), Zhang et al (1987), and Takamatsu et al. (2003) was and were developed in this study based on superposition, due to the limitations in the correlation for forced convective boiling of refrigerants in small channels.

Compared with conventional channels, evaporation in small channels may provide a higher heat transfer coefficient due to their higher contact area per unit volume of fluid. In evaporation within small channels, as reported by Bao et al. (2000), Zhang et al. (2004), Kandlikare-Steinke (2003), Tran et al. (2000), Pettersen (2004), Park and Hrnjak (2007), Zhao et al. [7], Yun et al. (2005), Yoon et al. (2004), Pamitran et al. (2008), and K.-I. Choi (2009), the contribution of nucleate boiling is predominant and laminar flow appears.

The study that was done by A.S. Pamitran et al (2007) and a/bK.-I.Choi et al. (2007), yielded a basic understanding about predicting pressure drop and heat transfer coefficients during refrigerants evaporation in minichannels. The research have established correlations and have been compared with other experimental correlations documented in the extensive literature on two-phase boiling heat transfer. The correlation creation approaches align well with the experimental results obtained utilizing refrigerants as working fluids.

2. METHODS

2.1 Experimental apparatus and method

Experimental facility for inner diameter 3 and 1.5 mm

The experimental facilities are schematically shown in Fig. 1(a) and (b). The test facilities were constructed by A.S. Pamitran et al. (2007) and K.-I. Choi et al (2007) and consisted of a condenser, a subcooler, a receiver, a refrigerant pump, a mass flow meter, a preheater, and test sections. For the test with 3.0 and 1.5mm inner diameter tubes, a variable AC output motor controller was used to control the flow rate of the refrigerant. A Coriolis-type mass flow meter was installed in a horizontal layout for the test with 1.5 and 3.0mm inner diameter tubes. A preheater or a cooler was installed to control the vapor quality of the refrigerant by heating or condensing the refrigerant before it entered the test section. For evaporation at the test section, a pre-determined heat flux was applied from a variable A.C voltage controller. The vapor refrigerant from the test section was then condensed in the condenser and subcooler, and then the condensed refrigerant was supplied to the receiver.

The test section was made of stainless steel circular smooth tubes with inner tube diameters of 3.0 and, 1.5 mm. The rate of input electric potential E and current I were adjusted in order to control the input power and to determine the applied heat flux, which was measured by a standard multimeter. The test parts were consistently heated by directly applying electric current to their tube walls. The test parts were adequately insulated with foam and rubber; hence, heat loss was disregarded in this study. The local saturation pressure of the refrigerant, utilized to ascertain the saturation temperature, was recorded utilizing Bourdon tube pressure gauges featuring a 0.005 MPa scale at both the input and outflow of the test sections. Differential pressure was assessed using Bourdon tube pressure gauges and a differential pressure transducer. Circular sight glasses matching the inner tube diameter of the test section were placed at both the inlet and outflow to visualize the flow. Each sight glass was held by flanges on both sides, as described in Fig. 1.

The temperature and flow rate measurements were recorded using the Darwin DAQ32 Plus logger R9.01 software program and version 2.41 of the Micro Motion ProLink Software package, respectively. The physical properties of the refrigerants were obtained from the REFPROP 8.0.

Experimental facility for inner diameter 0.5 mm

An additional experimental facility was constructed for tubes with a 0.5 mm inner diameter; this facility operates as an open-loop system. This method facilitates the transfer of refrigerant from high-pressure containers to refrigerant receivers. This system employed a needle valve to regulate the flow rate before to entering the test section constructed of stainless steel. A weighing balance was employed in the test utilizing a tube with a 0.5 mm inner diameter to quantify the refrigerant flow rate. The heating and measurements were analogous to those depicted in Fig. 1.The experimental conditions used in the studies of A.S Pamitran et al and K.-I. Choi et al. are listed in Table 1. Five refrigerants are use as the working fluid in the experiments; the studies express the effects of dimensional factors that are represented by the inner diameter of the tubes. The mass flux effect was detected by adjusting the fluid velocities and controlling the heat flux by electrical heating regulation.

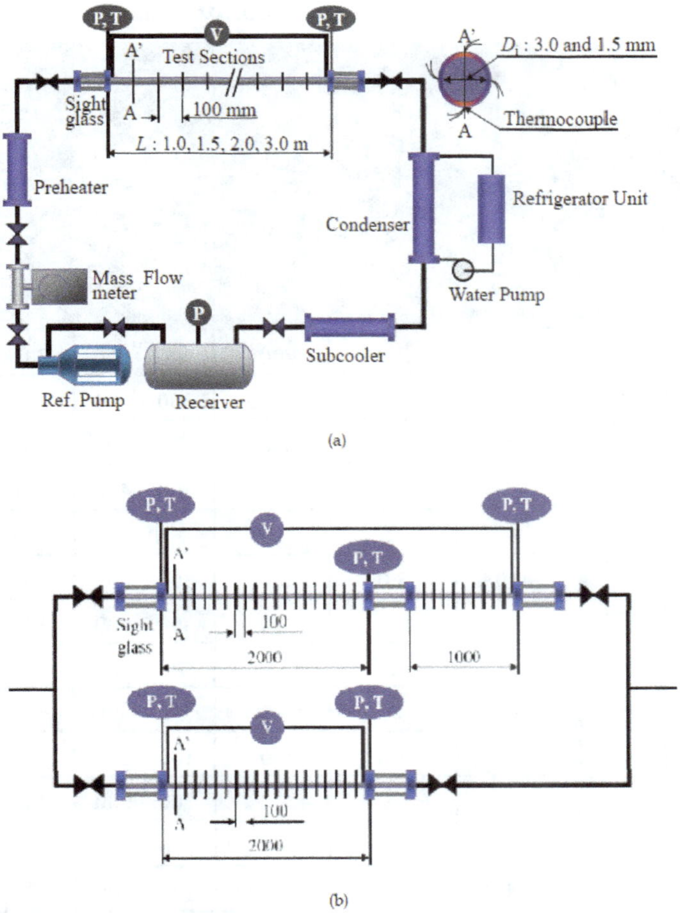

Fig. 1. Experimental test facility: (a) for test section with inner tube diameter of Di=3.0mm and D_i=1.5 mm

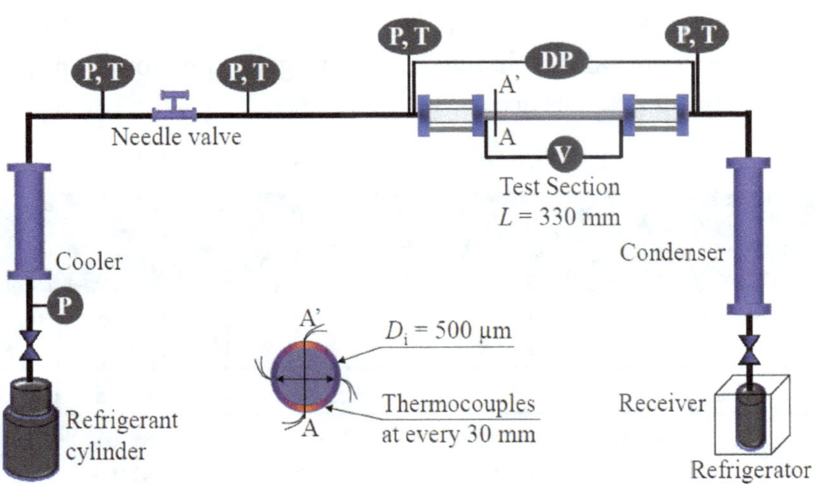

Fig. 2. For test section with inner tube diameter of D_i =0.5 mm

Table 1. Experiment conditions

Two-Phase Flow Boiling Heat Transfer for Evaporative Refrigerants in Diverse Circular Minichannels	Quality	Working refrigerant	Inner diameter (mm)	Tube length (mm)	Mass flux (kg/(m²·s))	Heat flux (kW/m²)	Inlet T_{sat} (ºC)
Horizontal circular smooth small tubes	up to 1.0	R-22	3.0	2000	400 – 600	20 – 40	10
			1.5	2000	300 – 600	10 – 20	10
		R-134a	3.0	2000	200 – 600	10 – 40	10
			1.5	2000	200 – 400	10	10
			0.5	330	100	5 – 20	6 – 10
		R-410A	3.0	3000	300 – 600	10 – 40	10 – 15
			1.5	1500	300 – 600	10 – 30	10 – 15
			0.5	330	70 – 400	5 – 20	1 – 11
		C_3H_8	3.0	2000	50 – 240	5 – 25	0 – 11
			1.5	2000	100 – 400	5 – 20	0 – 12
		CO_2	3.0	2000	200 – 600	20 – 30	1 – 10
			1.5	2000	300 – 600	10 – 30	0 – 11
		NH_3	3.0	2000	100 – 800	10 – 70	0 – 10
			1.5	2000	100 – 500	10 – 35	0 – 10

The experimental uncertainty associated with all the parameters is tabulated in Table 2. The uncertainties were obtained using both random and systematic errors, and these uncertainty values changed according to the flow conditions; their minimum to maximum ranges are shown.

Table 2. Summary of estimated uncertainty

Parameters	Uncertainty (%)				
	CO_2	C_3H_8	R-410A	R-134a	R-22
Twi (%)	±2.10 to ±4.56	±0.18 to ±5.58	±0.23 to ±6.53	±0.29 to ±3.92	±0.44 to ±3.90
P (kPa)	±2.5 kPa	±2.5 kPa	±2.5 kPa	±2.5 kPa	±2.5 kPa
G(%)	±1.85 to ±9.48	±3.24 to ±9.78	±1.84 to ±9.48	±1.85 to ±3.80	±1.84 to ±3.16

Parameters	Uncertainty (%)				
	CO_2	C_3H_8	R-410A	R-134a	R-22
q(%)	±1.67 to ±2.70	±2.07 to ±3.58	±1.67 to ±3.20	±1.79 to ±2.59	±1.78 to ±2.26
x(%)	±1.79 to ±9.71	±4.27 to ±9.82	±1.78 to ±9.85	±2.23 to ±4.19	±1.88 to ±3.39
h(%)	±4.46 to ±8.23	±1.78 to ±9.89	±2.59 to ±10.33	±2.75 to ±9.24	±2.74 to ±9.07

3. DATA REDUCTION

3.1 Pressure drop

The saturation pressure at the initial point of saturation was determined by interpolating the measured pressure and the subcooled length. The experimental two-phase frictional pressure drop can be obtained by subtracting the calculated acceleration pressure drop from the measured pressure drop.

$$\left(\frac{dp}{dz}\right) = \left(\frac{dp}{dz}F\right) + \left(\frac{dp}{dz}a\right) + \left(\frac{dp}{dz}z\right) \tag{1}$$

$$-\left(\frac{dp}{dz}a\right) = G^2\frac{d}{dz}\left(\frac{x^2 v_g}{\alpha} + \frac{(1-x)^2 v_f}{(1-\alpha)}\right) \tag{2}$$

$$-\left(\frac{dp}{dz}z\right) = g\,sin\,\theta\left[\frac{A_g}{A}\rho_g + \frac{A_f}{A}\rho_g\right] = g\,sin\,\theta\left[\alpha\rho_g + (1-\alpha)\rho_f\right] \tag{3}$$

The equation for the friction pressure gradient for horizontal tubes should be reduced to the static head factor, therefore the equation is

$$\left(\frac{dp}{dz}F\right) = \left(\frac{dp}{dz}\right) - \left(\frac{dp}{dz}a\right) = \left(\frac{dp}{dz}\right) - G^2\frac{d}{dz}\left(\frac{x^2}{\alpha\rho_g} + \frac{(1-x)^2}{(1-\alpha)\rho_f}\right) \tag{4}$$

The predicted void fraction with Steiner (1993), CISE (Premoli et al., 1971) and Chisholm (1972) are compared in the present study, as is shown in Fig. 3. The absolute deviations of the Steiner (1993), CISE (Premoli et al., 1971) and Chisholm (1972) void fractions from the homogenous void fraction are 7.11%, 3.72% and 11.49%, respectively. The predicted pressure drop for the present experimental data using some previous pressure drop correlations with the Steiner (1993) void fraction showed a better prediction than that using the CISE (Premoli et al., 1971) and Chisholm (1972) void fractions. Therefore, the void fraction in the present study was obtained from the Steiner (1993) void fraction.

$$\alpha = \frac{x}{\rho_g}\left[\left(1 + 0.12(1-x)\right)\left(\frac{x}{\rho_g} + \frac{(1-x)}{\rho_f}\right) + \left(\frac{1.18}{G^2}\right)\left(\frac{(1-x)}{\rho_f^{0.5}}\right)\left(g\sigma(\rho_f + \rho_g)\right)^{0.25}\right]^{-1} \tag{5}$$

The friction factor was determined from the measured pressure drop for a given mass flux by using the Fanning equation

$$f_{tp} = \frac{D\bar{\rho}}{2G^2}\left(-\frac{dp}{dz}F\right) \tag{6}$$

where the average density is calculated with the following equation:

$$\bar{\rho} = \alpha\rho_g + (1+\alpha)\rho_f \tag{7}$$

In order to obtain the two-phase frictional multiplier based on the pressure drop for the total flow assumed for the liquid, ϕ_{fo}^2, the calculated two-phase frictional pressure drop is divided by the calculated frictional two-phase pressure drop assuming the total flow to be liquid, as shown by

$$\phi_{fo}^2 = \frac{\left(-\frac{dp}{dz}F\right)_{tp}}{\left(-\frac{dp}{dz}F\right)_{fo}} = \frac{\left(-\frac{dp}{dz}F\right)_{tp}}{\left(-\frac{2f_{fo}G^2}{D\rho_f}F\right)} \tag{8}$$

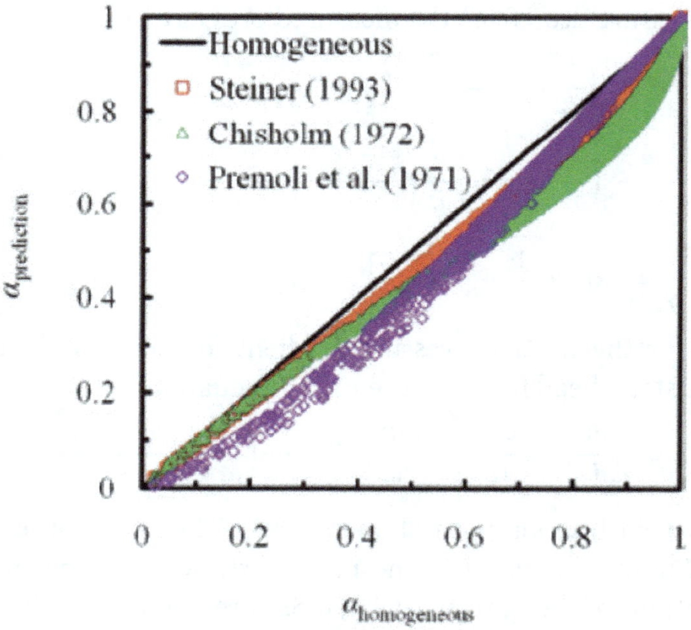

Fig. 3. Void fraction comparison

3.2 Heat transfer coefficient

The inside tube wall temperature, *Twi* was the average temperature of the top, both right and left sides, and bottom wall temperatures, and was determined using steady-state one- dimensional radial conduction heat transfer through the wall with internal heat generation.

The quality, x, at the measurement locations, z, were determined based on the thermodynamic properties, namely,

$$x = \frac{i - i_f}{i_{fg}} \tag{9}$$

The refrigerant flow at the inlet of the test section was not completely saturated. Even though it was just short of being completely saturated, it was necessary to determine the subcooled length to ensure reduction data accuracy. The subcooled length was calculated using the following equation to determine the initial point of saturation.

$$z_{sc} = L\frac{i_f - i_{fi}}{\Delta i} = L\frac{i_f - i_{fi}}{(Q/W)} \tag{10}$$

The outlet mass quality was then determined using the following equation:

4.1 Results and discussion

The flow with heat addition, or adiabatic flow, constitutes a linked thermodynamic issue. Heat transfer induces a phase change, resulting in alterations in phase distribution and flow patterns; concurrently, it modifies hydrodynamics, including pressure drop along the flow route, so influencing heat transfer characteristics. Additionally, a single-component, two-phase flow within a tube is unlikely to achieve full development at low pressure due to the morphological alterations in large bubbles and the intrinsic pressure variations along the tube, which perpetually modify the fluid state and consequently affect the phase distribution and flow configuration. This study delineates the attributes of a two-phase flow pattern and the associated pressure drop in tiny tubes for various refrigerants.

4.,2 Flow pattern

The hydrodynamic characteristic of two-phase flows, such as the pressure drop, void fraction, or velocity distribution, varies systematically with the observed flow pattern, just as in the case of a single-phase flow, whose behavior depends on whether the flow is in the laminar or turbulent regime. However, in contrast to a single-phase flow, liquid–vapor flows are difficult to describe by general principles, which could serve as a framework for solving practical work. The identification of a flow regime provides a picture of the phase boundaries.

The present experimental results were mapped on Wang et al. (1997) and Wojtan et al. (2005) flow pattern maps, which were developed for diabatic two-phase flows. The Wang et al. (1997) flow pattern map is a modified Baker (1954) map, developed using R-22, R-134a, and R-407C inside a 6.5mm horizontal smooth tube. Therefore, they claim that their flow pattern map is better for prediction of flow patterns in small tubes. The Wang et al. (1997) study showed that the flow transition for a mixture refrigerant showed a considerable delay compared with that of pure refrigerants. For R-410A, at the initial stage of evaporation, R-32 evaporated faster than R- 125. Therefore, R-32 increased the concentration of the vapor phase, and R-125 increased the concentration of the liquid phase throughout the evaporation process at the liquid–vapor interface. This resulted in a higher mean vapor velocity and a lower mean liquid velocity during evaporation. For the other working refrigerants, the physical properties of the refrigerant such as density, viscosity and surface tension have a strong effect on the flow pattern.

The predicted flow pattern for the selected current experimental data according to the existing flow pattern maps of Wojtan et al. (2005) and Wang et al. (1997) can be seen in Figs. 4(a) - (f) and 5(a) – (f), respectively. The Wang et al. (1997) map showed a better prediction of the flow pattern of the current experimental data for the beginning of annular flow than the Wojtan et al. (2005) map; however, the Wang et al. (1997) map could not show the prediction for dry-out conditions. The flow pattern prediction of the current test results for the all test conditions with the Wang et al. (1997) flow pattern map was for the intermittent, stratified wavy and annular flow. The stratified wavy flow appeared earlier for higher mass fluxes and its regime was longer for the low mass flux condition. The annular flow appeared earlier for higher mass fluxes.

The Wojtan et al. (2005) flow pattern map is a modified Kattan et al. (1998) map, developed using R-22 and R-410A inside a 13.6mm horizontal smooth tube. Kattan et al. (1998) used five refrigerants R-134a, R-123, R-402A, R-404A, and R-502 inside 12mm and 10.92mm (only for R-134a) tubes, which were heated by hot water flowing counter-currently to develop their flow pattern maps on the basis of the Steiner (1993) flow pattern map. The flow pattern prediction of the experimental results for all the test conditions with the Wojtan et al. (2005) flow pattern map was on intermittent, annular, dry-out, and mist flows. The flow pattern prediction as shown in Fig. 4(b)–(c), showed that the mass flux, heat flux and inner diameter had an effect on the flow-pattern. Since the Wojtan et al. (2005) map was developed using a conventional tube, the flow pattern transition of the present experimental data showed a delay on this map. However, the Wojtan et al. (2005) map could predict the dry-out condition. The Wojtan et al. (2005) map predicts the dry-out condition of the R-410A experimental data better than those of the other working refrigerants. Overall, the Wang et al. (1997) flow pattern map provides a better flow pattern prediction for the current experimental results than the Wojtan et al. (2005) flow pattern map.

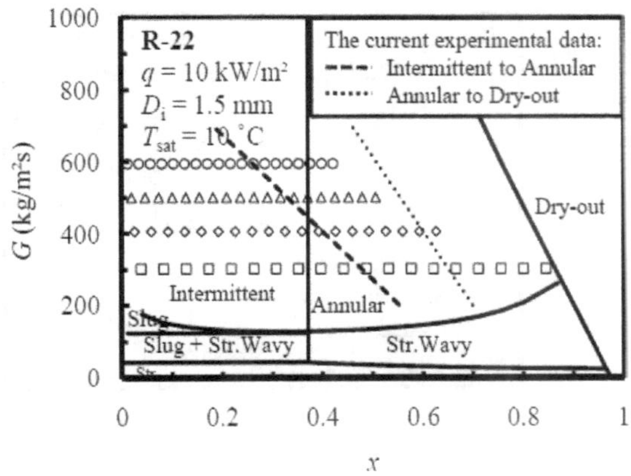

Fig. 4. (a) Present experimental results mapped on Wojtan et al. (2005) flow pattern map for R-22 at q = 10 kW/m², D_i = 1.5 mm, T_{sat} = 10°C

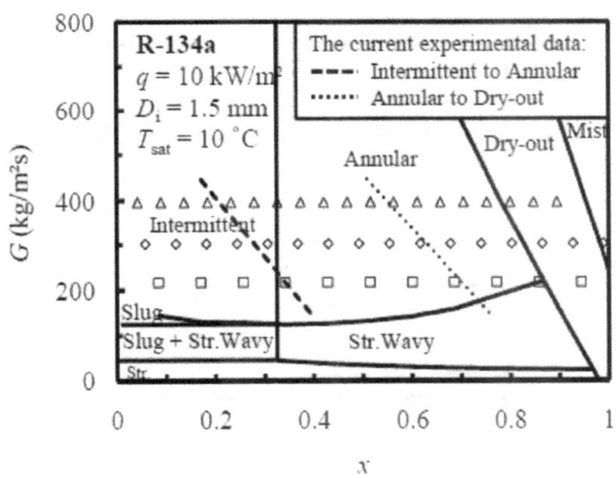

Fig. 4. (b) Present experimental results mapped on Wojtan et al. (2005) flow pattern map for R-134a at q = 10 kW/m², D_i = 1.5 mm, T_{sat} = 10°C

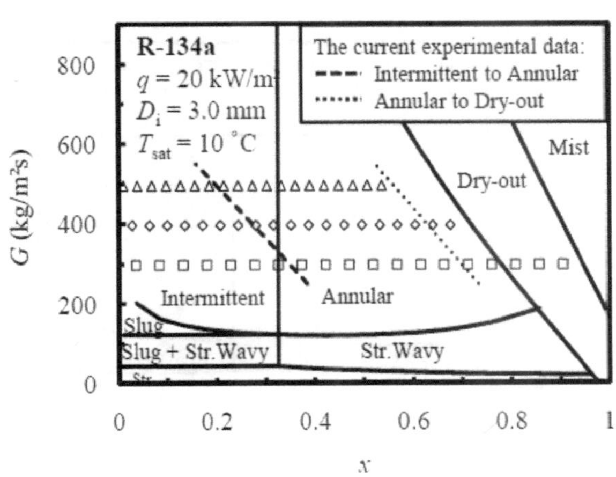

Fig. 4. (c) Present experimental results mapped on Wojtan et al. (2005) flow pattern map for R-134a at q = 20 kW/m², D_i = 3.0 mm, T_{sat} = 10°C

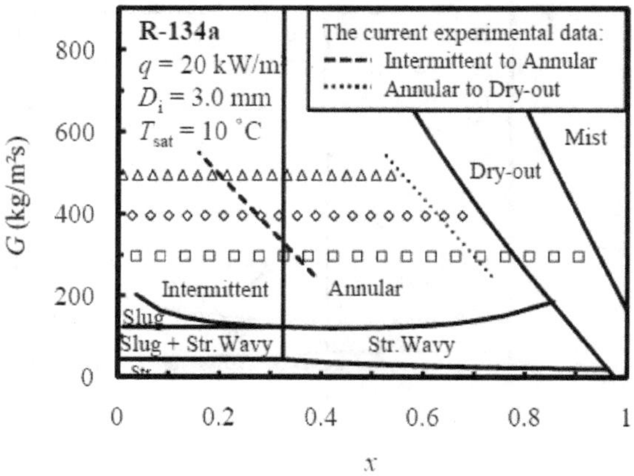

Fig. 4. (d) Present experimental results mapped on Wojtan et al. (2005) flow pattern map for R-410A at q = 20 kW/m^2, D_i = 3.0 mm, T_{sat} = 10°C

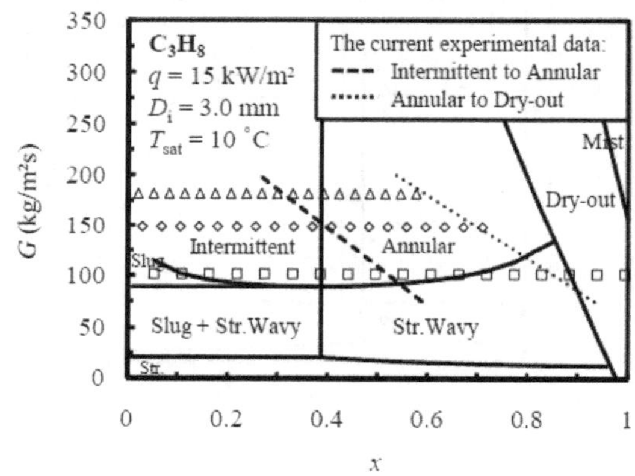

Fig. 4. (e) Present experimental results mapped on Wojtan et al. (2005) flow pattern map for C_3H_8 at q = 15 kW/m^2, D_i = 3.0 mm, T_{sat} = 10°C

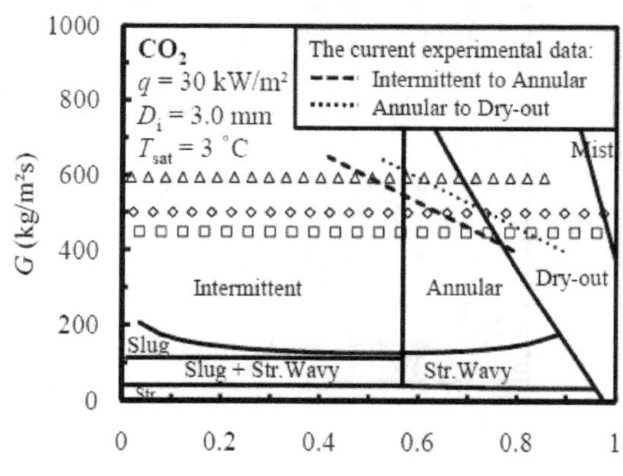

Fig. 4. (f) Present experimental results mapped on Wojtan et al. (2005) flow pattern map for CO_2 at q = 30 kW/m^2, D_i = 3.0 mm, T_{sat} = 3°C

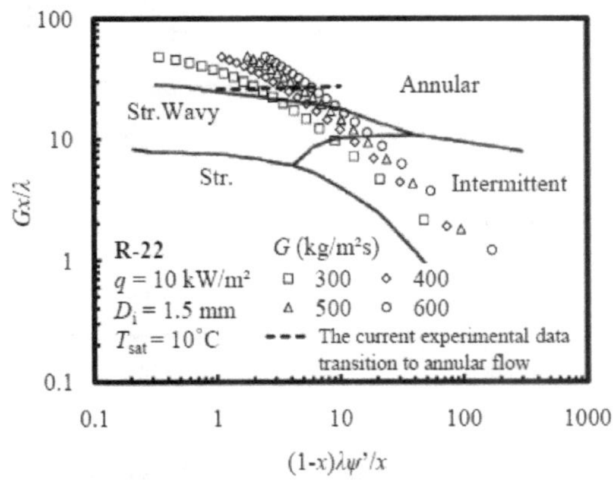

Fig. 5. (a) Present experimental results mapped on Wang et al. (1997) flow pattern map for R-22 at q = 10 kW/m², D_i = 1.5 mm, T_{sat} = 10°C

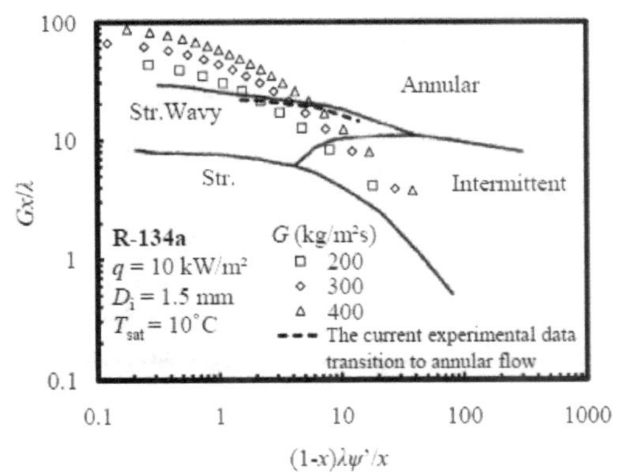

Fig. 5. (b) Present experimental results mapped on Wang et al. (1997) flow pattern map for R-134a at q = 10 kW/m², D_i = 1.5 mm, T_{sat} = 10°C

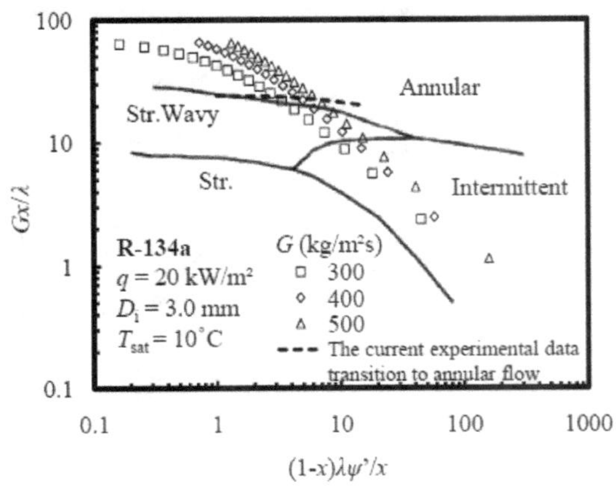

Fig. 5. (c) Present experimental results mapped on Wang et al. (1997) flow pattern map for R-134a at q = 20 kW/m², D_i = 3.0 mm, T_{sat} = 10°C

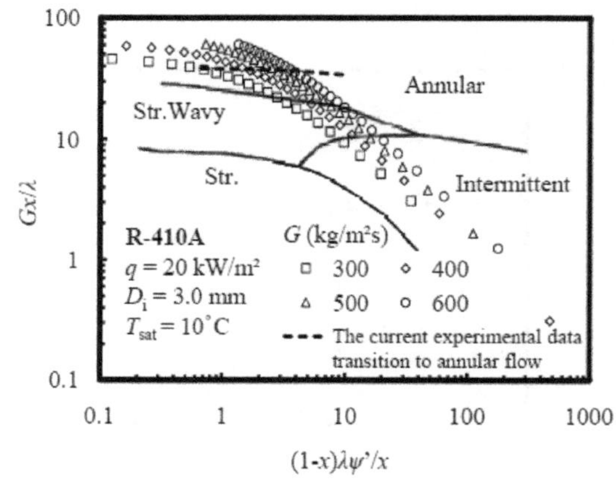

Fig. 5. (d) Present experimental results mapped on Wang et al. (1997) flow pattern map for R-410A at q = 20 kW/m², D_i = 3.0 mm, T_{sat} = 10°C

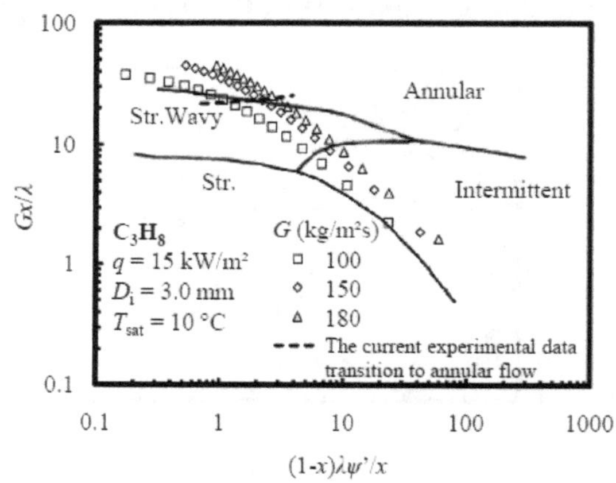

Fig. 5. (e) Present experimental results mapped on Wang et al. (1997) flow pattern map for C_3H_8 at q = 15 kW/m², D_i = 3.0 mm, T_{sat} = 10°C

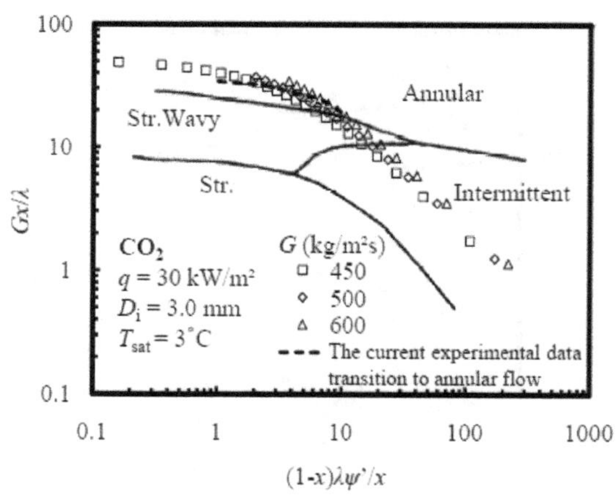

Fig. 5. (f) Present experimental results mapped on Wang et al. (1997) flow pattern map for CO_2 at q = 30 kW/m², D_i = 3.0 mm, T_{sat} = 3°C

4.2 Pressure drop

Fig. 6(a)-(f) shows that mass flux has a strong effect on the pressure drop. An increase in the mass flux results in a higher flow velocity, which increases the friction and acceleration pressure drops. This similar trend is shown by Zhao et al. (2000), Yoon et al. (2004), Park and Hrnjak (2007), Oh et al. (2008) and Cho and Kim (2007). Fig. 6 also illustrates that the pressure drop increases as the heat flux increases. It is presumed that the increasing heat flux results in a higher vaporization, which increases the average fluid vapor quality and flow velocity; this trend is similar to that shown by Zhao et al. (2000). The effect of the inner tube diameter on the pressure drop is also illustrated in Fig. 6(a)-(f). The pressure gradient in the 1.5 mm tube is higher than that in the 3.0mm tube. The explanation for this is that the smaller inner tube diameter results in a higher wall shear stress, wherein for a given temperature condition it results in a higher friction factor and flow velocity, and then results in both higher frictional and acceleration pressure drops.

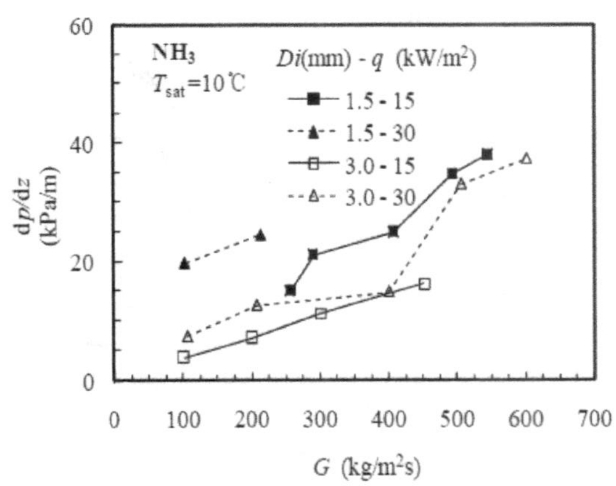

(a) Effect of mass flux, heat flux and inner diameter on pressure drop for NH_3 at T_{sat} = 10°C

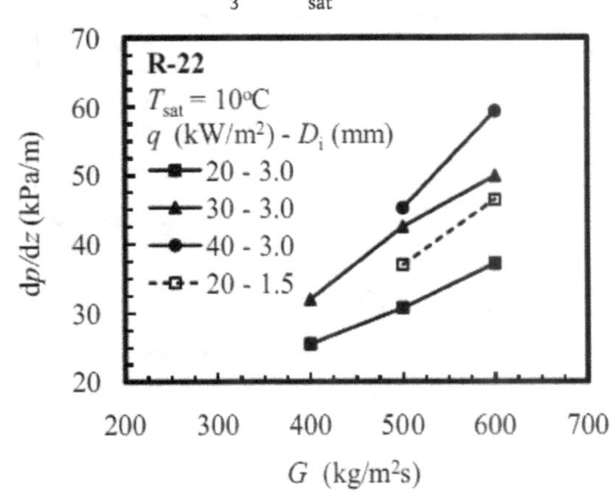

(b) Effect of mass flux, heat flux and inner tube diameter on pressure drop or R-22 at T_{sat} = 10°C

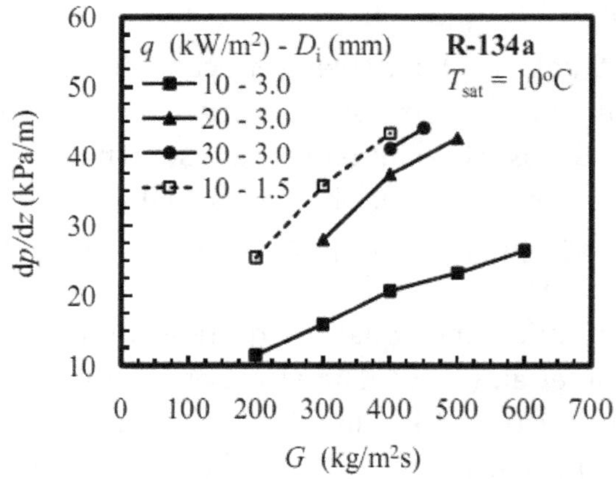

(c) Effect of mass flux, heat flux and inner tube diameter on pressure drop for R-134a at T_{sat} = 10°C

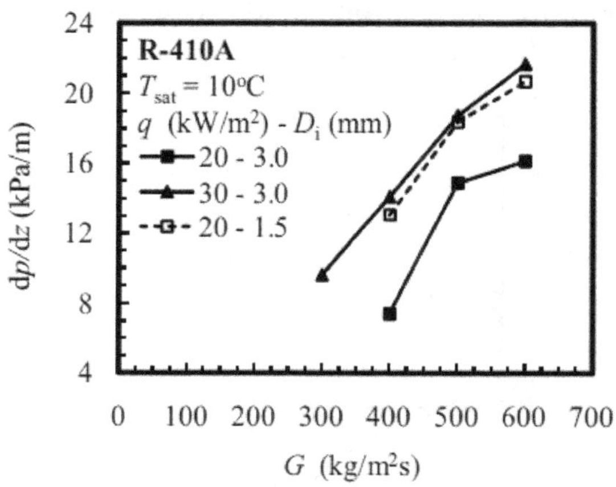

(d) Effect of mass flux, heat flux and inner tube diameter on pressure drop for R-410A at T_{sat} = 10°C

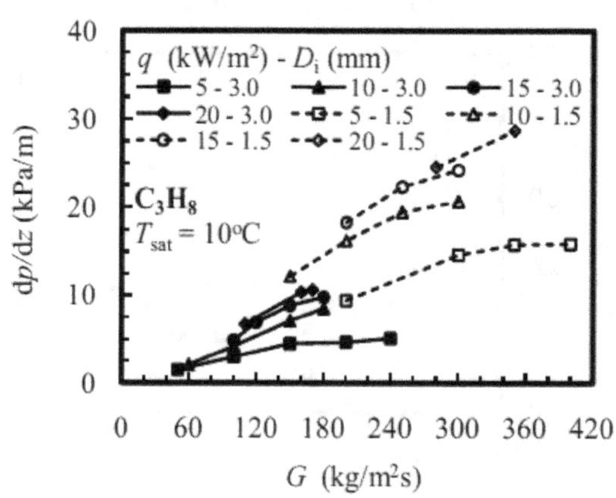

(e) Effect of mass flux, heat flux and inner tube diameter on pressure drop for C_3H_8 at T_{sat} = 10°C

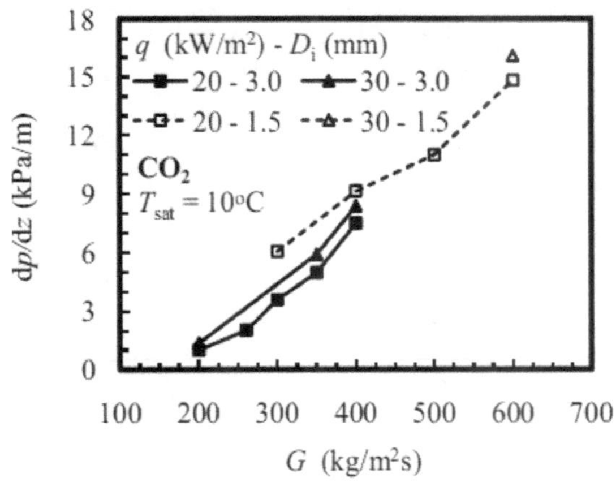

(f) Effect of mass flux, heat flux and inner tube diameter on pressure drop for CO_2 at T_{sat} = 10°C

Fig. 6. (a)-(f) The effects of mass flux, heat flux, inner tube diameter, and saturation temperature on pressure drop

Fig. 6 also depicts the effect of saturation temperature on pressure drop, where a lower saturation temperature results in a higher pressure drop. This can be explained by the effect of the physical properties of density and viscosity on the pressure drop at different temperatures. The liquid density, p_f, and liquid viscosity, μ_f, increase as the temperature decreases, whereas the vapor density, p_g, and vapor viscosity, μ_g, decrease as the temperature decreases. As the temperature decreases for a constant mass flux condition, the increasing liquid density and liquid viscosity result in a lower liquid velocity, whereas the decreasing vapor density and vapor viscosity are the result in a higher vapor velocity. It is clear that the pressure drop increases during evaporation, and this increasing of the pressure drop is higher when the saturation temperature is lower.

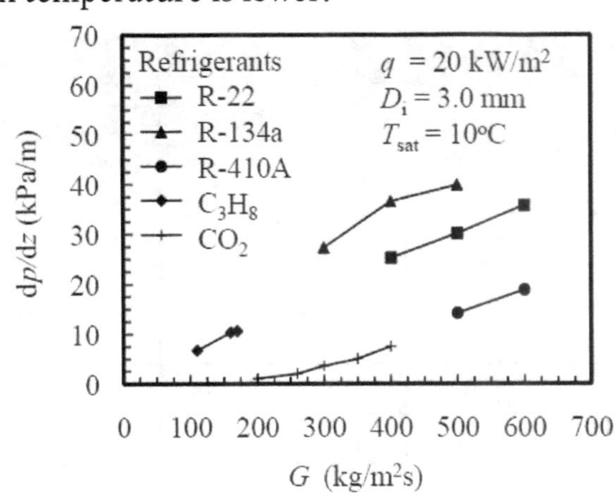

(a) Pressure drop comparison of the present working refrigerants at q = 20 kW/m², D_i = 3.0 mm, T_{sat} = 10°C

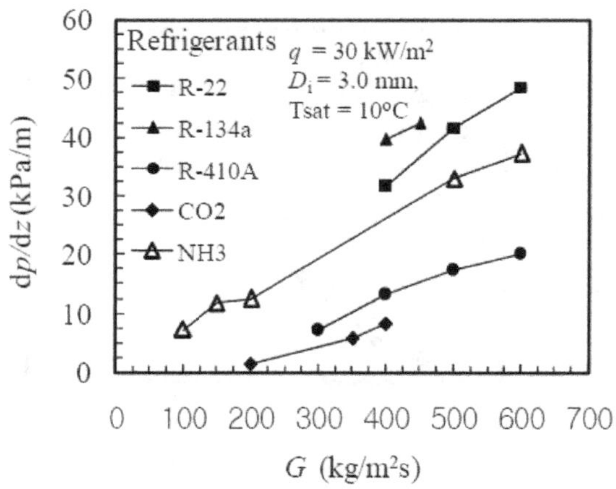

(b) Pressure drop comparison of the present working refrigerants at q = 30 kW/m², D_i = 3.0 mm, T_{sat} = 10°C

Fig. 7. Pressure drop comparison of the present working refrigerants

The pressure drop of the present working refrigerants is compared under several of the same experimental conditions. The pressure drop comparisons are illustrated in Fig. 7. The order of the pressure drop from the highest to the lowest is R-134a, R-22, NH_3, C_3H_8, R- 410A, and CO_2. The pressure drop is strongly affected by the physical properties of the working fluid such as density, viscosity, surface tension and pressure. As shown in Table 3, the working refrigerant with a higher pressure drop has a higher density ratio ρ_f/ρ_g, viscosity ratio μ_f/μ_g, surface tension, and a lower pressure, generally. Therefore, it is clear that in the present comparison that CO_2 has the lowest pressure drop, making CO_2 an effective future environmental friendly refrigerant.

Table 3. Physical properties of R-22, R-134a, R-410A, C_3H_8, CO_2 and NH_3 at 10 and 5°C

Refrigerant	T (°C)	σ (10-3 N/m)	μ_f/μ_g	μ_g (10-6 Pa s)	μ_f (10-6 Pa s)	ρ_f/ρ_g	ρ_g (kg/m³)	ρf (kg/m³)	P (MPa)
R-22	10	10.22	16.36	11.96	195.7	43.27	28.82	1247	0.681
R-134a		10.14	21.42	11.15	238.8	62.33	20.23	1261	0.415
R-410A		7.16	11.36	12.91	146.6	27.07	41.74	1130	1.085
C_3H_8		8.85	13.96	8.15	113.8	37.32	13.8	515	0.636
CO_2		2.77	5.59	15.46	86.37	6.41	134.4	861.7	4.497
NH_3		29.59	16.35	9.36	153.03	128.32	4.89	623.64	0.615

Refrigerant	T (°C)	σ (10-3 N/m)	μ_f/μ_g	μ_g (10-6 Pa s)	μ_f (10-6 Pa s)	ρ_f/ρ_g	ρ_g (kg/m^3)	ρ_f (kg/m^3)	P (MPa)
R-22		10.95	17.62	11.73	206.7	50.99	24.79	1264	0.584
R-134a	5	10.84	23.25	10.94	254.4	74.61	17.13	1278	0.35
R-410A		7.9	12.38	12.6	156	32.21	35.73	1151	0.934
C_3H_8		9.48	15.03	7.97	119.8	43.57	11.98	522	0.551
CO_2		3.64	6.46	14.83	95.84	7.85	114.1	896.7	3.965
NH_3		31.24	17.51	9.21	161.23	153.52	4.115	631.66	0.515

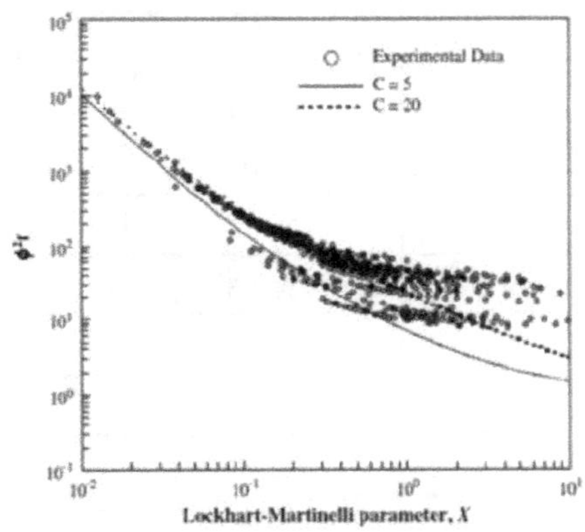

(a) Propane

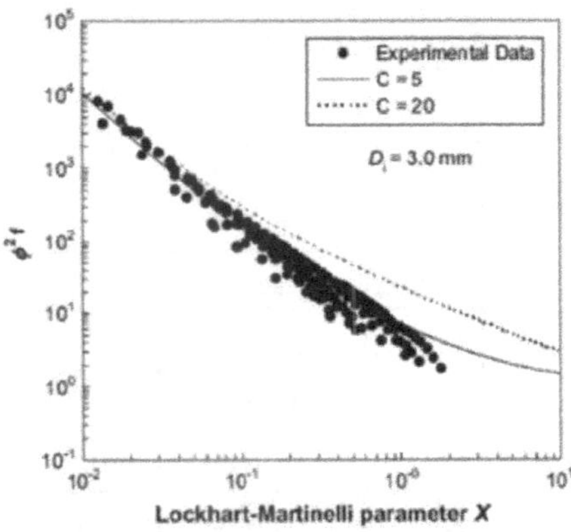

(b) R-410A

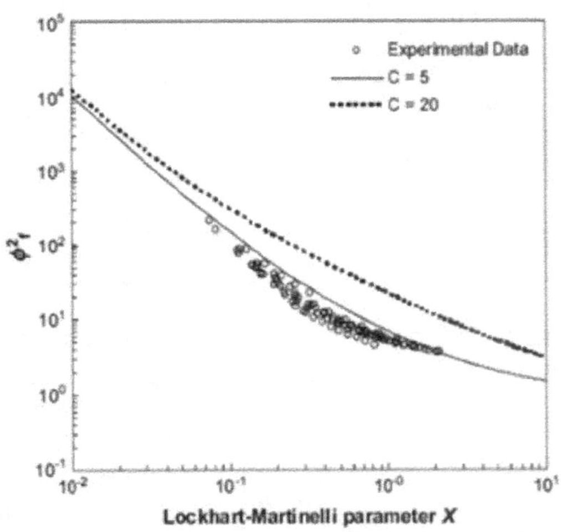

(c) CO_2

Fig. 8. Variation of the two-phase frictional multiplier data with the Lockhart–Martinelli parameter., (a) Propane, (b) R-410A and (c) CO_2

Fig. 8 (a)-(c) shows a comparison of the two-phase frictional multiplier data with the values predicted by the Lockhart–Martinelli correlation with C= 5 and C= 20, where the Cs are taken from Chisholm (1967). The figure shows that the present data are at the upper the baseline of C = 20, between the baseline of C = 20 and C = 5, and under the baseline of C = 5, which means that laminar, turbulent and the co-current laminar–turbulent flows exist in the present data.

Table 4. Deviation of the pressure drop comparison with some existing correlations.

	Previous correlations	Mean deviation(%)	Average deviation(%)
A.S. Pamitran et al. (2010)	Beattie and Whalley (1982)	23.35	15.91
	Cicchitti et al (1960)	23.51	2.9
	McAdams (1954)	25.25	21.75
	Chang et al. (2000)	26.41	22.65
	Dukler et al. (1964)	27.03	25.37
	Friedel (1979)	28.87	16.44
	Chisholm (1983)	34.65	3.89
	Tran et al. (2000)	42.48	33.05
	Zhang and Webb (2001)	49.04	22.4
	Mishima and Hibiki (1996)	50.71	39.89
	Lockhart and Martinelli (1949)	107	97.69
K.-I. Choi et al. (2009)	Mishima and Hibiki (1996)	35.37	25.08
	Friedel (1979)	38.79	21.91
	Chang at al. (2000)	38.86	21.99
	Lockhart and Martinelli (1949)	41.36	3.46
	Chisholm (1983)	45.24	18
	Zhang and Webb (2001)	46.82	7.58
	Chen et al. (2001)	63.67	63.6
	Kawahara et al. (2002)	73.04	72.92
	Tran et al. (2000)	77.99	35.6
	Homogeneous (Cicchitti et al., 1960)	48.67	-40.27
	Homogeneous (Beattie and Whalley, 1982)	54.68	-51.69
	Homogeneous (Mc Adam, 1954)	56.07	-53.73
	Homogeneous ((Dukler et al., 1964)	58.75	-57.32

	Previous correlations	Mean deviation(%)	Average deviation(%)
K.-I. Choi et al. (2008)	Chang and Ro (1996)	19.57	7.43
	Homogeneous (Beattie and Whalley, 1982)	20.04	7.73
	Homogeneous (Mc Adam, 1954)	20.08	13.31
	Homogeneous (Cicchitti et al., 1960)	20.12	8.11
	Homogeneous ((Dukler et al., 1964)	21.87	18.08
	Yu et al. (2002)	27.14	17.88
	Chen et al(2001)	29.85	7.25
	*Chen et al(2001)	14.4	1.43
	Zhang and Webb (2001)	30.82	30.77
	Friedel (1979)	39.21	39.18
	Chang et al. (2000)	39.43	39.41
	Kawahara et al. (2002)	44.63	44.63
	Mishima (1983)	67.81	66.11
	Chisholm et al. (2000)	98.54	97.25
	Tran et al. (2000)	133.97	133.97
	Lockhart and Martinelli (1949)	145.44	144.58
A.S. Pamitran et al. (2007)	Homogeneous (Dukler et al. (1964))	18.14	1.63
	Homogeneous (McAdams (1954))	18.33	0.53
	Homogeneous (Beattie and Whalley, 1982)	21.16	12.83
	Homogeneous (Cicchitti et al., 1960)	24.82	14.78
	Kawahara et al. (2002)	30.8	30.43
	Zhang and Webb (2001)	34.3	21.28
	Yu et al. (2002)	48.23	4.33
	Friedel (1979)	68.56	67.89
	Chang et al. (2000)	68.72	68.06
	Yoon et al. (2004)	101.18	93.59
	Chisholm (1983)	101.55	100.24
	Chen et al. (2001)	122.13	122.13
	Tran et al. (2000)	179.57	179.57

The current experimental two-phase pressure drop data were compared with the predictions from eleven existing correlations. The deviations from the correlation results from the previous experimental data are listed in Table 4, and the selected comparison figures are shown in figures 9(a)-(j). The pressure drop homogenous model of Beattie and Whalley (1982) provided the best prediction in the A.S.

Pamitran et al. (2010) study among the other methods, yielding a mean deviation of around 23%. Mishima and Hibiki (1996) is the best correlation to predict the pressure drop in K.-I Choi et al. (2009) with a mean deviation about 35%. The other presented homogenous models also provided relatively good predictions.

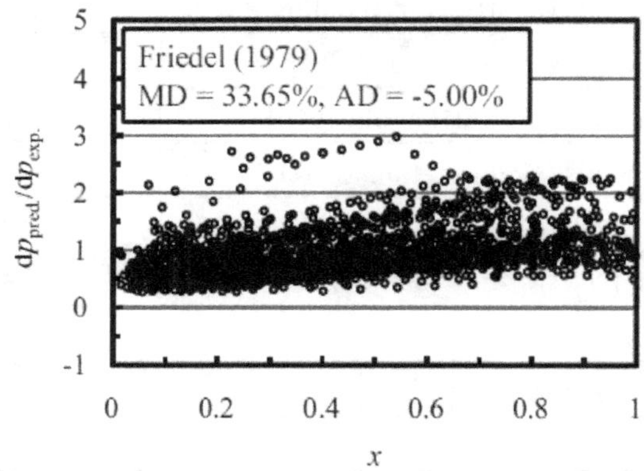

Fig. 9. (a) Comparison between the experimental data and the prediction pressure drop with Fridel (1979) correlation

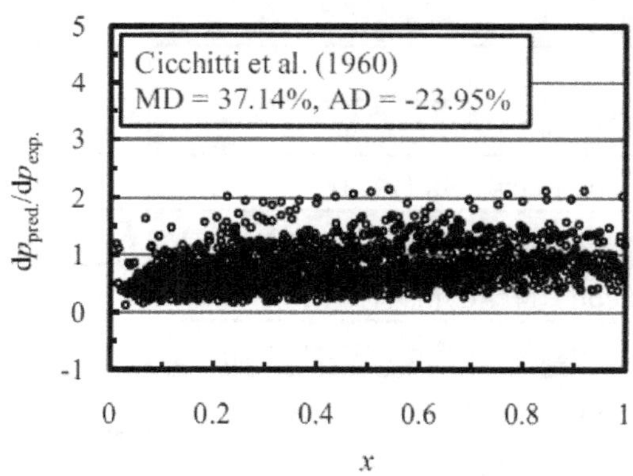

Fig. 9. (b) Comparison between the experimental data and the prediction pressure drop with Cicchitti et al (1960) correlation

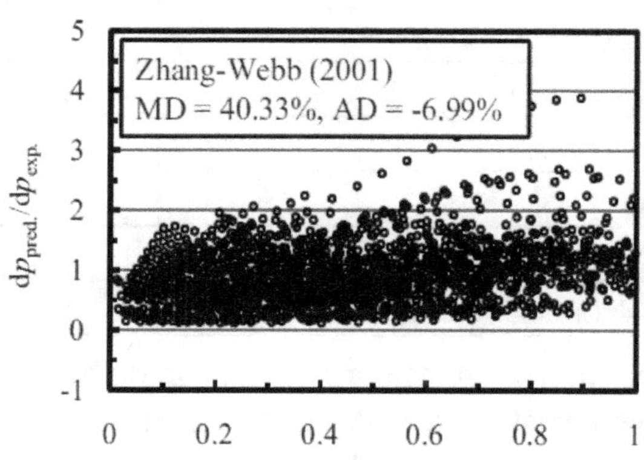

Fig. 9. (c) Comparison between the experimental data and the prediction pressure drop with Zang-Webb (2001) correlation

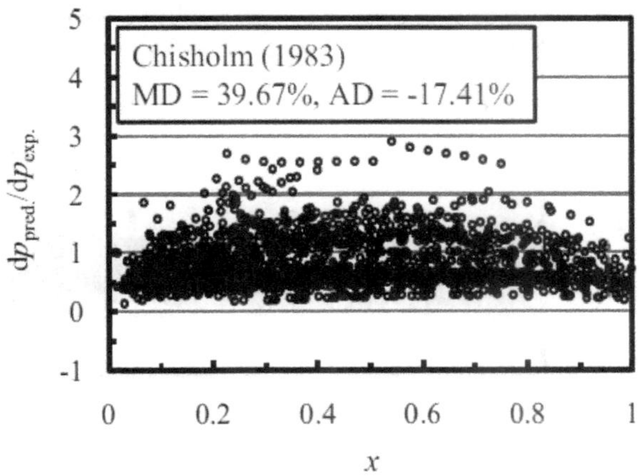

Fig. 9. (d) Comparison between the experimental data and the prediction pressure drop with Chisholm (1983) correlation

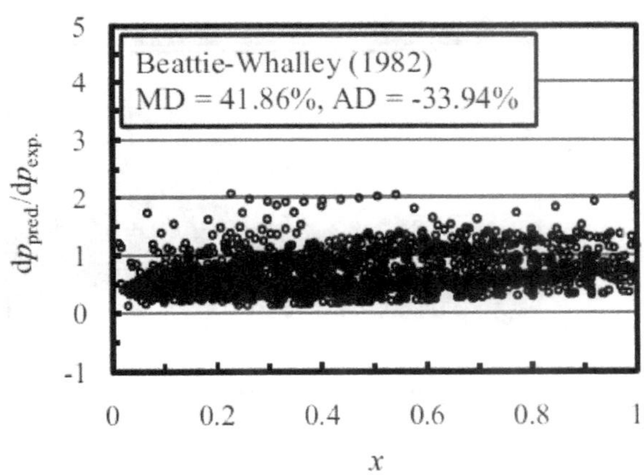

Fig. 9. (e) Comparison between the experimental data and the prediction pressure drop with Beattie-Whalley (1982) correlation

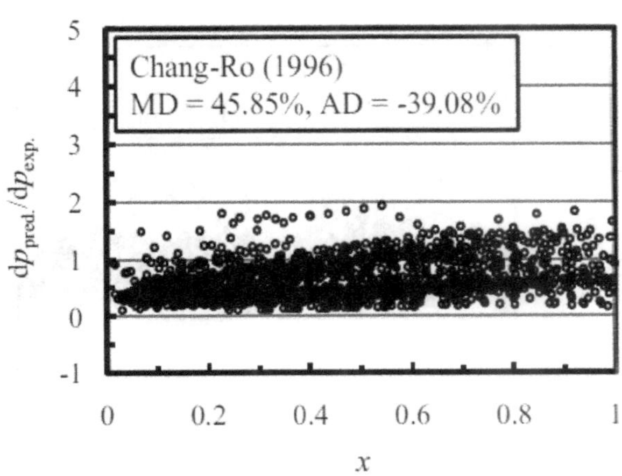

Fig. 9. (f) Comparison between the experimental data and the prediction pressure drop with Chang-Ro (1996) correlation

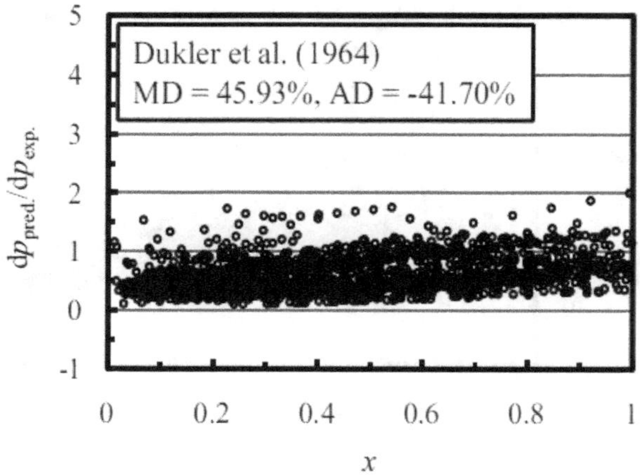

Fig. 9. (g) Comparison between the experimental data and the prediction pressure drop with Dukler et al. (1964) correlation

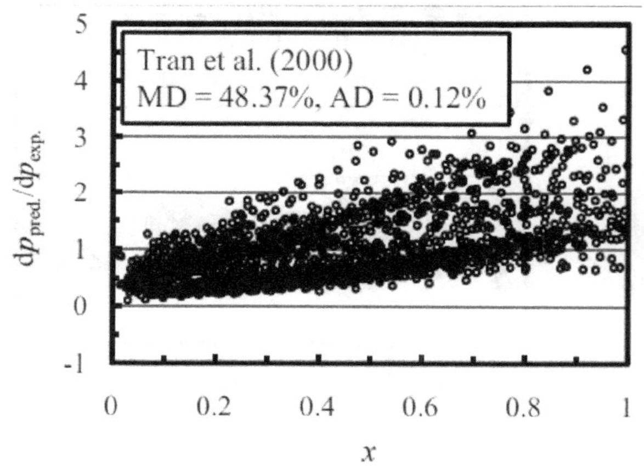

Fig. 9. (h) Comparison between the experimental data and the prediction pressure drop with Trans et al. (2000) correlation

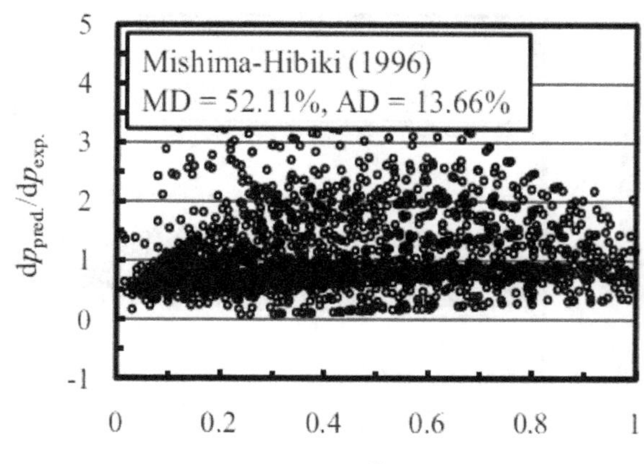

Fig. 9. (i) Comparison between the experimental data and the prediction pressure drop with Mishima – Hibiki (1996) correlation

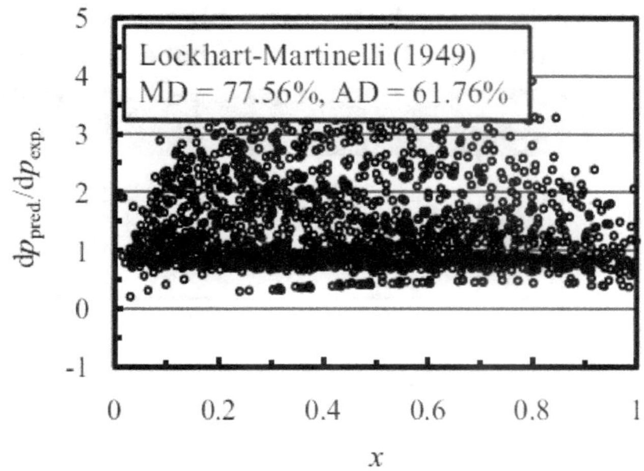

Fig. 9. (j) Comparison between the experimental data and the prediction pressure drop with Lochart-Martinelli (1949) correlation

The pressure drop homogenous model of Beattie and Whalley (1982) provided the best prediction in the A.S. Pamitran et al. (2010) study among the other methods, yielding a mean deviation of around 23%. Mishima and Hibiki (1996) is the best correlation to predict the pressure drop in K.-I Choi et al. (2009) with a mean deviation about 35%. The other presented homogenous models also provided relatively good predictions. The homogenous model assumes that vapor and liquid velocities are equal. The McAdams et al. (1954) correlation was developed based on a homogeneous model. The homogeneous model is a simplification model; therefore, the model is not an empirical model. For some conditions, the homogeneous model can predict the pressure drop well, but not for other conditions; it depends on how close the conditions are to the assumptions of the homogenous model. In this study we develop a new correlation based on the present experimental data. The prediction by the Chang et al.'s (2000) and Friedel's (1979) models yielded a deviation of lower than 28% in every study. The Chang et al. (2000) model was developed with R-410A and air–water in a 5 mm smooth tube. Friedel's (1979) model was developed using a large database; it was valid for horizontal flow and vertical upward flow. The prediction with the Chisholm (1983) method showed a fair result with a mean deviation of lower than 40%. The remaining predictions of Tran et al. (2000), Zhang and Webb (2001), Lockhart and Martinelli (1949) provided a large mean deviation of more than 40% in every study as mentioned in Table 4. The correlations have failed to predict the experimental data of A.S Pamitran et al. (2010) and (2007) and K.-I. Choi et al. (2009) and (2008). Table 5 shows the previous correlations as comparisons, predicting the pressure drop by using the experimental data.

Table 5. Some of previous correlations to predict pressure drop used by S. Pamitran et al. (2010) and (2007) and K.-I. Choi et al. (2009) and (2008)

Previous Correlations	
Homogeneous [Beattie and Whalley (1982)]	$\bar{\mu} = \mu_f\left(1 - \frac{x\rho_f}{x\rho_f + (1-x)\rho_g}\right)\left(1 - \frac{2.5x\rho_f}{x\rho_f + (1-x)\rho_g}\right) + \mu_g\frac{x\rho_f}{x\rho_f + (1-x)\rho_g}$
Homogeneous [Cicchitti et al (1960)]	$\bar{\mu} = x\rho_g + (1-x)\rho_f$
Homogeneous [McAdams (1954)]	$\bar{\mu} = \left(\frac{x}{\mu_g}\frac{1-x}{\mu_f}\right)^{-1}$
Chang et al. (2000)	$\phi_{fo}^2 = (1-x)^2 + x^2[(\rho_f f_{go})/(\rho_g f_{fo})] + 3.24A_2A_3\mathrm{Fr}^{-0.045}\mathrm{We}^{-0.035+\Phi}$ $A_2 = x^{0.78}(1-x)^{0.224}$, $A_3 = (\rho_f/\rho_g)^{0.91}(\mu_g/\mu_f)^{0.19}[1-(\mu_g/\mu_f)]^{0.7}$ $\Phi = e^{-\left(\frac{D_{i,ref}}{D_i}\right)^{1.7}}\log\left(\frac{350}{\mathrm{We}\left(\frac{Re_{fo}}{Re_{fo,ref}}\right)^{1.3}}\right)$
Homogeneous [Dukler et al. (1964)]	$\bar{\mu} = \mu_f\left(1 - \frac{x\rho_f}{x\rho_f + (1-x)\rho_g}\right) + \mu_g\frac{x\rho_f}{x\rho_f + (1-x)\rho_g}$
Previous Correlations Friedel (1979)	$\phi_{fo}^2 = (1-x)^2 + x^2[(\rho_f f_{go})/(\rho_g f_{fo})] + 3.24A_2A_3\mathrm{Fr}^{-0.045}\mathrm{We}^{-0.035}$ $A_2 = x^{0.78}(1-x)^{0.224}$, $A_3 = (\rho_f/\rho_g)^{0.91}(\mu_g/\mu_f)^{0.19}[1-(\mu_g/\mu_f)]^{0.7}$
Chisholm (1983)	$\phi_{fo}^2 = 1 + (\Gamma^2 - 1)[Bx^{0.875}(1-x)^{0.875} + x^{1.750}]$, where $\Gamma^2 = \frac{(dp/dz)_{go}}{(dp/dz)_{fo}}$ Γ / G (kg m⁻² s⁻¹) / B: ≤ 9.5: ≤ 500 → 4.8; $500 < G < 1900$ → $2400/G$; ≥ 1900 → $55/G^{0.5}$ $9.5 < \Gamma < 28$: ≤ 600 → $520/(\Gamma G^{0.5})$; > 600 → $21/\Gamma$ ≥ 28: - → $\Gamma^2 G^{0.5}$ $\phi_{fo}^2 = 1(4.3\Gamma^2 - 1)[\mathrm{N_{conf}}x^{0.875} + x^{1.75}]$ $\Gamma^2 = \frac{(dp/dz)_{go}}{(dp/dz)_{fo}}$ and $\mathrm{N_{conf}} = \frac{[\alpha/(g(\rho_f - \rho_g))]^{0.5}}{D}$

Previous Correlations	
Tran et al. (2000)	$\phi_{fo}^2 = 1(4.3\Gamma^2 - 1)[N_{conf}x^{0.875} + x^{1.75}]$ $\Gamma^2 = \frac{(dp/dz)_{go}}{(dp/dz)_{fo}}$ and $N_{conf} = \frac{\left[\alpha/\left(g(\rho_f - \rho_g)\right)\right]^{0.5}}{D}$
Zhang and Webb (2001)	$\phi_{fo}^2 = (1-x)^2 + 2.87x^2\left(\frac{P}{P_c}\right)^{-1} + 1.68x^{0.8}(1-x)^{0.25}\left(\frac{P}{P_c}\right)^{-1.64}$
Mishima and Hibiki (1996)	$\phi_f^2 = 1 - \frac{C}{X_{tt}} + \frac{1}{X_{tt}^2}$ $X_{tt} = \left(\frac{1-x}{x}\right)^{0.9}\left(\frac{\rho_g}{\rho_f}\right)^{0.5}\left(\frac{\mu_f}{\mu_g}\right)^{0.1}$ and $C(\text{turbulent} - \text{turbulent}) = 20$
Lockhart and Martinelli (1949)	$\phi_f^2 = 1 - \frac{C}{X_{tt}} + \frac{1}{X_{tt}^2}$ $X_{tt} = \left(\frac{1-x}{x}\right)^{0.9}\left(\frac{\rho_g}{\rho_f}\right)^{0.5}\left(\frac{\mu_f}{\mu_g}\right)^{0.1}$ and $C(\text{turbulent} - \text{turbulent}) = 20$

4.3 Heat transfer coefficient

Fig. 10(a) shows the effect of mass flux on heat transfer coefficient. Mass flux has an insignificant effect on the heat transfer coefficient in the low quality region. The insignificant effect of mass flux on the heat transfer coefficient indicates that nucleate boiling heat transfer is predominant. Several previous studies performed by Kew and Cornwell (1997), Lazarek and Black (1982), Wambsganss et al. (1993), Tran et al. (1996) and Bao et al. (2000) used small tubes and showed that, in small channels, nucleate boiling tends to be predominant. The significant nucleate boiling heat transmission is attributed to the physical qualities of refrigerants, specifically their surface tension and pressure, as well as the geometric influence of narrow channels. An elevated mass flux correlates with an increased heat transfer coefficient for moderate to high vapor quality, attributable to a rise in the convective boiling heat transfer contribution.

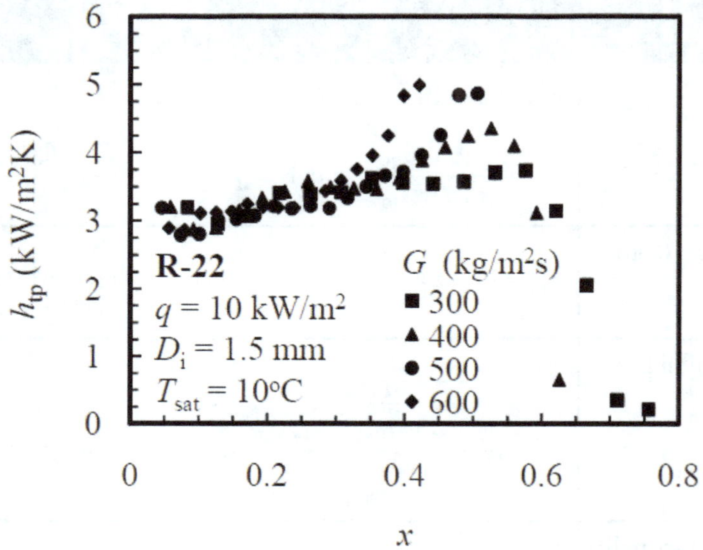

Fig. 10. (a) Effect of mass flux on heat transfer coefficient for R-22 at q = 10 kW/ m², D_i = 1.5 mm, T_{sat} = 10°C

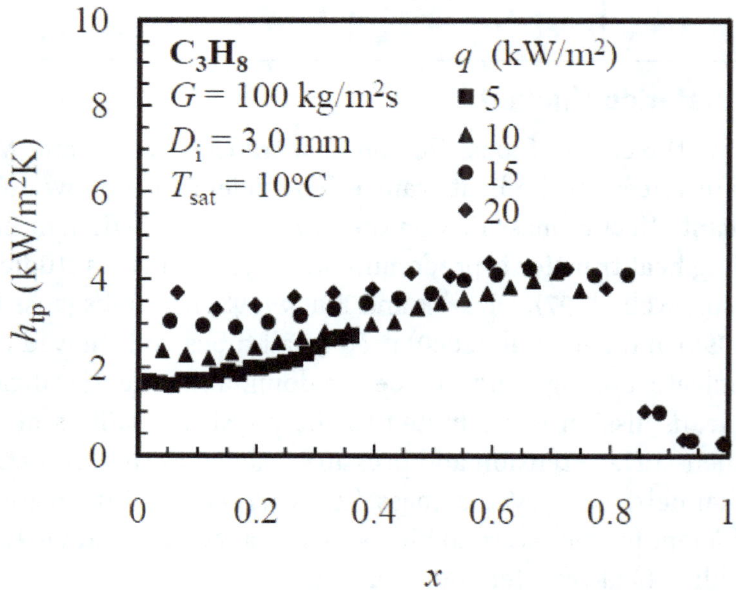

Fig. 10. (b) Effect of heat flux on heat transfer coefficient for C_3H_8 at G = 100 kg// (m²·s), D_i = 3.0 mm, T_{sat} = 10°C

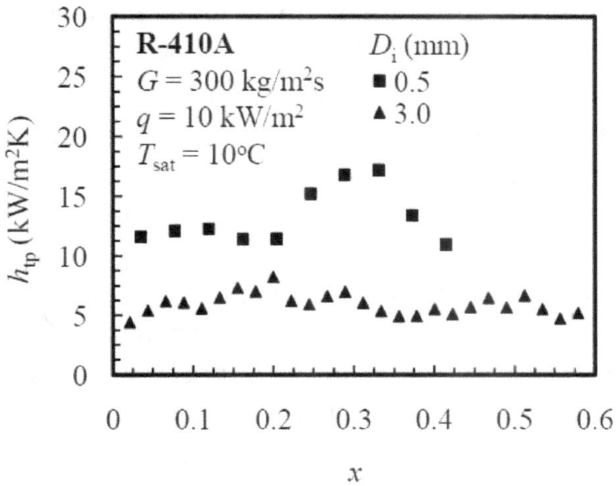

Fig. 10. (c) Effect of inner tube diameter on heat transfer coefficient for R-410A at G = 300 kg//(m^2·s), q = 10 kW/m^2, T_{sat} = 10°C

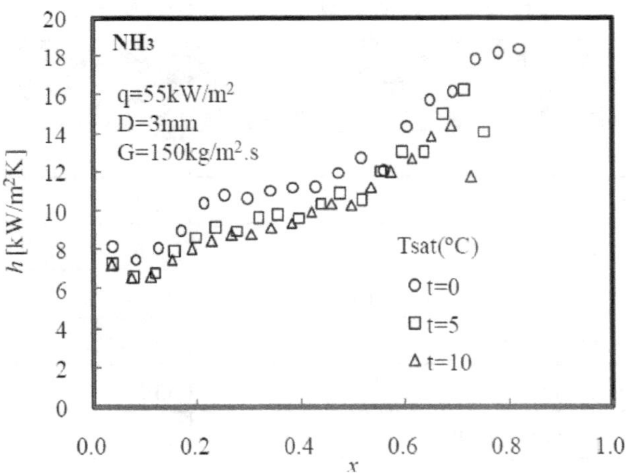

Fig. 10. (d) Effect of saturation temperature on heat transfer coefficient for NH_3 at G = 150 kg//(m^2·s), q = 55 kW/m^2, D_i = 3.0 mm

In the high quality region, a drop in the heat transfer coefficient occurs at lower qualities for a relatively higher mass flux. The steep decreasing of the heat transfer coefficient at high qualities is due to the effect of a small diameter tube on the boiling flow pattern because dry-patch occurs more easily in smaller diameter tubes and at a higher mass flux.

Fig. 10(b). depicts the dependence of heat flux on heat transfer coefficients in the low- moderate quality region. The large effect of heat flux on the heat transfer coefficient shows the domination of the nucleate boiling heat transfer contribution. At the higher quality region nucleate boiling is suppressed or convective heat transfer contribution is predominant; this is indicated by a low effect of heat flux on heat transfer coefficient.

Fig. 10 (c) shows that a smaller inner tube diameter has a higher heat transfer coefficient at low quality regions. This is due to a more active nucleate boiling in a smaller diameter tube. As the tube diameter gets smaller, the contact surface area for heat transfer increases. The more active nucleate boiling causes dry-patches to appear earlier. The quality for a rapid decrease in the heat transfer coefficient is lower for the smaller tube. It is supposed that the annular flow appears at a lower quality in the smaller tube and therefore, the dry-out quality is relatively lower for the smaller tube. The effect of saturation temperature on heat transfer coefficient is depicted in Fig. 10 (d). The heat transfer coefficient increases with an increase in saturation temperature, which is due to a larger effect from nucleate boiling. A higher saturation temperature provides a lower surface tension and higher pressure. The vapor formation in the boiling process explains that a lower surface tension and higher pressure provides a higher heat transfer coefficient.

Table 6. Mean deviation and average deviation calculated for the different pressure drop prediction methods

	Previous Correlation	Deviation	
		Mean (%)	Average (%)
K.-I. Choi et al. (2009)	Shah (1988)	19.21	3.55
	Tran et al. (1996)	21.18	-6.15
	Jung at al. (1989)	26.05	23.38
	Gungor and Winterton (1987)	28.44	23.78
	Takamatsu et al (1993)	32.69	32.15
	Kandlikar and Stainke (2003)	33.84	24.41
	Wattelet et al (1994)	48.28	48.28
	Chen (1966)	50.82	18.74
	Zhang et al. (2004)	79.21	77.89
K.-I. Choi et al. (2007)	Wattelet et al.	0.06	-3.03
	Gungore-Winterton	17.48	1.59
	Zhang et al	19.83	-13.35
	Kandlikare-Steinke	21.12	-15.43
	Tran et al.	22.31	-17.37
	Jung et al.	29.1	-23.44

	Previous Correlation	Deviation	
		Mean (%)	Average (%)
K.-I. Choi et al. (2007)	Wattelet et al.	19.09	12.22
	Jung et al.	23.48	9.53
	Kandlikare-Steinke	24.32	13.05
	Tran et al.	24.81	-10.51
	Shah	25.22	-12.29
	Gungore-Winterton	25.26	2.39
	Chen	36.13	-18.48
A.S. Pamitran et al (2007)	Zhang et al	27.45	-17.34
	Gungore-Winterton	31.07	-12.65
	Tran et al.	31.69	-28.52
	Takamatsu et al.	33.22	-20.65
	Jung et al.	33.39	-26.45
	Kandlikare-Steinke	38.59	-18.39
	Wattelet et al.	38.68	-31.48

Fig. 11 (a) and 11 (b) show the comparisons of the heat transfer coefficients of R-22, R-134a, R-410A, C_3H_8 and CO_2 at some experimental conditions. The mean heat transfer coefficient ratio of R-22, R-134a, R-410A, C_3H_8 and CO_2 was approximately 1.0 : 0.8 : 1.8 : 0.7 : 2.0.

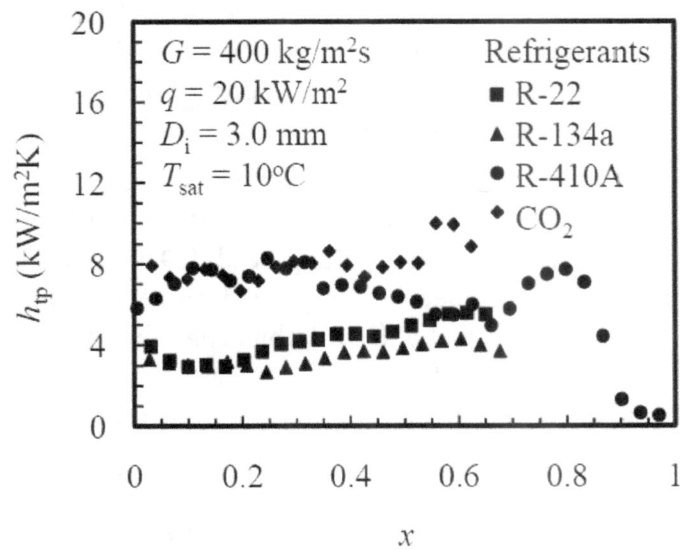

Fig. 11. (a) Heat transfer coefficient comparison of the present working refrigerants at G = 400 kg//(m²·s), q = 20 kW/m², D_i = 3.0 mm, T_{sat} = 10 °C

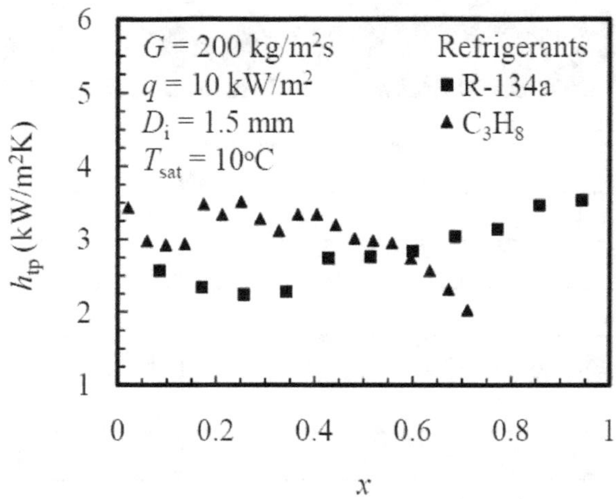

Fig. 11. (b) Heat transfer coefficient comparison of the present working refrigerants at G = 200 kg//(m²·s), q = 10 kW/m², D_i = 1.5 mm, T_{sat} = 10°C

The heat transfer coefficient of CO_2 was higher than that of the other working refrigerants during evaporation under all test conditions. The higher heat transfer coefficient of CO_2 is believed to be due to its high boiling nucleation. The CO_2 has much lower surface tension and applies much higher pressure than the other working refrigerants. The heat transfer coefficients of R-22, R-134a and C_3H_8 are similar due to their similar physical properties. The comparisons of the physical properties of the present working refrigerants are given in Table 3. The CO_2 has a much lower viscosity ratio μ_f/μ_g than the other working refrigerants, which means that the liquid film of CO_2 can break more easily than those of the other refrigerants. The CO_2 has also a much lower density ratio ρ_f/ρ_g than the other working refrigerants, which leads to a lower vapor velocity, which in turn causes less suppression of nucleate boiling, as shown.

The heat transfer coefficients of the present study are compared with the results given by several correlations for boiling heat transfer coefficients as shown in Table 6. The Gungor- Winterton (1987), Jung et al. (1989), Shah (1988) and Tran et al. (1996) correlations provided better predictions, with mean deviations of lower than 30% in A.S. Pamitran et al (2010), than the other correlations. Wattelet et al (1994) successively predict the experiment data for a,bK.I. Choi et al (2007). The Gungor-Winterton (1987) correlation was a modification of the superposition model; it was developed using fluids in several small and conventional tubes under various test conditions. The Jung et al. (1989) correlation was developed with pure and mixture refrigerants in conventional channels; its *F* factor contributed a big calculation deviation with the current experimental data. The Shah (1988) correlation was developed using a large data set for conventional tubes. The prediction with the Shah (1988) correlation was fair under conditions at the low quality region. The Tran et al. (1996) correlation was developed for R-12 and R-113 in small tubes. The correlations of Chen (1966) and Wattelet et al. (1994), which were developed for large tubes, have a high prediction deviation in K.I Choi et al.'s (2009) experimental data. The correlations of Kandlikar (1990) and Zhang et al. (2004) were developed for small tubes; however, the correlations could not predict well the present experimental data. The correlations of Wattelet et al. (1994), Kandlikar (1990) and Zhang et al. (2004) showed a large deviation in the prediction of the CO_2 data. The Kandlikar (1990) correlation failed to predict the heat transfer coefficient at the high quality region.

Table 7 shows the previous correlation use by A.S. Pamitran et al. (2007) and K.I Choi (2009), (2007)

Table 7. Some of previous correlations to predict heat transfer coefficient used by .S Pamitran et al. (2010) and (2007) and K.-I. Choi et al. (2009) and (2008)

Heat Transfer Coefficient Correlations	
Gungor and Winterton (1987)	$h_{tp} = E.h_l + S.h_{pb}$ $E = \text{fn}(Bo, X_{tt})$ and $S = \text{fn}(Bo, X_{tt}, Re_l)$ $h_l = h_{\text{Dittus-Boelter}}$, $h_{pb} = h_{\text{Cooper}}$
Jung et al. (1989)	$h_{tp} = \frac{N}{C_{un}} h_{un} + C_{me} F_p h_{lo}$ $N = \text{fn}(X_{tt}, Bo)$ $C_{un} = \text{fn}(X, Y, p, p_{cmvc})$ $h_{un} = \frac{h_1 h_2}{(h_1 X_2 + h_2 X_1)}$, h_1 and h_2 are $h_{sa} = h_{Stephan-Abdesalam}$ $C_{me} = \text{fn}(X, Y)$, $F_p = \text{fn}(X_{tt})$ $h_{lo} = h_{\text{Dittus-Boelter}}$
Shah (1988)	For horizontal tube, $Fr_f > 0.04$ $N = C_o = \left(\frac{1-x}{x}\right)^{0.8} \left(\frac{\rho_g}{\rho_f}\right)^{0.5}$ $h_{tp} = max(h_{nb}, h_c)$
Tran et al. (1996)	$h_{tp} = 8.4 \times 10^2 (Bo^2 We_f)^{0.3} \left(\frac{\rho_f}{\rho_g}\right)^{-0.4}$

Heat Transfer Coefficient Correlations	
Chen (1966)	$h_{tp} = S.h_{nb} + F.h_{lo}$ $h_{nb} = h_{Foster-Zuber}$ $h_{lo} = h_{Dittus-Boelter}$ $F = \text{fn}(X_{tt})$ and $S = \text{fn}(Re_{tp})$
Wattelet et al.(1994)	$h_{tp} = [h_{nb}^{n} + h_{cb}^{n}]^{1/2}$, $n = 2.5$ $h_{nb} = h_{Cooper}$ $h_{cb} = h_{Dittus-Boelter} \times F \times R$ $R = \text{fn}(X_{tt})$ and $R = \text{fn}(Fr_1)$
Kandlikar and Steinke (2003)	$\frac{h_{tp}}{h_l} = D_1 Co^{D_2}(1-x)^{0.8}\text{fn}(Fr_{lo}) + D_3 Bo^{D_4}(1-x)^{0.8} Fn$ $\text{fn}(Fr_{lo}) = 1$
Zang et al. (2004)	$h_{tp} = S.h_{nb} + F.h_{sp}$ for horizontal circular channel $h_{nb} = h_{Foster-Zuber}$ $h_{sp} = \max(h_{Dittus-Boelter})$, if $Re_f < 2300$ $h_{sp} = h_{Dittus-Boelter}$, if $Re_f \geq 2300$ $F = \text{fn}(\phi_f)$ and $S = \text{fn}(Re_f)$

5. DEVELOPMENT OF A NEW CORRELATION

5.1 Pressure drop

In this section, the effort of A.S. Pamitran et al and Choi et al. in developing a pressure drop correlation that is appropriate to use in specific refrigerants for the horizontal smooth minichannel is highlighted. The area average-gas fraction or void fraction denoted with α is defined as the ratio of gas phase cross-sectional area (Ag) to the total cross-sectional area. $A = A_g + A_f$

$$\alpha = \frac{A_g}{A} \tag{12}$$

In boiling and condensation it is often convenient to use the fraction of the total mass flow which is composed of vapor or liquid. The mass quality, *x*, is defined as

$$x = \frac{W_g}{W_g + W_f} = \frac{W_g}{W} \tag{13}$$

The rate of mass flow divided by the flow area is given in the name 'mass velocity' and the symbol *G*.

$$G = \frac{W}{A} = \rho . u = \frac{u}{v} \tag{14}$$

The mean velocity of the liquid phase is shown by

$$u_f = \frac{G(1-x)}{\rho_f . (1-\alpha)} \tag{15}$$

The mean velocity of the gas phase is shown by

$$u_g = \frac{Gx}{\rho_g . \alpha} \tag{16}$$

The total pressure drop consists of friction, acceleration and static head components as illustrated in the Equation below.

$$\left(\frac{dp}{dz}\right) = \left(\frac{dp}{dz}F\right) + \left(\frac{dp}{dz}a\right) + \left(\frac{dp}{dz}z\right)$$

For flow boiling in a horizontal layout test section, the static head pressure drop is excluded. Therefore, the experimental two-phase frictional pressure drop can be obtained by subtracting the calculated acceleration pressure drop from the measured pressure drop, as shown in Eq. (2), and may then be expressed in terms of the single-phase pressure drop for the liquid phase, considered to exist in the tube, as below.

$$\left(\frac{dp}{dz}F\right) = \left(\frac{dp}{dz}F\right)_f \phi_f^2 \tag{17}$$

The liquid frictional pressure drop is calculated from the Fanning equation. The frictional pressure drop can then be rewritten as Eq. (19) below.

$$\left(\frac{dp}{dz}F\right)_f = \frac{2f_f G^2(1-x)^2}{\rho_f D} \tag{18}$$

The friction factor in Eq. (7) was obtained by considering the flow conditions of laminar– turbulent flows, where f = 16Re-1 for Re < 2300 (laminar flow) and f = 0.079 Re-0.25 for Re > 3000 (turbulent flow). The laminar–turbulent transition Reynolds number was obtained from Yang and Lin (2007).

$$\left(\frac{dp}{dz}F\right) = \left[\frac{2f_f G^2(1-x)^2}{\rho_f D}\right] \phi_f^2 \tag{19}$$

A new modified pressure drop correlation was developed on the basis of the Lockhart– Martinelli method. The two-phase pressure drop of Lockhart–Martinelli consists of the following three terms: the liquid phase pressure drop, the interaction between the liquid phase and the vapor phase, and the vapor phase pressure drop. The relationship among these terms is expressed in Eq. (20).

$$\left(-\frac{dp}{dz}F\right)_{tp} = \left(-\frac{dp}{dz}F\right)_f + C\left[\left(-\frac{dp}{dz}F\right)_f \left(-\frac{dp}{dz}F\right)_g\right]^{1/2} + \left(-\frac{dp}{dz}F\right)_g \tag{20}$$

The two-phase frictional multiplier based on the pressure gradient for liquid alone flow, ¢2, is calculated by dividing Eq. (20) by the liquid phase pressure drop, as is shown in Eq. (21).

$$\phi_f^2 = \frac{\left(-\frac{dp}{dz}F\right)_{tp}}{\left(-\frac{dp}{dz}F\right)_f} = 1 + C\left[\frac{\left(-\frac{dp}{dz}F\right)_g}{\left(-\frac{dp}{dz}F\right)_f}\right]^{1/2} + \frac{\left(-\frac{dp}{dz}F\right)_g}{\left(-\frac{dp}{dz}F\right)_f} = 1 + \frac{C}{X} + \frac{1}{X^2} \tag{21}$$

The Martinelli parameter, X, is defined by the following equation:

$$X = \left[\frac{\left(-\frac{dp}{dz}F\right)_f}{\left(-\frac{dp}{dz}F\right)_g}\right]^{1/2} = \left[\frac{2f_f G^2(1-x)^2\rho_g/D}{2f_g G^2x^2\rho_f/D}\right]^{1/2} = \left(\frac{f_f}{f_g}\right)^{1/2}\left(\frac{1-x}{x}\right)\left(\frac{\rho_g}{\rho_f}\right)^{1/2} \tag{22}$$

The friction factor in Eq. (22) was obtained by considering the flow conditions of laminar (for Re < 1000, $f = 16\mathrm{Re}^{-1}$) and turbulent (for Re > 2000, $f = 0.079\mathrm{Re}^{-0.25}$).

The pressure drop prediction method can be summarized as follows:

$$\left(\frac{dp}{dz}\right) = \left(\frac{dp}{dz}F\right) + \left(\frac{dp}{dz}a\right)$$

$$\left(\frac{dp}{dz}\right) = \left(\frac{2f_f G^2(1-x)^2}{\rho_f D}\right)\varnothing_f^2 + G^2\frac{d}{dz}\left(\frac{x^2}{\alpha\rho_g} + \frac{(1-x)^2}{(1-\alpha)\rho_f}\right)$$

$$\left(\frac{dp}{dz}\right) = \left(\frac{2f_f G^2(1-x)^2}{\rho_f D}\right)\left(1 + \frac{C}{X} + \frac{1}{X^2}\right) + G^2\frac{d}{dz}\left(\frac{x^2}{\alpha\rho_g} + \frac{(1-x)^2}{(1-\alpha)\rho_f}\right)$$

$$\left(\frac{dp}{dz}\right) = \left(\frac{2f_f G^2(1-x)^2}{\rho_f D}\right)\left(1 + \frac{C}{\left(\frac{f_f}{f_g}\right)^{1/2}\left(\frac{1-x}{x}\right)\left(\frac{\rho_g}{\rho_f}\right)^{1/2}} + \frac{1}{\left(\frac{f_f}{f_g}\right)\left(\frac{1-x}{x}\right)^2\left(\frac{\rho_g}{\rho_f}\right)}\right) + G^2\frac{d}{dz}\left(\frac{x^2}{\alpha\rho_g} + \frac{(1-x)^2}{(1-\alpha)\rho_f}\right) \tag{23}$$

The calculated factor C is obtained from Chisholm (1967). For the liquid-vapor flow condition of turbulent-turbulent (tt), laminar-turbulent (vt), turbulent-laminar (tv) and laminar-laminar (vv), the values of the Chisholm (1967) parameter, C, are 20, 12, 10, and 5, respectively. The value of C in this thesis is obtained by considering the flow conditions of laminar and turbulent with thresholds of Re=2300 and Re=3000 for the laminar and turbulent flows, respectively. The laminar-turbulent transition Reynolds number was obtained from Yang and Lin (2007). Fig. 12 shows a comparison of the two-phase frictional multiplier data with the values predicted by the Lockhart-Martinelli correlation with C=5 and C=20. The figure shows that the present data are located at mostly between the baseline of C=5 and C=20, which means that laminar, turbulent and the co-current laminar-turbulent flows exist in the present data. For liquid-vapor flow conditions, the present experimental data shows 5.43% laminar-laminar, 28.46% laminar-turbulent, 3.78% turbulent-laminar, and.62.33% turbulent-turbulent.

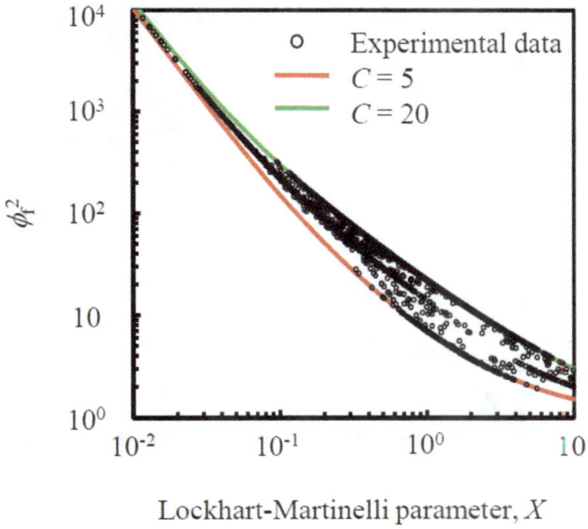

Fig. 12. Variation of the two-phase frictional multiplier data with the Lockhart-Martinelli parameter

The experimental result showed that the pressure drop is a function of mass flux, inner tube diameter, surface tension, density and viscosity, therefore the factor C in Eq. (21) and (23) will be developed as a function of the two-phase Weber number, We_{tp}, and the two-phase Reynolds number, Re_{tp}.

$$C = \left(\phi_f^2 - 1 - \frac{1}{X^2}\right) X = fn\left(We_{tp}, Re_{tp}\right) \tag{24}$$

Where We_{tp} and Re_{tp} are defined as:

$$We_{tp} = \frac{G^2 D}{\bar{\rho}\sigma} \tag{25}$$

$$Re_{tp} = \frac{GD}{\bar{\mu}} \tag{26}$$

A new factor C was developed using the regression method, as is shown in the Equations in the table below.

A new pressure drop correlation was developed on the basis of the Lockhart–Martinelli method as a function of the Weber number, Wetp, and the Reynolds number, Retp by considering the laminar–turbulent flow conditions. Using a regression method with 812 data points, a new factor C was developed by A.S. Pamitran et al. (2010) with mean and average deviations of 21.66% and -2.47%, respectively, based on comparison as shown in Fig. 13.

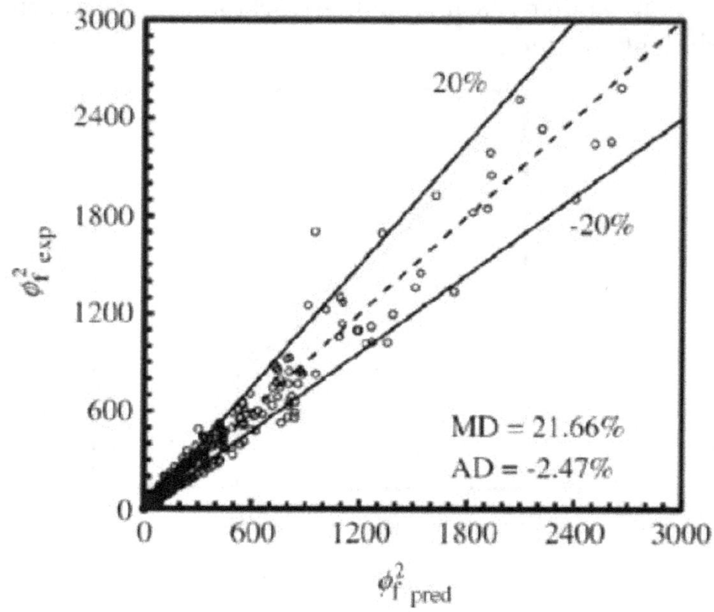

Fig. 13. Two-phase frictional multiplier comparison between the present experimental data ($\phi^2_{f\,exp}$) and the prediction with the newly developed correlation ($\phi^2_{f\,pred}$), A.S. Pamitran et al. (2010)

Table 8. Equations developed by A.S. Pamitran et al. and K.I. Choi et al

	Correlation	Deviation
Pamitran, A.S., et al. (2010)	$C = 3 \times 10^{-3} \times Re_{tp}^{1.23} We_{tp}^{-0.433}$	MD = 21.66% AD = -2.47%
Choi, K.-I., et al. (2009)	$C = 1732.953 \times Re_{tp}^{-0.323} We_{tp}^{-0.24}$	MD = 10.84% AD = 1.08%
Choi, K.-I., et al. (2008)	$C = 5.5564 \times Re_{tp}^{0.2837} We_{tp}^{-0.288}$	MD = 4.02 AD = -0.14%
Pamitran, A.S., et al. (2008)	$C = 1.2897 \times 10^{6} \times Re_{tp}^{0.5674} We_{tp}^{-3.3271}$	MD = 9.41% AD = -0.55%

This correlation will contribute to the design of heat exchangers with small tubes.

Fig. 14 illustrates the two-phase frictional multiplier comparison between the present experimental data and the prediction with the newly developed correlation. The comparison shows a mean deviation of 10.84% and an average deviation of 1.08%.

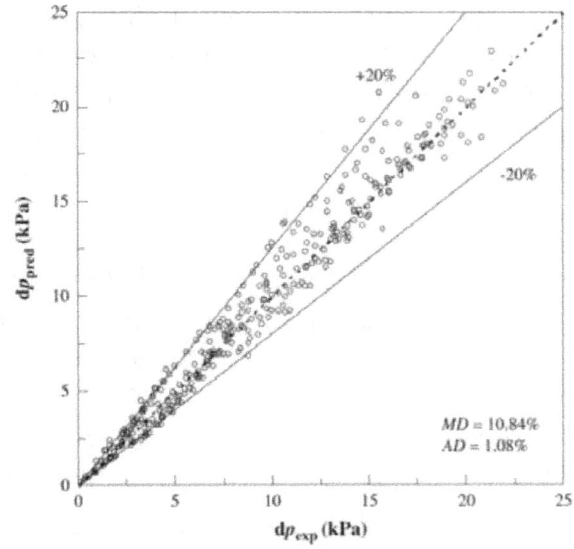

Fig. 14. Comparison of the experimental and predicted heat transfer coefficients using the new developed correlation. K.I. Choi et al (2009)

5.2 Heat transfer coefficient

It is well known that the flow boiling heat transfer is mainly governed by two important mechanisms, namely nucleate boiling and forced convective heat transfer.

$$h_{tp} = h_{nb} + h_c \tag{27}$$

In two-phase flow boiling heat transfer, the nucleate boiling heat transfer contribution is suppressed by the two-phase flow. Therefore, the nucleate boiling heat transfer contribution may be correlated with a nucleate boiling suppression factor, *S*. Another contribution of convective heat transfer may be correlated with a liquid single phase heat transfer. The *F* factor is introduced as a convective two-phase multiplier to account for enhanced convective properties due to co-current flow of liquid and vapor. A superposition model of heat transfer coefficient may be written as follows:

$$h_{tp} = S.h_{nbc} + F.h_f \tag{28}$$

The appearance of convective heat transfer for boiling in small channels occurs later than it does in large channels because of its high boiling nucleation.

The new heat transfer coefficient correlation in this study is developed by only using the experimental data prior to the dry-out. Chen (1966) introduced a multiplier factor, $F = fn(X_{tt})$, to account for the increase in the convective turbulence that is due to the presence of the vapor phase. The function should be physically evaluated again for flow boiling heat transfer in a minichannel that has a laminar flow condition, which is due to the small diameter effect. By considering the flow conditions (laminar or turbulent) in the Reynolds number factor, F, Zhang et al. (2004) introduced a relationship between the factor *F* and the two-phase frictional multiplier that is based on the pressure gradient for liquid alone flow, ϕ_f^2. This relationship is

$$F = fn(\phi_f^2) \tag{29}$$

Where ϕ_f^2 is a general form for four conditions according to Chisholm (1967), as is shown in Eq. (21). For the liquid–vapor flow condition of turbulent–turbulent (tt), laminar–turbulent (vt), turbulent–laminar (tv) and laminar–laminar (vv), the values of the Chisholm parameter, C, are 20, 12, 10, and 5, respectively. The value of C is found by an interpolation of the Chisholm parameter with thresholds of Re = 1000 and Re = 2000 for the laminar and turbulent flows, respectively.

The Martinelli parameter is defined as in equation (22) together with the Blasius equation of friction factors, f_f **and** f_g; the Martinelli parameter can be rewritten as

$$\left(\frac{f_f}{f_g}\right)^{1/8}\left(\frac{1-x}{x}\right)^{7/8}\left(\frac{\rho_g}{\rho_f}\right)^{1/2} \tag{30}$$

There is an important effect of quality, density ratio, p_f/p_g, and the viscosity ratio, μ_f/μ_g, on heat transfer coefficient. The liquid heat transfer is defined by existing liquid heat transfer coefficient correlations by considering flow conditions of laminar and turbulent. For laminar flow, Ref < 2300, where

$$Re_f = \frac{G(1-x)D}{\mu_f} \tag{31}$$

And the liquid heat transfer coefficient is obtained from the following correlation:

$$h_f = 4.36\frac{k_f}{D} \tag{32}$$

For flow with 3000 :: Ref :: 104, the liquid heat transfer coefficient is obtained from the Gnielinski (1976) correlation:

$$h_f = \frac{(Re_f-1000)Pr_f\left(\frac{f_f}{2}\right)\left(\frac{k_f}{D}\right)}{1+12.7\left(Pr_f{}^{2/3}-1\right)\left(\frac{f_f}{2}\right)^{0.5}} \tag{33}$$

where the friction factor is calculated from (for Re < 1000, $f = 16Re^{-1}$) and turbulent (for Re > 2000, $f = 0.079Re^{-0.25}$). For flow with 2300 :: Ref :: 3000, the liquid heat transfer coefficient is calculated by interpolation. For turbulent flow with 10^4 :: Re_f :: $5x10^6$, the liquid heat transfer coefficient is obtained from the Petukhov and Popov (1963) correlation:

$$h_f = \frac{Re_f Pr_f \left(\frac{f_f}{2}\right)\left(\frac{k_f}{D}\right)}{1 + 12.7\left(Pr_f^{2/3} - 1\right)\left(\frac{f_f}{2}\right)^{0.5}} \tag{34}$$

The Dittus Boelter (1930) correlation is used for turbulent flow with Re_f ;: $5x10^6$.

$$h_f = 0.023\frac{k_f}{D}Re_f^{0.8}Pr_f^{0.4} = 0.23\frac{k_f}{D}\left[\frac{G(1-x)D}{\mu_f}\right]^{0.8}\left[\frac{Cp_f\mu_f}{k_f}\right]^{0.4} \tag{35}$$

The F factor in this study is developed as a function of ϕ_f^2, F = fn(ϕ_f^2), where ϕ_f^2 is obtained from Eqs. (21) – (22). The liquid heat transfer is defined by the Dittus Boelter correlation and a new factor F, as is shown in Fig. 15(a)-(d), is developed using a regression method.

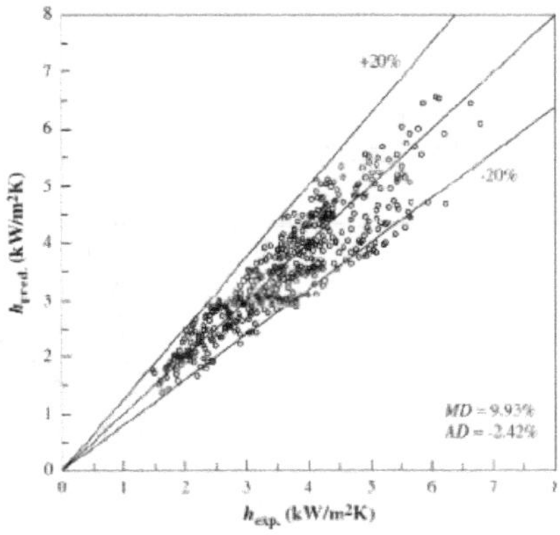

Fig. 15. (a) Comparison of the experimental and predicted heat transfer coefficients using the new developed correlation. K.I. Choi et al. (2009)

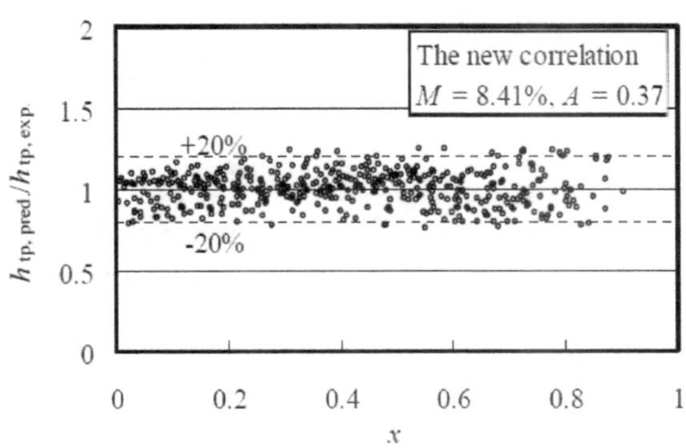

Fig. 15. (b) Diagram of the ratio of the experimental heat transfer coefficient, htp,exp, and the predicted heat transfer coefficient, $h_{tp,pred}$, vs. quality, x. [a]K.I. Choi et al. (2007)

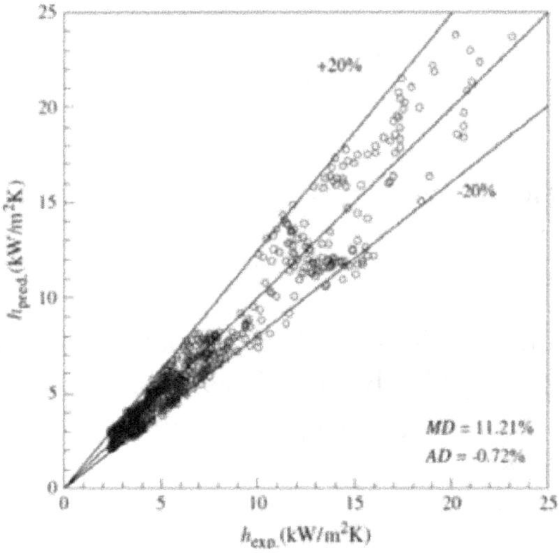

Fig. 15. (c) Diagram of the experimental heat transfer coefficient, $h_{exp.,}$ vs prediction heat transfer coefficient, $h_{pred.}$ [b]K.I. Choi et al. (2007)

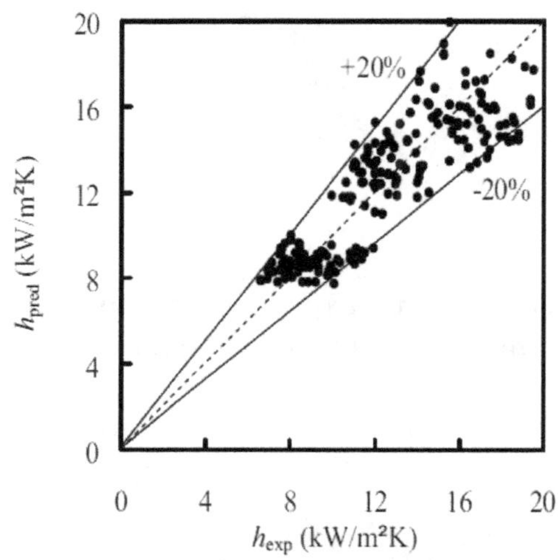

Fig. 15. (d) Diagram of experimental heat transfer coefficient h_{exp} vs prediction heat transfer coefficient h_{pred}. A.S. Pamitran et al (2007)

The prediction of the nucleate boiling heat transfer for the present experimental data used Cooper (1984), which is a pool boiling correlation developed based on an extensive study.

$$h = 55p_r^{0.12}(-0.4343 ln p_r)^{-0.55} M^{-0.5} q^{0.67} \tag{36}$$

Where the heat fluxes, q, is inWm^{-2}. Kew and Cornwell (1997) and Jung et al. (2003) showed that the Cooper (1984) pool boiling correlation best predicted their experimental data. Chen (1966) defined the nucleate boiling suppression factor, S, as a ratio of the mean superheat, ΔT_e, to the wall superheat, ΔT_{sat}. Jung et al. (1989)

proposed a convective boiling heat transfer multiplier factor, *N*, as a function of quality, heat flux and mass flow rate (represented by employing Xtt and Bo) to represent the strong effect of nucleate boiling in flow boiling as it is compared with that in nucleate pool boiling, h_{nbc}/h_{nb}. The Martinelli parameter, X_{tt}, is replaced by a two-phase frictional multiplier, ϕ_f^2, in order to consider laminar flow in minichannels. By using the experimental data of this study, a new nucleate boiling suppression factor, as a ratio of h_{nbc}/h_{nb}, is proposed as showen in table 9.

A.S. Pamitran et al. and K.I. Choi et al have developed a correlation based on the Zhang et al (2004) that modified the *F* factor and Chen (1966) introduced the suppression factor, *S*. The new heat transfer coefficient correlation is developed using a regression method with 461 data points for C_3H_8 in Choi, K.I et al (2009), 471 data points for CO_2 in aK.-I. Choi et al (2007), 681 data points for R-22, R-134a, and R-744 (CO_2) in bK.I. Choi et al.(2007), and 217 data points for R-410A in A.S. Pamitran et al. (2007).

The comparison of the experimental heat transfer coefficient, $h_{tp,exp}$, and the predicted heat transfer coefficient, $h_{tp,pred}$, for propane is illustrated in Fig. 15(a). The new correlation agrees closely for the comparison with a mean deviation of 9.93% and an average deviation of -2.42%. Fig. 15(b) illustrates CO_2 on the comparison with a mean deviation of 8.41% and an average deviation of 0.37%.

The comparison of the experimental heat transfer coefficient, h_{exp}, and the prediction heat transfer coefficient, h_{pred}, is shown in Fig. 15(c). The new correlation for three refrigerants R- 22, R-134a and R-744 (CO_2) showed good agreement with a mean deviation of 11.21% and an average deviation of -0.72%. Fig 14 (d), the new correlation for 410A also show a suitable range of mean deviation, 11.20% and an average deviation 0.09%; Table 9 gives a summary of the new correlation.

Table 9. Equations developed by A.S. Pamitran et al. and K.I. Choi et al

	F factor	Supression factor	Deviation
Choi, K.-I. et al. (2009)	F = AX(0 5 ϕ_f, 1)	$S = 181.458(\phi_f^2)^{0.002} Bo^{0.816}$	MD = 9.93% AD = 2,42%
aK.-I. Choi et al. (2007)	$F = 0.05\phi_f^2 + 0.95$	$S = 7.2694(\phi_f^2)^{0.0094} Bo^{0.2814}$	MD = 8.41% AD = 0.37%
bK.-I. Choi et al. (2007)	$F = 0.042\phi_f^2 + 0.958$	$S = 469.1689(\phi_f^2)^{-0.2093} Bo^{0.7402}$	Overall MD = 11.21% AD = -0,72%
A.S. Pamitran et al. (2007)	$F = 0.062\phi_f^2 + 0.938$	$S = 9.4626(\phi_f^2)^{-0.2747} Bo^{0.1285}$	MD= 11.20% AD = 0.09%

6. CONCISE SUMMARY

The pressure drop and heat transfer experiments in convective boiling performed with R-22, R-134a, R-410A, R-744 (CO_2), R-717 (NH3) and R-290 (C_3H_8) in horizontal smooth minichannels. It can be explained, that, the pressure drop is higher at the conditions of higher mass and heat fluxes, and for the conditions of smaller inner tube diameter and lower saturation temperature. The experimental results showed that pressure drop is a function of mass flux, inner tube diameter, surface tension, density and viscosity. The new pressure drop correlations were developed on the basis of the Lockhart–Martinelli method as a function of the two-phase Reynolds number, Retp, and the two-phase Weber number, Wetp. The new factor *C* was developed using a regression method Mass flux, heat flux, inner tube diameter and saturation temperature have an effect on the heat transfer coefficient. The heat transfer coefficient increases with a decreased inner tube diameter and with an increased saturation temperature. The geometric effect of the small tube must be considered to develop a new heat transfer coefficient correlation.

The two-phase flow pattern was mapped in the Wang et al (1997) and the Wotjan et al.(2005) flow pattern maps. The Wotjan et al. flow pattern map illustrates the formation of intermittent, annular, dry-out, and mist flows until the dry-out. The Wang et al (1997) flow pattern map illustrates the data start from annular flow, Str. wavy to intermittent flow.

Laminar flow appears for flow boiling in small channels, so the modified correlation of the multiplier factor for the convective boiling contribution, *F*, and the nucleate boiling suppression factor, *S*, is developed in the study using laminar and turbulent flows consideration. New correlations for the boiling heat transfer coefficient, derived from a superposition model for refrigerants in minichannels, were introduced. The development of the new correlations utilized 2,288 data points. This manuscript's documentation aims to elucidate two-phase flow boiling heat transfer and pressure drop in horizontal circular channels, serving as foundational knowledge for applications in refrigeration and similar domains involving two-phase flow. No singular correlation exists that can accurately forecast heat transfer and pressure drop in two-phase flow. Each individual or blended refrigerant may vary due to pressure drop correlations and heat transfer correlations. Consequently, the revision of the prior correlations or the emergence of a new development in this regard remains possible.

7. NOMENCLATURE

A Cross section area

AD Average Deviation, $AD = \left(\frac{1}{n}\right)\sum_{1}^{n}\left(\left(dp_{pred} - dp_{exp}\right)\times 100/dp_{exp}\right)$ *for pressure drop or* $AD = \left(\frac{1}{n}\right)\sum_{1}^{n}\left(\left(h_{pred} - h_{exp}\right)\times 100/h_{exp}\right)$ *for heat transfer coefficient*

Bo Boiling number, $Bo = q/Gi_{fg}$

C Correction factor for two-phase pressure drop

c_p **Specific heat capacity at constant pressure (kJ/(kg-K))**

D Diameter (m)

E Electric potential (V)

F Multiplier factor for convective heat transfer contribution

F Friction factor

G Mass flux ($kg/(m^2 \cdot s)$)

g Acceleration due to gravity (m/S_2)

h Heat transfer coefficient (kW/(m2·K))

I Electric current (A)

i Enthalpy (kJ/(kg·K))

k Thermal conductivity (kW/(m·K))

L Tube length (m)

M Molecular weight of the liquid

MD Mean Deviation, $MD = \left(\frac{1}{n}\right)\sum_{1}^{n}\left|\left((dp_{pred} - dp_{exp}) \times 100 / dp_{exp}\right)\right|$ *for pressure drop or* $MD = \left(\frac{1}{n}\right)\sum_{1}^{n}\left|\left((h_{pred} - h_{exp}) \times 100 / h_{exp}\right)\right|$ *for heat transfer coefficient*

n Number of data

P Pressure (N/m^2)

Pr Prandtl number, $Pr = c_p\mu/k$ Q Electric power (kW)

q Heat flux (kW/m^2)

$\dot{q}$ Heat generation (kW/m^3)

R Electrical resistance (ohm)

Re Reynolds number, Re = GD/μ

R_p Surface roughness parameter (μm)

S Suppression factor

T Temperature (K)

V Volume (m^3)

W Mass flow rate (kg/s)

We Webber number,

X Lockhart-Martinelli parameter

x Mass quality

z Length (m)

Greek letters

λ Correction factor on Baker (1954) flow pattern map, $\lambda = \left[\left(\rho_g/\rho_A\right)\left(\rho_f/\rho_W\right)\right]^{1/2}$

a Void fraction

μ Viscosity (Pa·s)

p Density (kg/m^3)

σ Surface tension (N/m)

φ^2 Two-phase frictional multiplier

ψ' Correction factor on Wang et al. (1997) flow pattern map, $\psi = (\sigma_W/\sigma)^{1/4}\left[(\mu_f/\mu_W)(\rho_W/\rho_f)^2\right]^{1/3}$

Gradients and differences

(dp/dz) Pressure gradient (N/m^{2}m)

(dp/dz F) Pressure gradient due to friction (N/m^{2}m)

(dp/dz a) Pressure gradient due to acceleration (N/m^{2}m)

(dp/dz z) Pressure gradient due to static head (N/m^{2}m)

Subscripts

A Air

c Convective

exp Experimental value

f Saturated liquid

g Saturated vapor

i Inner tube

fo Liquid only

nb Nucleate boiling

nbc Nucleate boiling contribution

o Outer tube

pb Pool boiling

pred Predicted value

red Reduced

sat Saturation

sc Subcooled

t Turbulent flow

tp Two-phase

v Laminar flow

W Water

w Wall

8. REFERENCES

Ali, M. I., Sadatomi, M. and Kawaji, M., 1993, "Two-phase flow in narrow channels between two flat plates", Can. J. Chem. Eng. 71, pp. 657-666.

Baker, O., 1954, "Design of pipe lines for simultaneous flow of oil and gas", Oil and Gas J. July, 26.

Bao, Z. Y., Fletcher, D. F., Haynes, B. S., 2000, "Flow boiling heat transfer of freon R11 and HCFC123 in narrow passages", Int. J. Heat and Mass Transfer 43;3347-3358.

Baroczy, C. J., 1963, "Correlation of liquid fraction in two-phase flow with application to liquid metals", NAA-SR-8171. Fluid Sci. 28;111-121.

Beattie, D. R. H. and Whalley, P. B., 1982, "A simple two-phase flow frictional pressure drop calculation method", Int. J. Multiphase Flow 8;83-87.

Chang Y. J., Chiang S. K., Chung T. W., Wang C. C., 2000, "Two-phase frictional characteristics of R-410A and air-water in a 5 mm smooth tube", ASHRAE Trans; DA-00-11-3;792-797.

Chang, S. D. and Ro, S. T., 1996, "Pressure Drop of Pure HFC Refrigerants and Their Mixtures Flowing in Capillary Tubes", Int J. Multiphase Flow 22(3); 551-561.

Chen I. Y., Yang K. S., Chang Y. J., Wang C. C., 2001, "Two-phase pressure drop of air-water and R-410A in small horizontal tubes", Int. J. Multiphase Flow 27;1293-1299.

Chen, J. C., 1966, "A correlation for boiling heat transfer to saturated fluids in convective flow", Industrial and Engineering Chemistry, Process Design and Development 5;322-329.

Chisholm, D., 1967, "A theoretical basis for the Lockhart-Martinelli correlation for two- phase flow", Int. J. Heat Mass Transfer 10;1767-1778.

Chisholm, D., 1968, "The influence of mass velocity on friction pressure gradients during steam-water flow", Paper 35 presented at Thermodynamics and Fluid Mechanics Convention, Institutes of Mechanical Engineers (Bristol), March.

Chisholm, D., 1972, "An equation for velocity ratio in two-phase flow", NEL Report 535.

Chisholm, D., 1983, "Two-phase flow in pipelines and heat exchangers", New York: Longman. Chisholm, D. and Sutherland, L. A., 1969, "Predicted of pressure gradients in pipeline systems during two-phase flow", Paper 4 presented at Symposium on Fluid Mechanics and Measurements in Two-phase Flow Systems, Leeds, 24-25, September.

Cho, J. M. and Kim, M. S., 2007, "Experimental studies on the evaporative heat transfer and pressure drop of CO2 in smooth and micro-fin tubes of the diameters of 5 and 9.52 mm", Int. J. Refrigeration 30;986-994.

Choi K-I A.S. Pamitran, Chun-Young Oh, Jong-Taek Oh, 2007, "Boiling heat transfer of R-22, R-134a, and CO2 in horizontal smooth minichannels", Int. J. of Refrigeration, Vol. 30, 1336-1346.

Choi K-I, A. S. Pamitran, Chun-Young Oh, Jong-Taek Oh, 2008 "Two-phase pressure drop of R-410A in horizontal smooth minichannels", Int. J. of Refrigeration, Vol. 31, 119-129.

Choi K-I. A.S. Pamitran, Jong-Taek Oh, 2007, "Two-phase flow heat transfer of CO2 vaporization in smooth horizontal Minichannels", Int. J. of Refrigeration, vol. 30, 767-777.

Choi K-I. A.S. Pamitran, Jong-Taek Oh, Kiyoshi Saito, 2009, "Pressure drop and heat transfer during two-phase flow vaporization of propane in horizontal smooth minichannels", Int. J. of Refrigeration, vol. 32, 837-845.

Cicchitti, A., Lombardi, C., Silvestri, M., Solddaini, G., Zavalluilli, R., 1960, "Two-phase cooling experiments- Pressure drop, heat transfer and burnout measurement", Energia Nucl 7(6);407-425.

Cooper, M. G., 1984, "Heat flow rates in saturated nucleate pool boiling–a wide-ranging examination using reduced properties", In: Advances in Heat Transfer. Academic Press 16;157-239.

Dittus, F. W. and Boelter, L. M. K., 1930, "Heat transfer in automobile radiators of the tubular type", University of California Publication in Engineering 2;443-461.

Dukler, A. E., Wicks, I. I. I. M., Cleveland, R. G., 1964, "Pressure drop and hold-up in two- phase flow", AIChE J. 10(1);38-51.

Friedel, L., 1979, "Improved friction pressure drop correlations for horizontal and vertical two-phase pipe flow", Presented at the European Two-phase Flow Group Meeting, Ispra, Italy, Paper E2, June.

Gnielinski, V., 1976, "New equations for heat and mass transfer in turbulent pipe and channel flow", International Chemical Engineering 16: 359-368

Grönnerud, R., 1979, "Investigation of Liquid Hold-Up, Flow-Resistance and Heat Transfer in Circulation Type Evaporators", Part IV: Two-phase flow resistance in boiling refrigerants, Annexe 1972-1, Bull. de l'Inst. du Froid, International Inst. of Refrigeration, Paris.

Gungor, K. E., Winterton, H. S., 1987, "Simplified General Correlation for Saturated Flow Boiling and Comparisons of Correlations with Data", Chem.Eng.Res 65; 148-156.

Jung, D. S., McLinden, M., Radermacher, R., Didion, D., 1989, "A Study of Flow Boiling Heat Transfer with Refrigerant Mixtures", Int J. Mass Transfer 32(9);1751-1764.

Jung, D., Kim, Y., Ko, Y., Song, K., 2003, "Nucleate boiling heat transfer coefficients of pure halogenated refrigerants", Int. J. Refrigeration 26;240-248.

Kandlikar, S. G., 1990, "A general correlation for saturated two-phase flow boiling heat transfer inside horizontal and vertical tubes", Journal of Heat Transfer 112;219-228.

Kandlikar, S. G., 2002, "Fundamental issues related to flow boiling in minichannels and microchannels", Experimental Thermal and Fluid Science 26;389-407.

Kattan N., 1996, "Contribution to the heat transfer analysis of substitute refrigerants in evaporator tubes with smooth or enhanced tube surfaces", PhD thesis No 1498, Swiss Federal Institute of Technology, Lausanne, Switzerland.

Kattan, N., Thome, J. R., Favrat, D., 1998, "Flow boiling in horizontal tubes: part 1 – development of a diabatic two-phase flow pattern map", J. Heat Transfer 120;140-147.

Kawahara, A., Chung, P. M. Y., Kawaji, M., 2002, "Investigation of two-phase flow pattern, void fraction and pressure drop in a microchannel", Int J. of Multiphase Flow 28;1411-1435.

Kew, P. A., Cornwell, K., 1997, "Correlations for the Prediction of Boiling Heat Transfer in Small-Diameter Channels", Applied Thermal Engineering 17(8-10);705-715.

Kuo, C. S. and Wang C. C., 1996, "In-tube evaporation of HCFC-22 in a 9.52 mm micro- fin/smooth tube", Int. J. Heat Mass Transfer 39;2559-2569.

Lazarek, G. M. and Black, S. H., 1982, "Evaporative heat transfer, pressure drop and critical heat flux in a small diameter vertical tube with R-113", Int. J. Heat Mass Transfer 25;945-960.

Lockhart, R. W. and Martinelli, R. C., 1949, "Proposed correlation of data for isothermal two- phase, two-component flow in pipes", Chem. Eng. Prog. 45;39-48.

McAdams, W. H., 1954, "Heat transmission", third ed. New York: McGraw-Hill.

Mishima, K. and Hibiki, T., 1996, "Some characteristics of air-water two-phase flow in small diameter vertical tubes", Int. J. Multiphase Flow 22;703-712.

Müller-Steinhagen, H. and Heck, K., 1986, "A simple friction pressure drop correlation for two-phase flow in pipes", Chemical Engineering and Processing 20(6);297-308

Oh, H. K., Ku, H. G., Roh, G. S., Son, C. H., Park, S. J., 2008, "Flow boiling heat transfer characteristics of carbon dioxide in a horizontal tube", Applied Thermal Engineering 28;1022-1030.

Ould Didi, M. B., Kattan, N., Thome, J. R., 2002, Prediction of Two-phase Pressure Gradients of Refrigerants in Horizontal Tubes, Int J. Refrig. 25;935-947.

Pamitran A.S, Kwang-Il Choi, Jong-Taek Oh, Hoo-Kyu Oh, 2008, "Two-phase pressure drop during CO2 vaporization in horizontal smooth minichannels" Int. J. of Refrigeration, Vol. 31, 1375–1383.

Pamitran A.S. Kwang-Il Choi, Jong-Taek Oh, Hoo-Kyu Oh, 2007, "Forced convective boiling heat transfer of R-410A in horizontal minichannels", Int. J. of Refrigeration, Vol. 30,155-165.

Pamitran A.S. Kwang-Il Choi, Jong-Taek Oh, Pega Hrnjak., 2010, "Characteristics of two- phase flow pattern transitions and pressure drop of five refrigerants in horizontal circular small tubes", Int. J. of Refrigeration, vol. 33, 578-588.

Park, C. Y. and Hrnjak, P. S., 2007, "CO2 and R410A flow boiling heat transfer, pressure drop, and flow pattern at low temperatures in a horizontal smooth tube", Int. J. Refrigeration 30;166-178.

Peng, X. F. and Peterson, G. P., 1996, "Forced convective heat transfer of single-phase binary mixtures through microchannels", Experimental Thermal and Fluid Science 12;98-104.

Pettersen, J., 2004, "Flow vaporization of CO2 in microchannels tubes", Exp. Therm.

Petukhov, B. S. and Popov, V. N., 1963, "Theoretical calculation of heat exchanger in turbulent flow in tubes of an incompressible fluid with variable physical properties", High Temp. 1(1);69-83.

Shah, M. M., 1988, "Chart correlation for saturated boiling heat transfer: equations and further study", ASHRAE Trans 2673;185-196.

Smith, S. L., 1969, "Void fractions in two phase flow: a correlation based upon an equal velocity head model", Proc. Inst. Mech. Engrs, London 184;647–657 Part 1.

Steiner, D., 1993, "Heat transfer to boiling saturated liquids", VDI-Wärmeatlas (VDI Heat Atlas), Verein Deutcher Ingenieure, ed., VDI-Gessellschaft Verfahrenstechnik und Chemieinge- nieurwesen (GCV), Düsseldorf, Germany, (J.W. Fullarton, translator).

Tran, T. N., Chyu, M. C., Wambsganss, M. W., France, D. M., 2000, "Two-phase pressure drop of refrigerants during flow boiling in small channels: An experimental investigation and correlation development", Int. J. Multiphase Flow 26;1739-1754.

Tran, T. N., Wambsganss, M. W., France, D. M., 1996, "Small circular- and rectangular-channel boiling with two refrigerants", Int. J. Multiphase Flow 22(3);485-498.

Wambsganss, M. W., France, D. M., Jendrzejczyk, J. A., Tran, T. N., 1993, "Boiling Heat Transfer in a Horizontal Small-diameter Tube", AMSE Trans 115;963-975.

Wang, C. C., Chiang, C. S., Lu, D. C., 1997, "Visual observation of two-phase flow pattern of R- 22, R-134a, and R-407C in a 6.5-mm smooth tube", Exp. Therm. Fluid Sci. 15;395-405.

Wattelat, J. P., Chato, J. C., Souza, A. L., Christoffersen, B. R., 1994, "Evaporative Characteristics of R-12, R-134a, and a Mixture at Low Mass Fluxes", ASHRAE Trans 94-2-1;603-615.

Wojtan, L., Ursenbacher, T., Thome, J. R., 2005, "Investigation of flow boiling in horizontal tubes: part I – a new diabatic two-phase flow pattern map", Int. J. Heat Mass Transfer 48;2955-2969.

Yan, Y. Y., Lin, T. F., 1998, "Evaporation Heat Transfer and Pressure Drop of Refrigerant R- 134a in a Small Pipe", Int J. of Heat and Mass Transfer 41;4183-4194.

Yang, C. Y. and Lin, T. Y., 2007, "Heat transfer characteristics of water flow in microtubes", Experimental Thermal and Fluid Science 32(2);432-439.

Yoon, S. H., Cho, E. S., Hwang, Y. W., Kim, M. S., Min, K., Kim, Y., 2004, "Characteristics of evaporative heat transfer and pressure drop of carbon dioxide and correlation development", Int. J. Refrigeration 27;111-119.

Yu, W., France, D. M., Wambsganss, M. W., Hull, J. R., 2002, "Two-phase Pressure Drop, Boiling Heat Transfer, and Critical Heat Flux to Water in a Small-diameter Horizontal Tube", Int J. of Multiphase Flow 28;927-941.

Yun, R., Kim, Y., Kim, M. S., 2005, "Convective boiling heat transfer characteristics of CO2 in microchannels", Int. J. Heat. Mass Transfer 48;235-242.

Zhang, L., Hihara, E., Saito, T., Oh, J. T., 1997, "Boiling Heat Transfer of a Ternary Refrigerant Mixture inside a Horizontal Smooth Tube", Int J. Mass Transfer 40(9);2009-2017.

Zhang, M. and Webb, R. L., 2001, "Correlation of two-phase friction for refrigerants in small- diameter tubes", Experimental Thermal and Fluid Science 25;131-139.

Zhang, W., Hibiki, T., Mishima, K., 2004, "Correlation for flow boiling heat transfer in mini- channels", Int. J. Heat and Mass Transfer 47;5749-5763.

Zhao, Y., Molki, M., Ohadi, M. M., Dessiatoun, S. V., 2000, "Flow boiling of CO_2 in microchannels", ASHRAE Trans DA-00-2-1;437-445.

Zivi, S. M., 1964, "Estimation of Steady-State Steam Void-Fraction by Means of the Principle of Minimum Entropy Generation", J. Heat Transfer 86;247-252.

11
Chapter

ANALYSIS OF THE IMPACT OF AIRFLOW AND PRODUCT CONFIGURATION ON THE FREEZING PROCESS VIA CONVECTIVE HEAT TRANSFER COEFFICIENT ASSESSMENT

1. INTRODUCTION: AN OVERVIEW

The energy level can influence the ultimate cost of food items, making the reduction of freezing time a primary objective for enterprises and researchers. The thermal load of the product often represents the highest demand on the refrigeration system, with the majority of heat removal occurring during the initial hours of the process, particularly during the freezing phase. Food goods are mostly frozen in air blast freezing tunnels. Air blast systems operate on the principle of fluid convection, wherein cooling air circulates around the product requiring temperature reduction. Convection is the transmission of thermal energy through the movement of molecules inside a substance. The rate of heat exchange between the bulk of a fluid and a known solid is given by (1):

$$\frac{dQ}{dt} = h_c A(T_b - T_\infty) \tag{1}$$

where Q is the amount of energy (J) drawn back per time t (s); hc is the convection heat transfer coefficient ($W/m^2{}^\circ C$); A is the heat transfer area (m^2); Tb is the product temperature (oC) and T_∞ is the air temperature (oC).

The convective heat flow is enhanced as the fluid motion increases. Air blast cooling processes are majorly ruled by the convective heat transfer, which relates the amount of transferred energy from the product surface to the cooling air. Wide variations in convective heat transfer coefficients may occur in different positions.

Determination of this parameter can be a useful tool to measure the efficiency of the process.

The freezing process of substantial quantities of food goods in commercial contexts can be highly intricate. Items requiring cooling impede airflow, hence compromising heat transfer. An exhaustive examination of product attributes and processing parameters is essential, since improper sample placement and insufficient air circulation may result in extended cooling durations. If the system is improperly constructed, issues concerning product or process quality may arise, including insufficient cooling or freezing rates, microbial proliferation, modifications in consistency, and compositional changes due to chemical reactions during subsequent storage. It is essential to examine appropriate packaging materials and the potential presence of air voids within the package, together with unsuitable product layout, which may result in inadequate heat transfer and ineffective temperature reduction. To achieve uniform heat transfer, it is essential to ensure appropriate sample placement inside the system and sufficient air circulation. Consequently, utilizing exhausted air may constitute a cost-effective option for air blast systems, since it reduces air short-circuiting, resulting in more consistent cooling. Numerous implementations of forced air devices to facilitate cooling and freezing processes have been documented. The predominant technique involves utilizing a forced-air freezing tunnel that improves air circulation around the items. Consequently, recent reports include detailed investigations into objective approaches for identifying optimal designs for cooling and freezing systems. Air velocity and air temperature are the characteristics that receive the most attention from designers of air blast systems. It is essential to evaluate the airflow pathways, as product dispersion may directly influence the effectiveness of cooling and freezing processes. Evaluating the efficacy of freezing methods is a challenging endeavor. Recent study indicates that the convective coefficient, among other factors, might serve as an effective metric for comparing various processing conditions. This chapter aims to examine the impact of key parameters on the freezing time of food samples by measuring the convective heat transfer coefficient, given the intricacies of the industrial process. The discussion will focus on results from comparative experiments regarding various airflow configurations and sample arrangements inside the system, illustrating how minor modifications to product placement and airflow direction can influence heat transfer.

2. REVIEW OF LITERATURE

2.1 Cooling and freezing processes for food products

In light of the swift rise in the production and consumption of frozen foods, there is an increasing focus on assessing the thermal properties of food during the freezing process to enhance system development and optimize processing equipment (Scott et al. 1992). Temperature lowering procedures attempt to inhibit microbial growth, hence prolonging the shelf life of perishable food items. The processes of freezing and cooling are governed by the thermal exchange between the product and the cooling medium (Welti-Chanes et al., 2005). Certain scholars have proposed models to evaluate the expenses associated with various food freezing techniques (Becker & Fricke, 1999; Chourot et al., 2003). The procedures of chilling and freezing food are intricate. The freezing of food mostly relies on the quantity of water contained within the meal that will undergo freezing during the process. Before freezing, sensible heat must be extracted from food until it attains the freezing temperature. Supercooling transpires when the temperature falls below the freezing point, initiating crystal nucleation. In pure water, nucleation releases heat, resulting in a temperature increase to the initial freezing point, which remains constant until all the water transforms into ice. Nevertheless, it diminishes marginally in meals, owing to the rising concentration of solutes in the unfrozen water component. Subsequently, during the phase transition of water to ice, latent heat is extracted from the solidified substance. The process commences when the majority of freezable water has transformed into ice and concludes when the temperature is lowered to the storage temperature (Zaritzky, 2000).Among most popular frozen food products are fruits and fruit pulps, normally used as raw material for processing industries as ice cream, yogurt, jams and others. Such products can be frozen in batch or continuous processing (Salvadori & Mascheroni, 1996). In order to make cooling processes financially affordable, studies based on information from manufacturers and users are necessary to design refrigeration equipment in accordance with the demanded application.

2.2 Heat transfer for cooling process

Cooling time is directly influenced by the ratio between the resistances to heat transfer inside and outside the sample. This ratio is called the Biot number, defined by Equation 1:

$$Bi = \frac{hL_c}{k_b} \tag{2}$$

where h is the convective heat transfer coefficient (W/m^2 °C), Lc is the characteristic length of the body (m), usually defined as the volume of the body divided by the surface area in contact with the cooling medium, and kb is the thermal conductivity of the body (W/m °C). Values of Biot number close to zero (Bi → 0), imply that the heat conduction inside the body is much faster than the heat convection away from its surface, and temperature gradients are negligible within the body

(Heldman, 1992). This condition allows for the applicability of certain methods of solving transient heat transfer problems. Within this condition, it can be assumed a lumped-capacitance model of transient heat transfer, leading to Newtonian cooling behaviour since the amount of thermal energy in the body is directly proportional to its temperature, which can be assumed uniform throughout the body. This leads to a simple first-order differential equation which describes heat transfer in these systems.

When the Biot number is very large ($Bi \to \infty$), the internal resistance to heat transfer is much larger and it only can be assumed that the surface temperature is equal to the cooling medium, but not the interior of the body. For this situation, solutions of the equation of Fourier heat transfer are useful. When the Biot number is within the range of $0.1 < Bi < 40$, both internal and external resistance should be considered.

The intrinsic factors of the products, such as thermal conductivity and diffusivity, are immutable. Therefore, the minimization of cooling and freezing duration must be accomplished by modifying system factors, including temperature, cooling air velocity, and product configuration.

2.3 Factors influencing cooling duration

The cooling or freezing rate is a critical parameter in the design of freezing systems, as it directly influences product quality. In industrial applications, it is the most critical parameter for comparing various types of systems and equipment. In addition to product qualities, the duration of temperature removal is influenced by various elements in the heat transfer process, including the dimensions and apertures of the packaging and the properties of the cooling medium. The expense of the cooling procedure is contingent upon the cooling rate, which is directly influenced by the packaging's opening area for air circulation, bed depth, temperature, and the velocity of the cooling air (Baird et al., 1988). The freezing systems can be categorized based on various heat removal techniques. Forced air freezers are among the most utilized systems. This technology has numerous advantages, including straightforward installation and operation, process efficiency, and a diverse range of items that can be cooled using this equipment. The optimal cooling method selection is contingent upon the intended use and influenced by various aspects, including the needed cooling rate, subsequent storage conditions, and the expenses associated with equipment and operation. Well-designed systems can enhance efficiency and save operational costs (Talbot & Chau, 1998; ASHRAE, 2002).Talbot & Fletcher (1996) compared the efficiency between an air blast system and a storage chamber. During the cooling process of grapes, there was a reduction of 6.7°C in one hour and 14.6°C after 2.5 hours, compared to a decrease of only 2 °C in one hour and 3.5°C in 2.5 hours in the storage chamber. Experiments carried out using a prototype portable forced-air device offered promising results (Barbin et al., 2009). The system was designed to be used inside cooling and freezing chambers, aiming to improve heat flow rates.

2.4 Air flow

Cooling time by forced air systems is determined by airflow and product thermal load, which affects the amount of energy to move the air around the product and inside the system. The most common industrial applications use direct cold air insufflations inside the system. The airflow varies according to the speed and amount of air flowing through products and its variation results in longer or shorter freezing time. A correct orientation of the air flow inside the equipment and around the product can significantly reduce processing times (Cortbaoui et al., 2006). Surface area of contact between products and cooling air and products arrangement are other parameters that affect forced air cooling (Baird et aL, 1988; Fraser, 1998; Laguerre et al., 2006).

In industrial plants, air flow is highly turbulent due to the fans movement and to wakes originated from upstream obstacles. Resende & Silveira Jr. (2002b) showed that the air velocity in forced air tunnels are strongly influenced by any changes in the amount of product inside the system, causing the air to flow through preferential paths, leading to increased freezing time and poor heat transfer coefficients. Results show that variations in heat transfer coefficients may occur according to the product positioning inside the equipment. Vigneault et al. (2005) studied how gravity influences air circulation in horizontal air flow, showing that low levels of air flows may be more affected, causing temperature variation of the cooling air and reducing the flow to the upper chamber. This could be an important parameter to consider when designing cooling and freezing systems. Exhausting air is more appropriate to avoid air to flow through preferential corridors, leading to more uniform heat exchange when compared to insufflations processes (Fraser, 1998).

Conventional cooling methods by forced air are an efficient alternative to removing the heat load of fruits and vegetables during post-harvest cooling. Air exhaustion is widely used for this purpose, as it improves the air distribution in the products surrounding. This system is usually used inside cooling chambers. With the fan in operation, it creates a low pressure region surrounding the products. The cooling air flows through this region between the small opening areas, reducing the product temperature (Talbot & Fletcher, 1996; Talbot & Chau, 1998; Fraser, 1998; Abrahao, 2008). The possibility of adapting a cold room for use as a system for forced air represents an economical advantage of this process (Talbot & Fletcher, 1996).

Baird et al. (1988) showed that the velocity of cooling air influences directly the operational cost of cooling systems, as it can change with the increase of air velocity in the system. The lowest costs were obtained with air velocities between 0.1 and 0.3 m/s. To study the cooling of plastic balls filled with a solution of carrageen, Allais et al. (2006) showed that increasing the speed of air flow, ranging from 0.25 m/s to 6 m/s, reduced the half-time cooling of samples from 800 s to 500 s. But this variation is exponential, and the reduction tends to be smaller from speed of 2 m/s. Results obtained by Vigneault et al. (2004a, b) for cooling process using forced air show that air flows above 2 l/s.kg and air velocities of insufflations higher than 0.5 m/s cause no influence on half-cooling time of the samples.

2.5 Devices and techniques for quantifying air flow velocity

Numerous methodologies for quantifying air flow velocity are documented in the literature, each based on distinct concepts, and the precision of the sensors employed in these techniques differs considerably, rendering them appropriate for specific applications. The hot-wire anemometer is a widely utilized instrument due to its extensive applicability. Owing to their compact dimensions and rapid response times, these instruments are ideal for the meticulous examination of fluid dynamics and are frequently employed to assess airflow in ventilation and air conditioning systems. The hot-wire anemometer quantifies the instantaneous velocity of fluids. The anemometer's core consists of an exposed, extremely fine hot wire heated by a steady current to a temperature exceeding the ambient level. The airflow surrounding the wire has a cooling influence on it. By assessing the variation in wire temperature under a constant current, a correlation can be established between the wire's resistance and the fluid flow velocity, given that the electrical resistance of most metals is contingent upon their temperature. This instrument possesses superior spatial resolution relative to alternative measuring techniques, making it suitable for the meticulous examination of turbulent flows or scenarios where rapid velocity fluctuations are pertinent.Resende & Silveira Jr. (2002b) and Nunes et al. (2003) suggest that several measurements in the cross-section of the air flow could lead to a more accurate determination of the air velocity. The average velocity is therefore used for determination of the air flow in the selected position, according to (3):

$$\dot{V} = \int_S \overline{v} dS \tag{3}$$

In this equation, V is the flow (m^3/s), S is the total area (m^2) and v is the velocity vector (m/s).

2.6 Packaging and storage

Packaging influences the heat transfer coefficients of food products in multiple ways. It serves as a barrier to energy transmission from food by functioning as insulation, hence reducing the heat transfer coefficient. Packaging can generate air-filled holes and bubbles surrounding the food item, which enhances insulation and reduces the heat transmission coefficient (Becker and Fricke, 2004). The findings of Santos et al. (2008) indicated that the freezing process of beef in cardboard boxes is undervalued, and the duration of processing is occasionally insufficient for all samples to attain the requisite temperature. Substituting cardboard boxes with metal perforated boxes resulted in a reduction of up to 45% in freezing time for this product.Becker and Fricke (2004) developed an iterative algorithm to estimate the surface heat transfer coefficients of irregularly shaped food items based upon their cooling curves, considering the density of the food item and the packaging. This algorithm extends to irregularly shaped food items existing techniques for the calculation of the surface heat transfer coefficient previously applicable to only regularly shaped food items, taking into account the concept of equivalent heat transfer dimensionality. In this method, the density used to calculate the heat transfer coefficient is affected by the packaging, as it is calculated from the mass of the food item plus the packaging and the outside dimensions of the package around the food item, generating results for the heat transfer coefficient for the food within its packaging.

An important parameter for improved performance of an air blast cooling system is the apertures and gaps that the packages and pallets must have to allow the circulation of the cold air through the packed product in order to achieve rapid and uniform heat transfer between the cooling air and the product (Vigneault et al. 2004a; Zou et al., 2006a, b).

Results obtained by Talbot & Fletcher (1996) and Abrahao (2008) showed the importance of proper cooling system design, proving that the larger the opening area in the packaging, the lower the requirement on refrigeration and air circulation systems to obtain a more uniform cooling rate. Meana et al. (2005) showed that the empty regions between the plastic containers that are used in the cooling of strawberries by forced air influence significantly the cooling time of the products. According to Baird et al. (1988), opening areas smaller than 10% of the total area of the box can significantly increase the cost of cooling processes. Castro et al. (2003) suggest that an opening area of 14% is appropriate for a rapid and uniform cooling process.

Extensive openings may result in inadequately built containers that are unsuitable for industrial processing. The primary objective is to achieve an appropriate opening area of the boxes to facilitate a reduced freezing time while preserving the mechanical integrity of the boxes.

2.7 Convective heat transfer coefficients (hc)

Convective heat transfer is related to the amount of energy transferred from the product surface when it is in contact with the refrigerating fluid (Welty et al, 2000).

Dincer (1995a) determined the experimental heat transfer coefficient with data obtained during forced air cooling of figs in air blast systems, with results varying from 21.1 to 32.1 Wm-2°C-1 for air velocities of 1.1 to 2.5 ms-1.

Experiments carried out in a forced air room with air velocities in the range of 1 to 2 ms-1 resulted in hc values varying from 28 Wm-2°C-1 up to 52 Wm-2°C-1 for cylindrical products (cucumber) during cooling (Dincer and Genceli, 1994). Mohsenin (1980) obtained hc values in the range of 20 to 35 Wm-2°C-1 for forced air systems with air velocity from 1.5 to 5.0 ms-1. Dussán Sarria et al. (2006) studied the influence of the air velocity in a cooling tunnel. According to the authors, air velocities greater than 2.0 ms-1 did not affect the convective coefficients (hc), as results obtained were not greater than 23.8 Wm-^{2}°C-1.

Considering the complexity of freezing processes and the recent results presented, many parameters influence the experimental results for heat transfer coefficients, thus varying according to the flow characteristics of the cooling medium and the products involved.

Accurate descriptions of the boundary conditions are rather difficult for industrial air blast systems, and software solutions such as CFD will not be effective in solving the momentum and heat transport equation without precise information (Mohamed, 2008). Regarding the wide range of convective heat transfer coefficients (hc) reported, it is important to calculate this coefficient in order to understand different operating conditions of distinct cooling systems and compare to any new systems developed. Several methods for convective heat transfer measurements are reported. The most common are those involving temperature measurements in permanent and transient state (Cleland, 1990).

2.7.1 Temperature measurements in steady state

This method involves establishing a steady thermal load in the system, such as an electrical heating probe. The heat transfer coefficient can be determined using the surface area of the heating probe, the energy input, and the temperatures of the cooling medium and the probe. Nonetheless, maintaining a steady temperature and velocity of the cooling medium poses significant challenges under experimental conditions, hence constraining the applicability of this approach.

2.7.2 Temperature measurements in transient state

Temperature measurement in transient state consists of a metallic test body with a known high thermal conductivity being used to minimize the temperature gradient during the heat exchange between the cooling medium and the product ($Bi<0.1$), allowing the test body to have an almost uniform temperature during the cooling process. When the internal resistance of the test body to heat transfer is neglected, an energy balance conducts to the convective heat transfer coefficient.

By Newton's cooling law, the rate of heat transfer in a given volume of control is given by equation 1.

The variation of energy in a metal body with constant properties is given by the equation:

$$\frac{dQ}{dt} = \rho_m V c_{pm} \frac{dT}{dt} \tag{4}$$

where ρm is the density, V is the volume and cpm is the specific heat of the metallic body, respectively. Combining equations 1 and 4, integrating and adopting the initial boundary condition T (t = 0) = T_i, leads to the solution for the temperature variation as a function of time:

$$\frac{T_b - T_\infty}{T_i - T_\infty} = e^{\frac{-hcAt}{\rho_m c_{pm} V}} \tag{5}$$

Equation 5 proves that the cooling process has an exponential behaviour, as verified by several authors for horticultural products (Mohsenin, 1980; Dincer, 1995a).

In practice, this method consists of using a test body made of some material with high thermal conductivity, so that tests are carried out without phase change and assuming the constant thermal properties within temperature variation range. Le Blanc et al. (1990a, b), Resende et al. (2002), Mohamed (2008) and Barbin et al. (2010) reported experiments using the described method for obtaining convective coefficients from the cooling curves obtained for a metallic test body, indicating the capability of the present method in handling complex boundary situation such as encountered in industrial systems. Results for convective heat transfer coefficients were reported by Barbin et al. (2010), comparing two air flow direction in the same equipment, concluding that this is a useful method for studying temperature reduction processes.

According to Resende et al. (2002), some points arise when using this method. If the test body consists of a metal block, there may be heat transfer through the edges of the material, affecting the values of hc calculated. Furthermore, condensation can occur in experiments with cooling air, causing changes to the measurements. Thus, the positioning of the test body must be carefully chosen in order to prevent any condensation of water during the tests, and the edges of the body or other parts that may interfere with the temperature measurements during the process must be perfectly insulated to avoid heat transfer through these regions.

3.METHODS

3.1 Samples for simulation of thermal load

Food model system with 15% (weight / weight of solution) of sucrose and 0.5% (weight / weight of solution) of carboxyl-methyl cellulose (Carbocel AM, Arinos, SP, Brazil) was packed in polyethylene bags (0.1 kg) with similar dimensions (0.095 m x 0.07 m x 0.015 m) to pulp fruit products in the market. The samples were stored in 35 plastic boxes (Figure 1a), with external dimensions of 0.6 m x 0.4 m x 0.12 m, which were stacked on a commercial pallet (1.00 m x 1.20 m, Figure 1b) and kept inside the freezing room. The boxes had an opening area of 21% of the total area, accounting for more than the minimum values recommended for proper air flow (Castro et al., 2003).

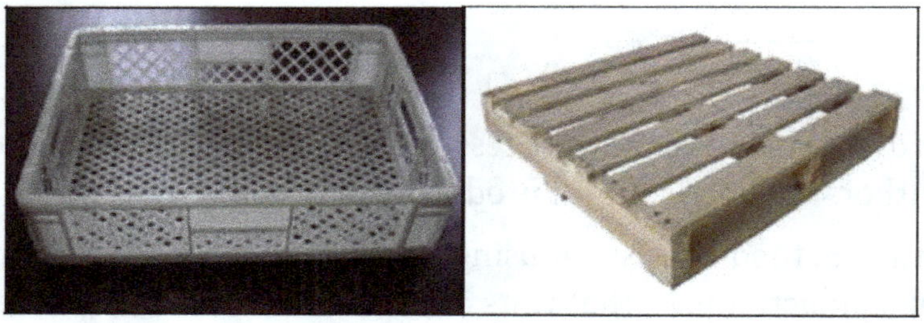

Fig. 1. (a) Plastic box for freezing products; (b) Commercial pallet

3.2 Product arrangement

Two arranges of samples were tested to determine the influence of opening areas to the refrigeration process. Using the industrial arrangement of samples, ninety six packages of sample were allocated in each box, in three layers (top, middle and bottom), with thirty two packs in each layer, corresponding to about 9.6 kg of product, similar to the amount used in the industrial process. This assembly is shown in Figure 2 (Arrangement 1). The boxes were piled in six layers with five boxes per layer, totalling thirty boxes, simulating a commercial assembly of a pallet used regularly in the process (Figure 3, Arrangement 1).

A second distribution of packages inside the boxes was tested, with larger distance between the packages inside the boxes in order to improve the circulation of air around the samples. In this assembly, eighty four packs of sample were allocated in each box, within five layers three layers with twenty units, and two layers of twelve units, distanced from each other for air circulation, totaling 8.4 kg of product per box.

Figure 2 (Arrangement 2) shows the new distribution of packaging inside the boxes. The second arrangement had a smaller amount of samples in each box. Hence, it was added another layer of boxes in the system in order to have the same amount of product and the same thermal load for all the tests (Figure 3, Arrangement 2).

Fig. 2. Schematic diagram for sample distribution used for packaging boxes.

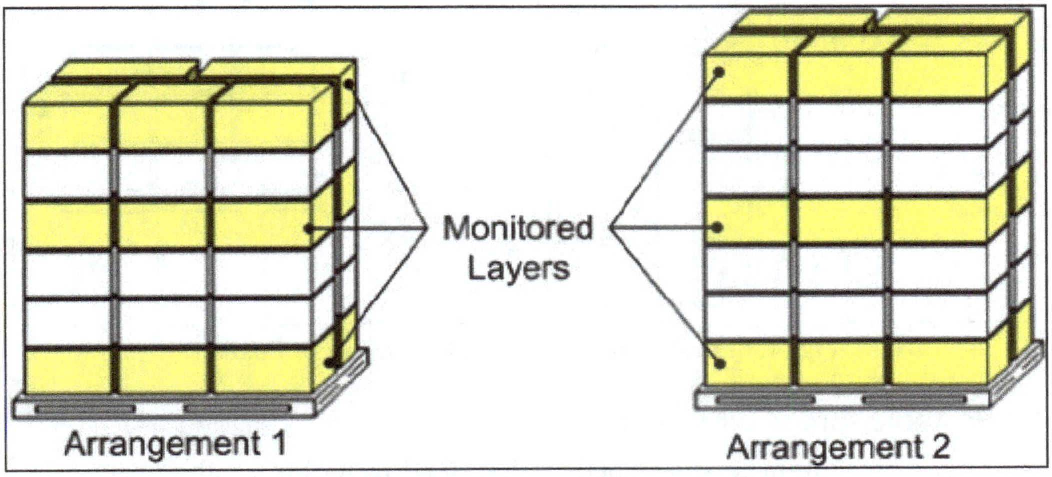

Fig. 3. Pallet with boxes and layers that were monitored

3.3 Thermal assessment

Insufflations and exhaustion air tests were run in triplicate. The velocity of the cooling air was measured for comparison with the convective coefficients. Three layers of boxes had its temperature monitored until the centre of the samples reached -18ºC. Each of the monitored layers had four thermocouples in the corners of the layers and one in the middle, as shown in Figure 4. The thermocouples were inserted inside the samples in the plastic bags to measure the samples temperatures variation during the freezing process.

The monitoring system used for temperature acquisition is composed by an automatic channel selector system (Scanner 706, Keithley Instruments Inc.). Samples temperatures were monitored using T-type thermocouples (copper-constantan). The thermocouples were calibrated using a controlled temperature bath with a propylene glycol solution and a standard thermometer as reference. Five different temperature values were chosen (-19°C, - 10°C, 0°C, 10°C and 20°C).

The average temperatures measured in the water bath (10 measurements for each one of the five different chosen values) were plotted against the corresponding thermocouple (mV) values (ASTM, 1989). The difference for the correlation coefficient of the curve-fitted line (R2) were not lower than 0.99.

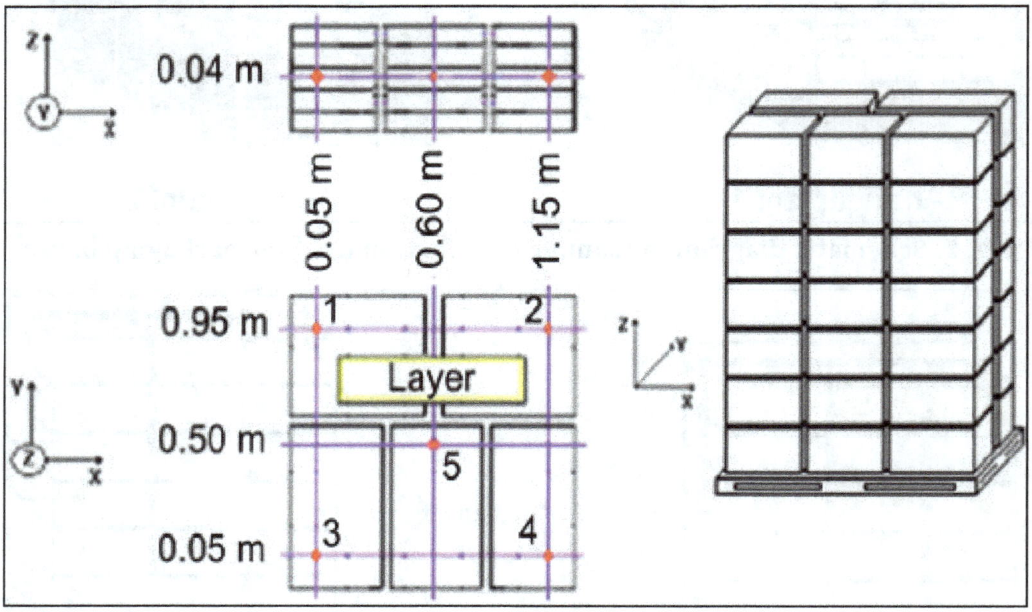

Fig. 4. Samples monitored in the layers of boxes

3.4 Portable forced air system

The forced air system was designed as described in Barbin et al. (2009), with a plastic sheet cover connected to a flexible duct and a fan that insufflates or exhausts the air inside the system. The plastic covers the boxes that contain the product, stacked on a commercial pallet. The portable tunnel fan used has axial airscrews with a tri-phase induction engine (Weg, Brazil, model 71586, 0.5 hp). The device was placed inside a freezing storage room (Recrusul, Brazil), with internal dimensions of 3 m x 3 m x 2.3 m (20.7m3) and walls made of 0.01 m aluminium panels filled with expanded polystyrene as insulation.

The cooling method involves circulating the interior air of the storage room through the open areas of the boxes and around the product samples. During the exhaustion process, the system is linked to the fan's suction, allowing air to move from the lower section of the system into the boxes and then back to the room via the fan. During the insufflation procedures, the airflow is modified, directing ambient cooled air directly onto the product. The forced air circulation is angled vertically during both the exhaust and insufflation processes. During exhaustion the flow proceeds from the base to the apex of the pallet; conversely, in the blowing process, it traverses from the apex to the base (Figure 5).

3.5 Air flow measurement

A hot-wire anemometer (Tri-Sense, model EW-37000-00, Cole-Parmer Instrument Company, IL, USA) was used for measurements of air velocity. The sensor was inserted through openings in the air diffuser for measuring air velocity in different positions of the area normal to the air flow. The measurement points were aligned and positioned at regular distances. The sensor was introduced for measuring the air speed with different depth of insertion, providing fixed points in the surface area perpendicular to the airflow.

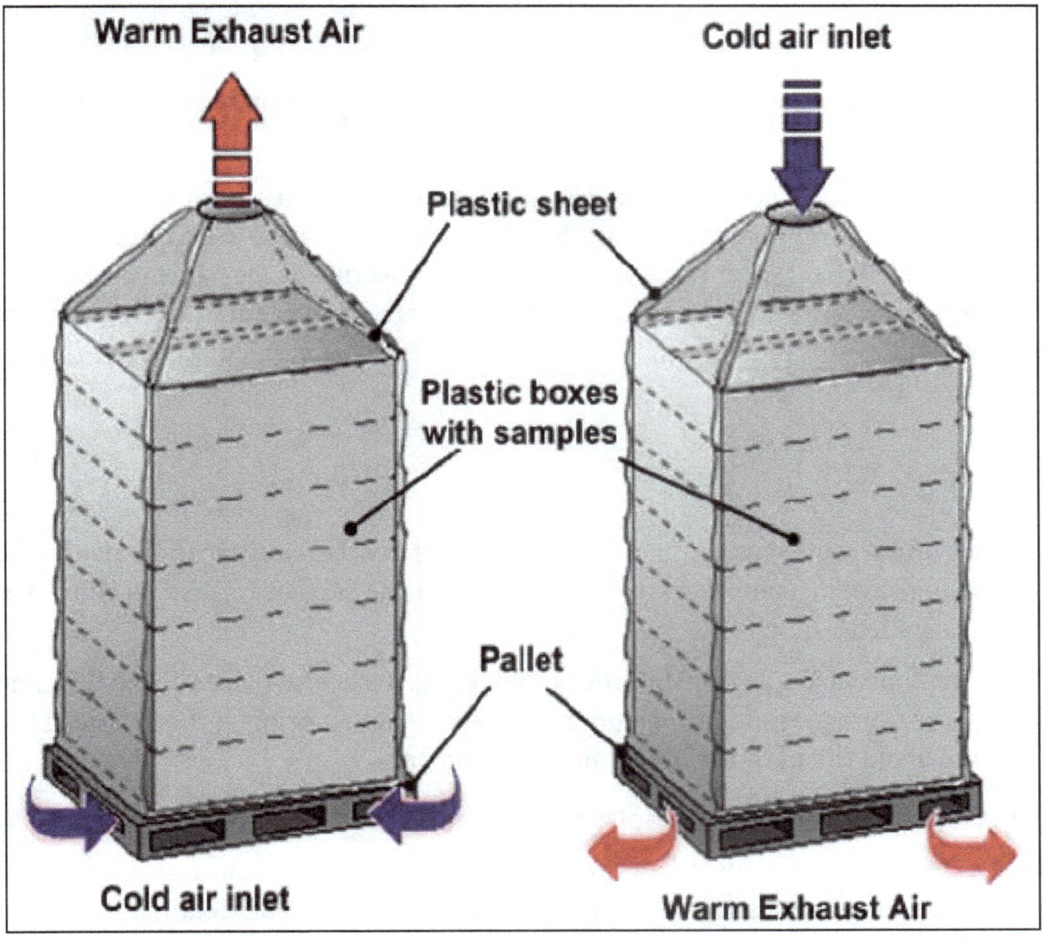

Fig. 5. Portable tunnel with boxes stacked on a commercial transport pallet covered with plastic, and air flow orientation during the exhaustion and insufflations processes.

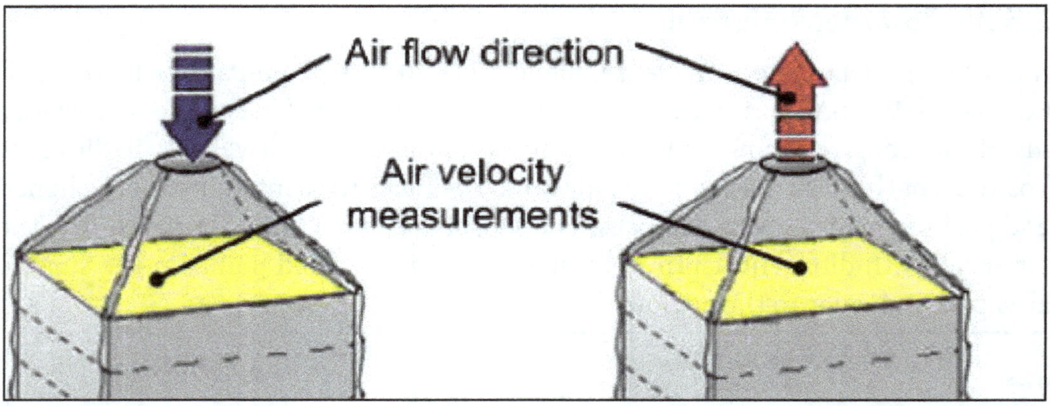

Fig. 6. Surface area for air velocity measurements during insufflations and exhaustion processes.

Velocities were measured for comparison between the exhaustion and insufflations with the fan operating at steady state and no obstructions to the air flow. In this study, the area that the air flows through is a cross-section of the pallet represented by the five boxes. The greater the number of velocity measurements, the more accurate the result of air flow. Thus the new equation for calculating the flow in the tunnel is:

$$\dot{V} = \int_{x_0}^{x_1} \int_{y_0}^{y_1} \overline{v}(x,y)dydx \tag{6}$$

where x and y represent the coordinates of the cross-section perpendicular to the air flowing stream, comprising the dimensions of the surface formed by the boxes from the pallet.

Common approach is to measure the air velocities in several points of the flow and obtain one average result, for a more consistent representation of the profile of the flow and avoid the high variability of measurements.

3.6 Measurement of Convective Heat Transfer

The experiments for determining the convective heat transfer coefficients during freezing processes were carried out according to procedures described by Le Blanc et al. (1990a, b) and Resende et al. (2002), using a specimen of high thermal conductivity metal. The method consisted of measuring the temperature variation of a test body with high thermal conductivity during cooling. The high conductivity is necessary to minimize the temperature gradient formed during the heat transfer process between the sample and the cooling medium. The test body shown in Figure 4a is an aluminium brick with known dimensions (0.10 m x 0.07 m x 0.025 m), with perforations for insertion of thermocouples for temperature measurement. Empty spaces around the thermocouples were filled with thermal paste to prevent formation of air pockets within the holes that could affect the measurements. Aluminium thermo physical properties (as a metallic test body) used for the determination of the convective heat transfer coefficients at 20°C are: density (ρ_{Al}=2701.1 kgm^{-3}), specific heat (C_{pAl}=938.3 Jkg^{-1}°C^{-1}), thermal conductivity(k_{Al}=229 Wm^{-1}°C^{-1}) (Welty et al., 2000).

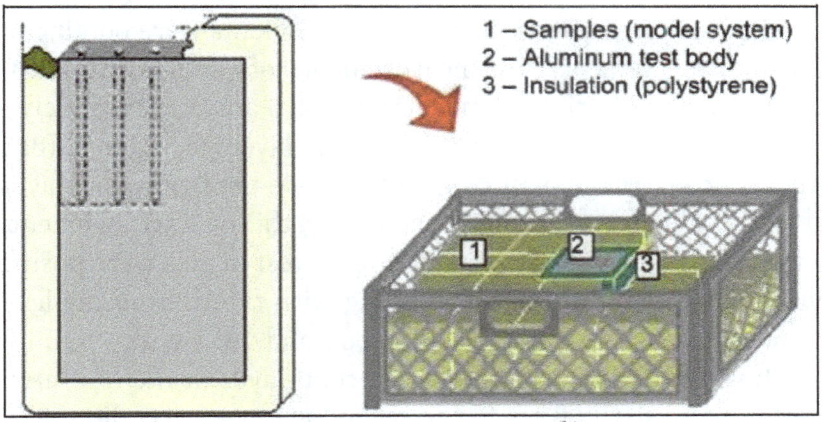

Fig. 7. (a) Aluminium test body insulation and (b) positioning inside the box.

Regarding heat flow analysis, polystyrene was used as insulation around the test body to keep only one surface exposed in contact with the cooling air. The test body had all of its edges insulated, except the upper face which was kept exposed to the cold air flow (Figure 4). This procedure was adopted to avoid edge effects and generate a one-dimensional heat flow. The test bodies were assembled inside the boxes and over the packages of samples, as shown in Figure 4b.

The fast cooling curve can be described by Equation 7, which is a simplification of Equation 6:

$$\frac{(T - T_\infty)}{(T_i - T_\infty)} = e^{S_2 t} \tag{7}$$

where T is the average value obtained from three thermocouple temperature measurements inside the test body during the cooling process, T_i is the initial temperature of the aluminium plate and T_∞ is the cooling medium temperature. The S_2 parameter represents the cooling coefficient, defined as the test body temperature change per time for each temperature degree difference between the product and the cooling medium, and is expressed in s^{-1}.

The values of S_2 were used for the calculation of the heat transfer coefficients for the period of sensible heat removal of the samples during the cooling process, as Equation 8:

$$h_c = \frac{-\rho_m V c_{pm}}{A} S_2 \tag{8}$$

Values for the average dimensionless temperature [(T - T1)/ (Ti - T1)] logarithm were calculated for every test body monitored during chilling period, with values obtained plotted versus cooling time. The angular coefficients (S_2) were calculated from these graphs. Convective heat transfer coefficients were obtained according to Equation 8, using temperature measurements of the 5 identified (T1 to T5) aluminium test bodies, distributed in layers 1, 3, 4, 5 and 6, respectively, including both the extreme layers (1 and 7) and the central layers (3, 4 and 5) (Barbin et al., 2010). With the new arrangement of samples inside the boxes, one layer of boxes was added to the pallet, namely layer number 7. In this new arrangement, the layer number 6 was not monitored. All the aluminium test bodies were positioned over the samples in the centre of the boxes along with the thermocouples identified with number 5 as last algorism (15, 35, 45, 55 and 75, Figure 3) and in contact with the cooling air (Figure 4b). Only the second layer in the first arrangement, and both second and sixth layers in the second arrangement, did not have a test body. The main objective was to analyze the air circulation in the central layers in comparison to the top and bottom layers, and to determine the local convection coefficient of heat transfer. After obtaining S_2 values (according to equation 7), the effective heat transfer coefficients were calculated for the samples in the sensible heat loss phase, as shown in Equation 8.

4. RESULTS

4.1 Determination of air flow velocity

The airflow in the vicinity of the product was examined, and findings were previously documented by Barbin et al. (2009). The results are presented here for comparison with the heat transfer coefficients. The flow direction and turbulence severity are typically not precisely understood in practical scenarios. The forthcoming alteration of the heat transfer coefficient for a product of specific dimensions is predominantly influenced by airflow characteristics, including velocity and turbulence, while being minimally impacted by the product's shape and orientation relative to the flow. Because of the significant turbulence in airflow within industrial applications, wind tunnel tests are ineffective for determining heat transfer values (Kondjoyan, 2006); thus, understanding airflow behavior for these studies is crucial.

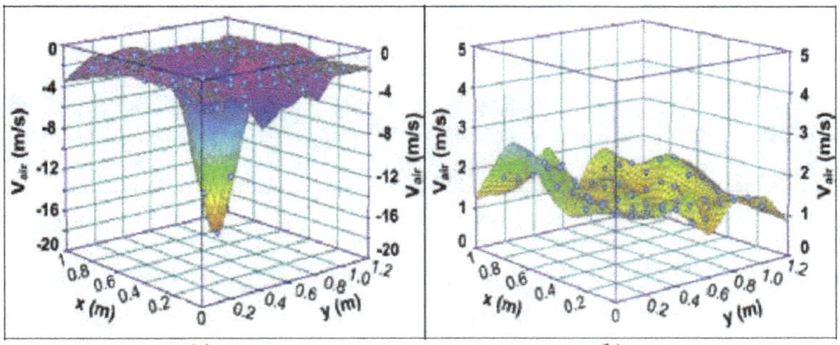

Fig. 8. Results for air flow velocity obtained in the product surface for: (a) insufflation process, and (b) exhaustion process.

The graphs shown in Figure 8 illustrate the results of measurements of air velocity for each process (Barbin et al., 2009). Figure 8a shows the experimental values for the air velocities measured during the process of insufflations, while Figure 8b shows the results for the exhaustion process. The dimensions x (m) and y (m) represent the surface area perpendicular to the air flow, where velocities were measured. The vertical axis shows the velocity results (m/s), containing a response surface for visualization of the airflow.

High values for the air velocity were observed in the insufflation process, reaching above 15 m/s in the central region of measurement, while in the edge and corner positions the velocity results are lower, ranging between 1 and 2 m/s. This difference or lack of uniformity in this distribution of velocities is not observed in the exhaustion process, as it can be seen in Figure 5. This phenomenon may be consequence of the fact that the exhaustion is performed more uniformly, causing the distribution and movement of air to be more even inside the tunnel. Another reason could be that the air had not enough space to expand properly along the short region between the duct exit and the uppermost layer of samples. This makes it a difficult task to overcome because it needs more space for the system according to the appropriate technique for the design of air blast systems (ASHRAE, 1977). However, given the dimensions of the chamber and set of boxes with the product, the assembly was done with the maximum extension possible to fit inside the cold room.

Another factor that may cause interference measures in this place of assembly is that the anemometer does not indicate the direction of air flow during the measurements. Table 1 shows the values obtained and the average velocity of air flow calculated for each flow.

Table 1. Average values obtained for air velocity and air flow during insufflation and exhaustion processes.

Test	Average air velocity (m/s)	Air flow (m^3/s.kg)
Insufflation	- 3.05 ± 0.2	3.7 x 10^{-3}
Exhaustion	1.88 ± 0.2	2.3 x 10^{-3}

The negative sign of the inflation rate indicates a flow direction contrary to exhaustion, as depicted on the zaxis of the graph in Figure 8, and it did not influence air flow calculations, as absolute values of air velocity were employed in both scenarios.

4.2 Freezing time

A freezing process without the portable tunnel was carried out as a reference test to be compared to the experiments using the tunnel device. This reference freezing process consisted in leaving the boxes inside the cold room until all of the samples reached the final freezing temperature.

The average freezing time for the pallet with the industrial sample arrangement was 47 hours, for the reference test, without the use of the portable tunnel. It was also observed the lack of uniformity of process during the freezing process, where the samples located in the upper and lower layers reached -18 ° C in about 42 hours, while it took more than 52 hours for all the samples in the central layer to reach the final freezing temperature.

The top layer of boxes (layer 6) was quickly frozen in the insufflation test, as all the samples reached -18 ° C after 35 hours. The samples in the bottom and middle layers were frozen after about 45 hours, showing that the cooling air was not flowing properly in the bottom part of the pallet.

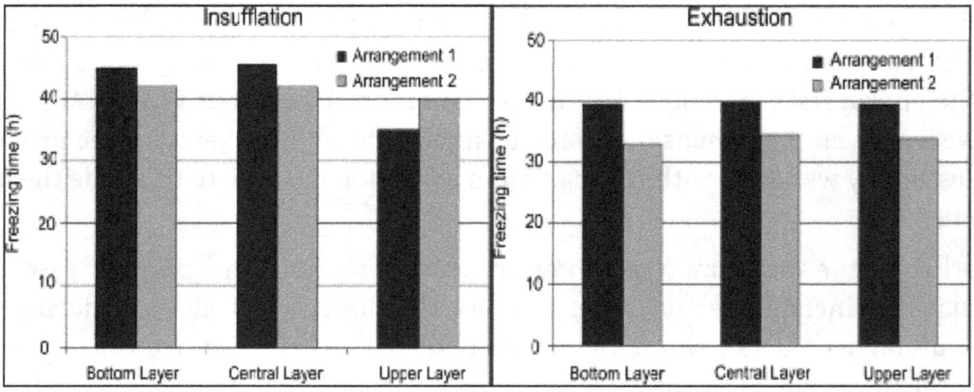

Fig. 9. Comparison of freezing time between two arrangements of samples for insufflation and exhaustion processes.

During the exhaustion test it was observed a difference for the freezing time between the central and peripheral layers of about 1 to 2 hours; however, all samples reached the temperature of -18°C after 40 hours of processing.

Regarding the new arrangement of samples, the freezing time for the exhaustion process was 38 hours. It took 43 hours for the samples to be frozen in the insufflation process. The comparison between the freezing time for industrial and new arrangement can be seen in Figure 9.

The heat transfer analysis comparing the air distribution performed shows that, although showing values around 2 m/s for air velocity results, the exhaustion process reduced up to 14% of the freezing time compared to the blowing system, where air velocity were up to 19m/s.

The top layer of the assembly showed an increase in freezing time for the new configuration of the samples for the testing of inflation (Figure 9). This may have occurred because the industrial configuration might block air circulation for the other layers, reducing the efficiency of heat exchange between air and the other layers of product assembly.

4.3 Convective heat transfer coefficients

The lumped-capacitance method was applied to obtain the heat transfer coefficients for comparison between two different arrangements of products inside the cold chamber. Results of Table 2 show that there is an increase in the convective coefficient for the all the layers of boxes in the pallet, except the test body number 4, located on the second layer from the top.

In the insufflation process for the industrial arrangement, the results were in the range from 3 to 6 W/m^2°C for the three lower layers increasing to 15 W/m^2°C for the test body 4 and reaching 30 W/m^2°C the top layer. The values for the new proposed arrangement reached 10 W/m^2°C for the first two bodies and were greater than 30 W/m^2°C the top layer.

In the exhaustion process, results were nearly twice as high as the industrial process for the first two test bodies. The top layer had also higher values compared to the industrial arrange. For the central layer, results were similar, around 10 W/m^2°C. Only the test body in the second layer from the top showed higher results for the industrial arrange. Thus, results show that enlarging opening space for air flow increases the convective heat transfer, even if there is more product or layers in the pallet.

Table 2. Results for convective coefficient measured between two arrangements of samples in the boxes and two different air flow orientation

Convective coefficient (W/m^2°C)				
	Exhaustion		Insufflation	
Test body	Arrangement 1	Arrangement 2	Arrangement 1	Arrangement 2
T1 lower	5.79	10.74	3.58	9.90
T2	5.25	12.88	6.13	9.92
T3 middle	10.41	10.76	5.77	6.38
T4	14.97	11.06	15.41	13.04
T5 upper	5.92	7.73	29.46	31.72

The variation of heat transfer coefficients with location within a refrigerated room is a critical factor, as it directly influences cooling duration. Given that energy expenses are the primary cost for most refrigerated warehouses, this factor is crucial for managerial decisions, encompassing both early investments and long-term operational costs.

5. CONCISE SUMMARY

This study was conducted to address shortcomings in temperature reduction methods by examining the convective heat transfer coefficient data for food cooling and freezing processes, aiming to aid the food refrigeration industry in enhancing the efficiency of heat transfer processes through straightforward solutions. A literature review was performed to gather supporting material and compare it with the acquired results. Results indicate that augmenting the cooling air velocity within the system or employing more powerful equipment to bring cooler air is not the most effective method for enhancing cooling rates. Enhanced outcomes can be attained by implementing straightforward modifications to the procedure, such as reorganizing the products or samples designated for cooling or freezing, and altering the airflow direction around the products, which can expedite freezing operations. The calculation of the convective heat transfer coefficient helps elucidate the efficacy of cooling air in specific areas of the system or samples, facilitating more accurate methods to enhance the cooling process. The experimental data generated from this effort will be utilized by designers of food cooling and freezing equipment. This information will enable a more precise assessment of convective heat transfer coefficients within these devices, resulting in enhanced system efficiency and decreased cooling and freezing durations. This information is crucial for the establishment and functioning of cooling and freezing facilities and will be immediately beneficial to engineers engaged in the design and operation of these systems. Incorporating these aspects into the equipment's design will significantly enhance energy savings.

6. REFERENCES

Abrahao, R. F.; Correia, E.; Teruel, B. J. (2008). Computational simulation for the development of packages for bananas [Simulação computacional aplicada ao desenvolvimento de embalagens para bananas]. Revista Brasileira de Fruticultura, 30 (1). p. 79-87.

Allais, I.; Alvarez, G.; Flick, D. (2006). Modeling cooling kinetics of spheres during mist chilling. *Journal of Food Engineering*. V. 72, p. 197-209.

ASHRAE Handbook. *Fundamentals*. (1977) Atlanta, Georgia: American Society of Heating, Refrigerating and Air Conditioning Engineers, Inc. 1977.

ASHRAE. *Refrigeration*. (2002). Atlanta, Georgia: American Society of Heating, Refrigerating and Air Conditioning Engineers, Inc. 2002.

ASTM. (1989). *Manual on the use of thermocouples in temperature measurement*. American society for testing and materials.

Baird, C. D.; Gaffney, J. J.; Talbot, M. T. (1988). Design criteria for efficient and cost effective forced air cooling systems for fruits and vegetables. *ASHRAE Transactions*, Atlanta, v.94, p.1434-1453.

Barbin, D. F.; Neves Filho, L. C.; SIlveira Junior, V. (2009). Processo de congelamento em túnel portátil com convecção forçada por exaustão e insuflação para paletes. [Freezing process evaluation using a portable forced air system with air evacuation and air blowing in pallets]. *Ciencia e Tecnologia de Alimentos*, 29, (3), p.1-9.

Barbin, D. F.; Neves Filho, L. C.; Silveira Junior, V. (2010). Convective heat transfer coefficients evaluation for a portable forced air tunnel. *Applied Thermal Engineering*, 30. p.229–233.

Becker, B. R.; Fricke, B. A. (1999). Food thermophysical property models. *International Communications in Heat and Mass Transfer*, v.26, n.5, p.627-636.

Becker, B. R.; Fricke, B. A. (2004). Heat transfer coefficients for forced-air cooling and freezing of selected foods. *International Journal of Refrigeration*, v. 27, (5). p. 540-551.

Castro, L. R.; Vigneault, C.; Cortez, L. A. B. (2003). Container opening design for horticultural produce cooling efficiency. *International Journal of Food Agriculture and Environment*, v.2, n.1, p.135-140.

Chourot, J. M.; Macchi, H.; Fournaison, L.; Guilpart, J. (2003). Technical and economical model for the freezing cost comparison of immersion, cryomechanical and air blast freezing processes. *Energy Conversion and Management*, n.44, p.559-571.

Cleland, A. C. (1990). *Food Refrigeration Process. Analysis, Design and Simulation*. Elsevier Applied Science. London and New York. 1990. 284p.

Cortbaoui, P.; Goyette, B.; Gariepy, Y.; Charles, M. T.; Raghavan, V. G. S.; Vigneault, C. (2006). Forced air cooling system for Zea mays. *Journal of Food Agriculture and Environment*. V. 4, p. 100-104.

Dincer, I. (1995a). Thermal cooling data for figs exposed to air cooling. *International Communications Heat Mass Transfer*. v.22, n.4. p.559-566.

Dincer, I. (1995b). Transient heat transfer analysis in air cooling of individual spherical products. *Journal of Food Engineering*, v. 26, p. 453-467.

Dincer, I.; Genceli, F. (1994). Cooling process and heat transfer parameters of cylindrical products cooled both in water and air. International Journal Heat Transfer, v. 37, n. 4, p. 625-633.

Dussán Sarria, S.; Honório, S.L.; Nogueira, D.H. (2006). Precooling parameters of 'Roxo de Valinhos' figs (Ficus carica L.) packed in a carton box. *Fruits*. v. 61, Issue 6, Nov. 2006, p 401-406.

Fraser, H. (1998). Tunnel Forced-Air Coolers for Fresh Fruits & Vegetables. Ministry of Agriculture, Food and Rural Affairs, Government of Ontario, Canada, 1998. Available: <http://www.omafra.gov.on.ca/english/engineer/facts/98-031.htm>.

Heldman, D.R. Food Freezing. In: Heldman, D. R.; Lund, D. B. *Handbook of Food Engineering*. New York: Dekker, 1992. 277-315.

Kondjoyan, A. (2006). A review on surface heat and mass transfer coefficients during air chilling and storage of food products. *International Journal of Refrigeration*, v.29, p.863–875.

Laguerre, O.; Ben Amara, S.; Flick, D. Heat transfer between wall and packed bed crossed by low velocity airflow. *Applied Thermal Engineering*, v. 26, p. 1951-1960, 2006.

Le Blanc, D. I.; Kok, R.; Timbers, G. E. (1990a). Freezing of a parallelepiped food product. Part 1: Experimental determination. *International Journal of Refrigeration*, v.13, p.371- 378.

Le Blanc, D. I.; Kok, R.; Timbers, G. E. (1990b). Freezing of a parallelepiped food product. Part 2: Comparison of Experimental and Calculated Results. *International Journal of Refrigeration*, v.13, p.379-392.

Meana, M. B.; Chau, K. V.; Emond, J. P.; Talbot, M. T. Forced-air cooling of strawberries in reusable plastic containers. *Proceedings of Florida State Horticultural Society*. v. 118, p. 379-382, 2005.

Mohamed, I. O. (2008). An inverse lumped capacitance method for determination of heat transfer coefficients for industrial air blast chillers. *Food Research International*, 41. p.404–410.

Mohsenin, N. N. (1980). *Thermal properties of foods and agricultural materials.* New York: Gordon and Breach. p. 198-224. 1980.

Nunes, M.; Nader, G.; Jesus, F. B. G.; Cardoso, M.; Jabardo, P. J. S.; Pereira, M. T. (2003). Estudo sobre técnicas de medição de vazão em bocas de insuflamento e de exaustão de ar. Metrologia-2003 – Metrologia para a Vida Sociedade Brasileira de Metrologia (SBM) Setembro 01-05, 2003, Recife, Pernambuco - BRASIL

Resende, J. V.; Silveira Jr, V. (2002a). Medidas da Condutividade Térmica Efetiva em Modelos de Polpas de Frutas no Estado Congelado. *Ciência e Tecnologia de Alimentos*, v.22, n.2, p.177-183.

Resende, J. V.; Silveira Jr. V. (2002b). Escoamento de ar através de embalagens de polpa de frutas em caixas comerciais: Efeitos sobre os perfis de velocidade em túneis de congelamento. *Ciência e Tecnologia de Alimentos*, v.22, n.2, p.184-191.

Resende, J. V.; Neves Filho, L. C.; Silveira Jr, V. (2002). Coeficientes de Transferência de Calor Efetivos no Congelamento com Ar Forçado de Modelos de Polpas de Frutas em Caixas Comerciais. *Brazilian Journal of Food Technology*, v.5, p.33-42.

Salvadori, V. O.; Mascheroni, R. H. (1996). Freezing of strawberry pulp in large containers: experimental determination and prediction of freezing times. *International Journal of Refrigeration*, v.19, n.2, p.87-94.

Santos, C. A.; Laurindo, J. B.; Silveira Junior, V.; Hense, H. (2008). Influence of secondary packing on the freezing time of chiken meat in air blast freezing tunnels [Influência da embalagem secundária sobre o tempo de congelamento de carne de frango em túneis de circulação de ar forçada]. *Ciencia e Tecnologia de Alimentos*, 28 (SUPPL.). p.252-258.

Scott, E. P.; Beck, J. V.; Heldman, D. R. (1992). Estimation of Time Variable Heat Transfer Coefficients in Frozen Foods during Storage. *Journal of Food Engineering*, v.15, p.99- 121, 1992.

Talbot, M. T.; Chau, K. V. (1998). Precooling Strawberries. Agricultural and Biological Engineering Department, Florida Cooperative Extension Service, Institute of Food and Agricultural Sciences, University of Florida, pub.CIR942/AE136, 1998. Available in: <http://edis.at.ufl.edu/AE136>.

Talbot, M. T.; Fletcher, J. H. A (1996). Portable Demonstration Forced-Air Cooler. Agricultural and Biological Engineering Department, Florida Cooperative Extension Service, Institute of Food and Agricultural Sciences, University of

Florida, pub.CIR1166/AE096, 1996. Available in:<http://edis.at.ufl.edu/AE096>.

Vigneault, C.; Goyette, B.; Markarian, N. R.; Hui, C. K. P.; Cote, S.; Charles, M. T.; Emond,J.-P. (2004a). Plastic container opening area for optimum hydrocooling. *Canadian Biosystems Engineering*, v.46, p.41-44.

Vigneault, C.; De Castro, L. R.; Gautron, G. (2004b). Effect of open handles on packages during precooling process of horticultural produce. *ASAE Annual International Meeting.* p. 6901-6908.

Vigneault, C.; De Castro, L. R.; Cortez, L. A. B. (2005). Effect of Gravity on Forced-air Precooling. *IASME Transactions* Vol. 2(3):459-463.

Welty, J. R.; Wicks, C. E.;, Wilson, R. E.; Rorrer, G. (2000). *Fundamentals of Momentum, Heat, and Mass Transfer*. 4th Edition, John Wiley & Sons: New York, 2000.

Welty-Chanes, J.; Vergara-Balderas, F.; Bermudez-Aguirre, D. (2005). Transport phenomena in food engineering: Basic concepts and advances. *Journal of Food Engineering* V. 67, (1-2), March 2005, Pages 113-128.

Zaritzky, N. E. (2000). Factors affecting the stability of frozen foods. In: KENNEDY, C. J. *Managing frozend foods*. Cambridge: Woodhead Publishing Limited, 2000.

Zou, Q; Opara, L. U.; Mckibbin, R. (2006a). A CFD modeling system for airflow and heat transfer in ventilated packaging for fresh foods: I. Initial analysis and development of mathematical models. *Journal of Food Engineering*. v. 77, Issue 4, p. 1037-1047.

Zou, Q; Opara, L. U.; Mckibbin, R. (2006b). A CFD modeling system for airflow and heat transfer in ventilated packaging for fresh foods: II. Computational solution, software development, and model testing. *Journal of Food Engineering*. v. 77, Issue 4, p. 1048-1058.

12 Chapter HEAT TRANSFER FROM CONDENSATION IN MARANGONI

1. INTRODUCTION: AN INTRODUCTION

The Marangoni condensation phenomena, which exhibit an extremely high heat transfer coefficient in the dropwise condensation regime, are the result of the condensate's surface tension instability in the condensation of a binary vapor mixture of a positive system. This system is characterized by a negative gradient in the surface tension of the mixture with respect to the mass fraction of the volatile component, as seen in water-ethanol and water-ammonium mixtures. Marangoni dropwise condensation is distinct from the so-called dropwise condensation, which is limited to the lyophobic surface and may occur readily on the wetting surface. This phenomenon was first reported by Mirkovich & Missen (1961) for a binary mixture of organic vapors. Ford & Missen (1968) demonstrated that the criterion for instability of a condensate liquid film. Fujii et al. (1993) experimentally investigated the condensation of water - ethanol mixtures in a horizontal tube and found that several different condensation modes such as dropwise and rivulets occur depending on concentration. Morrison & Deans (1997) measured the heat transfer characteristics of a water - ammonium vapor mixture and found that it exhibited enhanced heat transfer.

Utaka & Terachi (1995a, 1995b) measured the condensation characteristic curves and clarified that surface subcooling is one of the dominant factors in determining the condensate and heat transfer characteristics of Marangoni condensation although the effect of concentration of binary mixture was major factor deciding the condensation modes in most experimental reserches. Moreover, the effects of external conditions such as the vapor mass fraction (Utaka & Wang, 2004) and the vapor velocity (Utaka & Kobayashi, 2003) were investigated. Heat transfer was significantly enhanced at low mass fractions of ethanol in a water - ethanol mixture. Murase et al. (2007) studied Marangoni condensation of steam - ethanol mixtures using a horizontal condenser tube. Their results showed similar trends as those of (Utaka & Wang, 2004) for vertical surfaces.

On the other hand, the mechanisms of Marangoni condensation have also been studied. Hijikata et al. (1996) presented a theoretical drop growth mechanism for Marangoni dropwise condensation. They found that the Marangoni effect that occurs due to a difference in the surface tension plays a more important role than the absolute value of the surface tension in Marangoni condensation. Utaka et al. (1998) investigated the effect of the initial drop distance, which is the average distance of the initially formed drops grown from a thin flat condensate film that appears immediately after a drop departs. They clarified that the initial drop distance is closely related to the heat transfer characteristics of Marangoni condensation. Further, Utaka & Nishikawa (2003a, 2003b) measured the thickness of condensate films on the tracks of departing drops and between drops by applying the laser extinction method. The condensate layer thickness was determined to be around 1 μm, strongly correlating with the initial drop distance and the heat transfer parameters. This work elucidates the mechanics and heat transport properties of Marangoni condensation phenomena based on existing research.

2. MECHANISMS AND CHARACTERISTICS OF MARANGONI CONDENSATION PHENOMENA

2.1 Outline of mechanisms and characteristics of Marangoni condensation phenomena

The factors influencing the condensate modes in the condensation of binary mixes are contingent upon the phase equilibrium relationship between liquid and vapor, as well as the comparative magnitudes of the surface tensions of the two liquids. Marangoni condensation appears typically for the case of so-called positive system, in which the surface tension of more volatile component σ_L is lower than that of non-volatile component σ_H. Such a instability in the system of liquid evaporation was shown by Hovestreijit (1963) for the first time. The mechanism of Marangoni condensation was explained by Fujii et al. (1993) from the similar point of view as shown in Fig. 1. The thicker condensate liquid (point B in Fig. 1) pull the thinner condensate (point A) due to the surface tension difference occurred by the distributions of surface temperature and concentration of liquid. As a result, the irregular condensate thickness augmented and the irregular modes of condensate such as dropwise appear by the surface instability. The phase equilibrium ralation and the variation of surface tension against mass fraction for water – ethanol binary mixture, which is major test material as a positive system in this study, is shown in Fig. 2.

The major dominant parameters in Marangoni condensation are the concentration of vapor and the surface subcooling of condensing surface. Althoug the concentration of vapor was a main factor determining the condensation mode, in investigating the heat transfer characteristics of Marangoni condensation, surface subcooling was found to be the fundamental factor controlling the condensate modes by Utaka-Terachi (1995a, 1995b). In other words, the change in heat transfer coefficient showed strong non-linearity with condensate mode transition because the Marangoni dropwise condensation formed in the vapor-side appears over the wide range of surface subcooling, in addition to the change in diffusion resistance, which is inherent in the condensation of binary mixtures. The features of the condensation characteristic curves are summarized schematically in Fig. 3, in which the variations of heat flux q and heat transfere coefficient α aginst surface subcooling ΔT are shown. Table 1 shows the characters in the change in heat transfer codfficient. Alphabetical symbols in the figure and table denote characteristic points in the condensation characteristic curves. Points B and D are the steep increase point of heat transfer and the maximum heat transfer point, respectively. The condensate shows the dropwise mode in a wide range of surface subcooling, encompassing B and D, and the dropwise mode appears when approaching point D from B. With further increase of the surface subcooling, the minimum point of the heat flux and the point of inflection of the heat transfer coefficient (E) appear as the end point of the transition region.

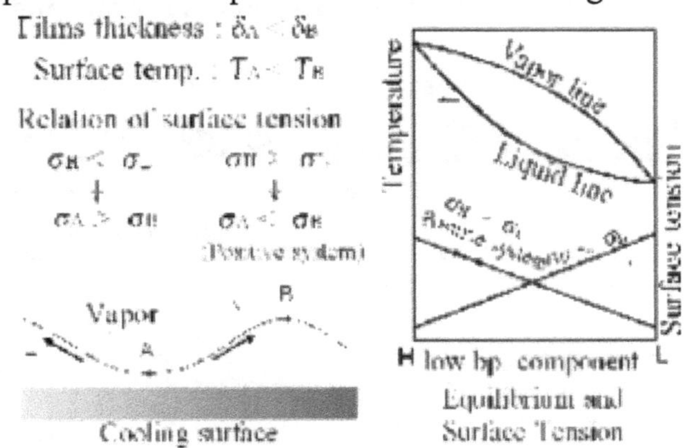

Fig. 1. Mechanisms of Marangoni condensation phenomena

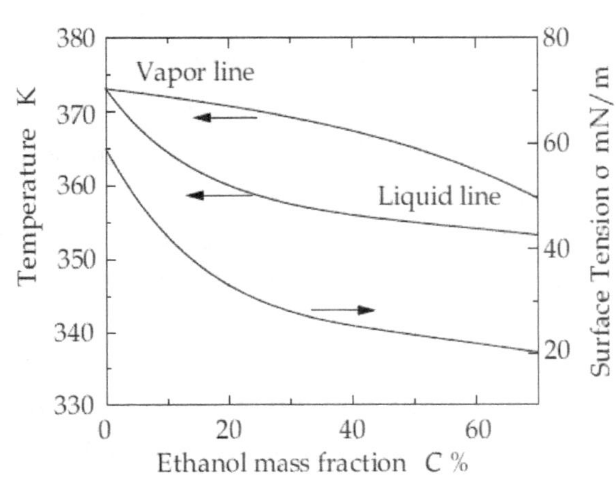

Fig. 2. Phase equilibrium relation and surface tension variation for water-ethanol mixtures

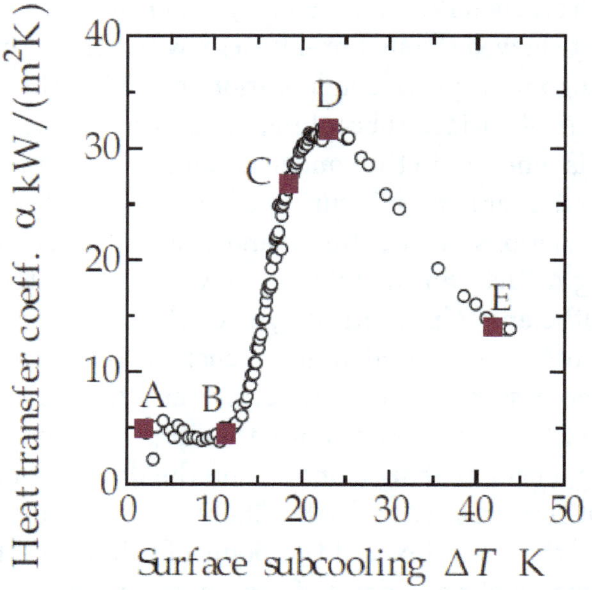

Fig. 3. Nature of condensation characteristic curves for Marangoni condensation

Table 1. Nature of Marangoni condensation heat transfer

Domain	Characteristics
Smaller ΔT than steep increase pt. (B) (Vapor-side dominant region, A -B)	Small α Diffusion resistance dominant
Linear increase of α from pt. (B) to pt. (C) (Steep increase region, B -C)	Reduction of diffusion resistance Dropwise condensation
Departing from linear increase Maximum α pt. (D) Negative gradient region (Transition region, C -E)	Reduction of α
(Film region (E -))	Film condensation

2.2 Marangoni dropwise condensation process

Marangoni dropwise condensation exhibits drop cycles analogous to those observed on a lyophobic surface. As the detailed components of condensate variation are addressed in subsequent sections, just the fundamental items are discussed here. Initially, following the passage of a leaving droplet in a standard condensation process, a slender film of liquid condensate persists. The creation of first droplets begins concurrently with ongoing condensation. Subsequently, the droplets increase in size through condensation and the merging of droplets. In the final phase of the cycle, some of the larger droplets start to detach due to external factors, including gravity and vapor flow, resulting in the formation of a new thin liquid layer. The cycles occur irregularly.

3. STANDARDIZED EXPERIMENTAL APPARATUS AND METHODOLOGIES UTILIZED IN MEASUREMENTS

The standard experimental apparatus and methodology utilized in Marangoni condensation experiments are presented here. A copper heat transfer block devised specifically for investigating phenomena with large heat flux and high heat transfer coefficients (Utaka & Kobayshi, 2003) was shown in Fig. 4. The heat transfer block having a cross-section of trapezoidal shape with notches was constructed in order to realize uniformity of surface temperature and large heat flux. The condensing surface had an area of 10 mm×20 mm. The copper block of one-dimensional rectangular column was also utilized for the cases realizing moderate heat flux. Oxidized titanium was applied to the condensing surface in order to achieve a wetting surface. In addition, impinging water jets from a bundle of thin tubes were used so as to provide high and uniform cooling intensity. A schematic diagram of the leak-tight experimental apparatus, intended to minimize the effects of non- condensing gas, is shown in Fig. 6 (Utaka & Wang, 2004). After passing through the condensing chamber (Fig. 5) in which the heat transfer block is placed, the vapor generated in the steam generator is condensed almost entirely in the auxiliary condenser. The condensate is returned to the vapor generator by the plunger pump via the flow measurement equipment. The vapor flow is in the same direction in which gravity acts, through a duct. Non-condensing gas is continuously extracted by the vacuum pump near the outlet of the auxiliary condenser. The inlet of the vacuum pump is cooled by an electronic cooler to maintain a constant concentration in the vapor mixture, by maintaining low vapor pressure. The loop was divided into a high-pressure part and a low-pressure part bounded by the pressure adjusting valve and the return pump. The vapor pressure of the high-pressure-side is maintained at approximately 1 kPa above atmospheric pressure. The concentration of non-condensing gas in the vapor mixture is measured before and after the experiment. Another heat transfer block shown in Fig. 6 for the vapor concentration measurement is attached in the condensing chamber located downstream of the main heat transfer block.

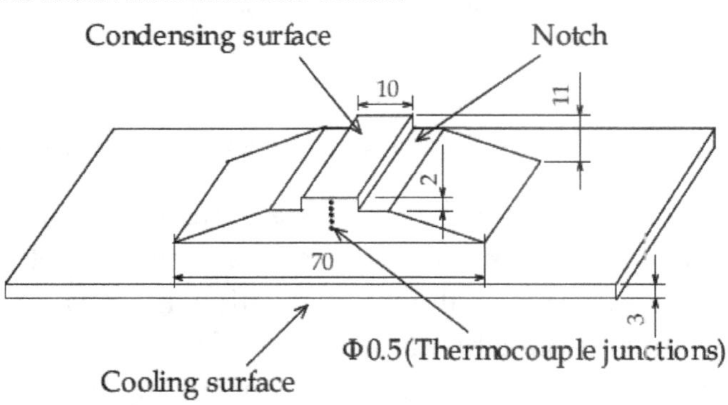

Fig. 4. Heat transfer block for large heat flux

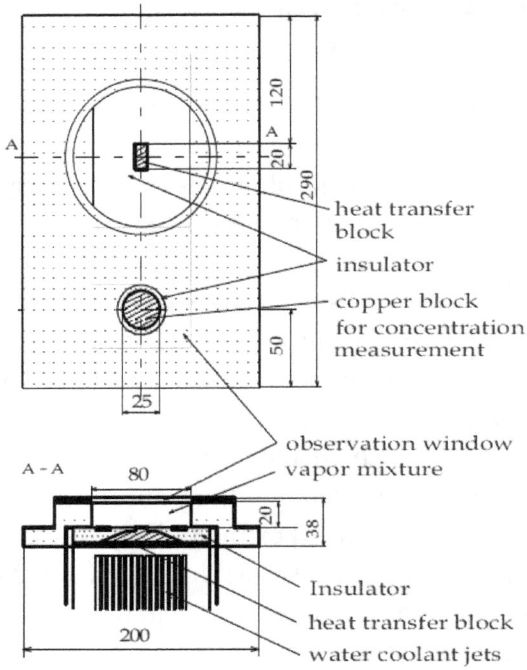

Fig. 5. Condensing chamber

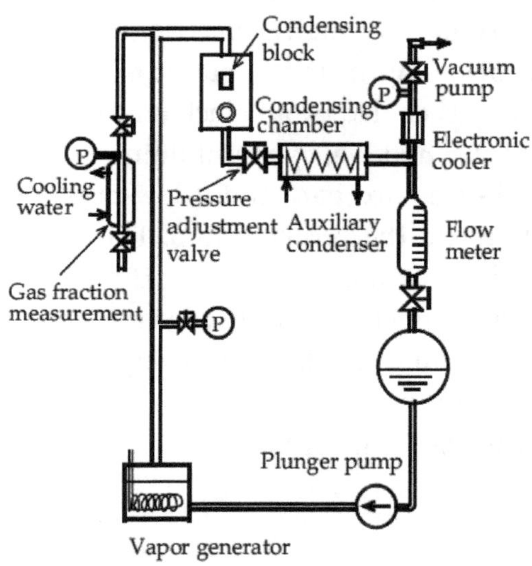

Fig. 6. Schematic diagram of experimental apparatus

Once the vapor condition attained a stable state, the condensation characteristic curves were continually measured using a quasi-steady method, wherein the cooling water temperature was altered gradually while maintaining a constant vapor concentration and velocity. The features of the condensate were seen and documented through the glass window of the condensing chamber utilizing a high-speed camera..

4. FACTORS RELATED TO THE MECHANICS OF MARANGONI CONDENSATION

As described in Section 2, the irregurarity of condensate appears due to surface tenstion instability of the condensation system for binary mixtures of positive system in Marangoni condensation. Hijikata et al. (1996) performed the instability analyses and gave proof of Maragoni condensation phenomena theoretically. In this section, the dominant factors concerning the mecanism of heat transfer in Marangoni condensation is discussed.

4.1 Relation between initial drop distance and heat transfer

(a) Observation and measurements of drop formation

The observation and measurement of the process of formation and growth of condensate drops commenced after the sweeping action by the departing drops were carried out for four ethanol mass fractions of vapor c, i.e., c = 0.07, 0.17, 0.37 and 0.52 under constant vapor velocity at atmospheric pressure (Utaka, et al. 1998). Figure 7 (a) - (c) shows the photographs taken by the high speed camera. Figure 8 (a) shows the condensation heat transfer characteristic curves measured simultaneously with the taking photographs. The sizes of local maximum drops diminished as the distance from the up-flow side peripherals of the departing drops increased. This phenomenon is illustrated by the largest drops in the images, as the time elapsed from the sweep motions of the departing drops increased with distance due to the constrained velocity of drop departure. It is observed that nearly no droplets were present in the rather thin belt-like regions adjacent to the up-flow side peripheries of the leaving drops. The minute droplets manifested in the subsequent sections. Consequently, the initial droplets originate from the thin liquid coating and then increase in size by condensation and coalescence, as documented in the observations of Hijikata et al. (1996), by uniformly sweeping across the entire surface. The drop diameters adjacent to the peripheries of the leaving drops fluctuated with the surface subcooling. The initial drop distance di, defined as the distance between adjacent drops formed after sweeping, was utilized as a metric to assess the characteristics in relation to variations in ethanol concentration and surface subcooling. The rationale for selecting the beginning drop distance to determine the parameters of drop formation is as follows. Determining the minimal drop size is challenging due to the indistinguishability between the irregularities of the thin condensate film and the shape of the drop. Conversely, the initial drop distance remains constant until the occurrence of the first coalescence, allowing for a definitive definition. Figure 8(b) illustrates the variability in the measured initial drop distances. As the elapsed time following the conclusion of the sweep varies along the trajectory of the leaving drop, the diameters of the drops differ in the direction of departure. Consequently, the distance between adjacent drop centers in the direction of the departing drop periphery was utilized.

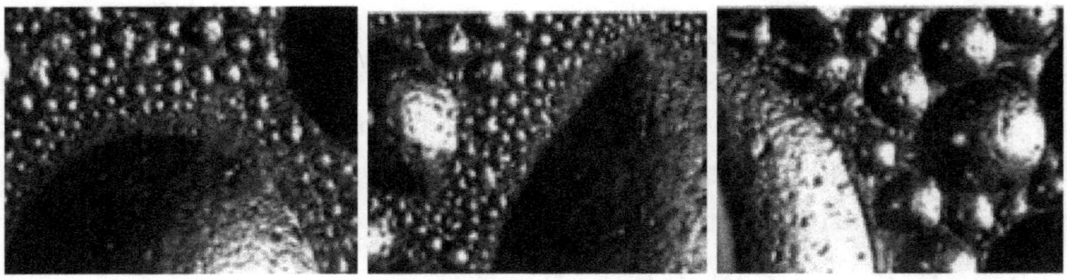

(a) ΔT = 15.2 K (b) ΔT = 18.3K (c) ΔT = 29.5 K

Fig. 7. Aspect of condensate (*c* =0.37) 0.2 mm

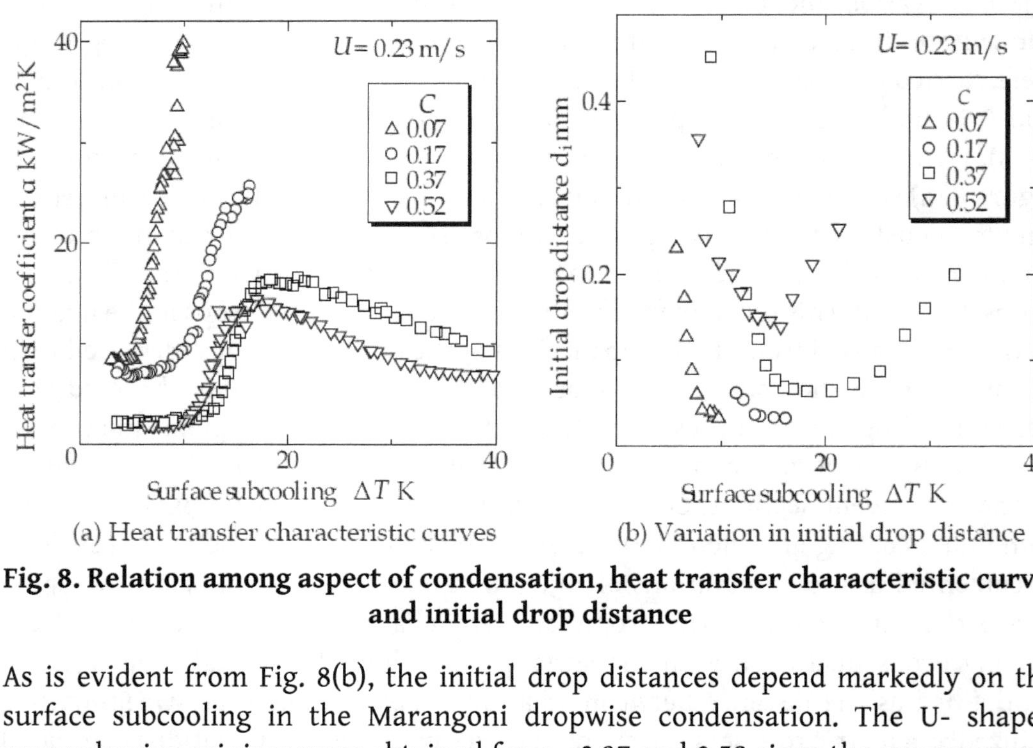

(a) Heat transfer characteristic curves (b) Variation in initial drop distance

Fig. 8. Relation among aspect of condensation, heat transfer characteristic curve, and initial drop distance

As is evident from Fig. 8(b), the initial drop distances depend markedly on the surface subcooling in the Marangoni dropwise condensation. The U- shaped curves having minima were obtained for c = 0.37 and 0.52 since the measurement of the whole regions of the surface subcoolings were achieved. The curves until the minima were obtained for c = 0.07 and 0.17 as same subcooling regions as the characteristic curves. The ratio of least to greatest initial drop distances for each concentration changed widely and was in the range of 2 to 7. It is notable that the surface subcooling at the minimum of the initial drop distances nearly coincides with those at the maximum heat transfer coefficient for each concentration.

The smallest drops, which is nearly equal to the sizes of initial drop distances, were quite larger than those of dropwise condensation on a lyophobic surface and were

0.03 mm to 0.15 mm in this experimental range. Thus, the initial drop distance depended considerably on the ethanol concentration and surface subcooling. That is, the initial drop distance decreased with decreasing ethanol concentration and the maximum is about 5 times larger than the minimum. The minimum diameters of drops measured by Fujii et al (1993) were 0.2 mm and 0.05 mm for $c = 0.48$ and 0.34, respectively, and are ranged in those measurements. However, since their conditions of the surface subcooling were unknown as well as the concentrations are different from this measurement, it is difficult to get detailed comparison. Therefore, it indicates the possibility that the reduction of the thickness of the thin condensate film among drops together with the decrease of the initial drop distance and the approaching of the spherical shape is given by the increase of the driving force of the Marangoni effect.

4.2 Condensate film thickness

4.2.1 Apparatus for measuring condensate film thickness

The entire experimental system, composed of the vapor loop, condensing chamber, cooling system, which are similar to those shown in Section 3, and laser extinction measurement system, is shown in Fig. 9 (Utaka & Nishikawa 2003a, 2003b).

Lambert's law (Eq. 1) was used for determining the condensate liquid thickness. Where, A is the laser exstinction coefficient, $I0$ and I are the intensities of incident light and of transmission light, respectively, and δ is the optical path length.

$$\delta = -A^{-1} \times \log\left(I / I_0\right) \tag{1}$$

Since the extinction coefficient A is unknown and depends upon the mass fraction of the liquid mixture through which the laser beam passes, and was measured prior to the measurement of film thickness. The helium-neon laser beam, having 3.39 **μm** of 2.6 mm in diameter, is transformed to parallel light of 53 mm in diameter by the beam collimator lens. The beam is concentrated by passing through a condenser lens made from optical silicon. The beam waist is defined at the thin liquid layer of the mixture. The transmitted laser beam, which is partially absorbed by the test liquid enclosed in a quartz glass gap, is converged by the quartz glass lens. The transmitted light intensity I is measured by the detector (light reception area of 3 × 3 mm2, response time of 1 ~ 3 μs) made of lead selenide. The incident light intensity I0 is measured under the same conditions of transmitted light intensity without the test liquid. The diameter of the laser beam is approximately 30 μm at the test liquid.

The aspect of the condensate was photographed by a high-speed digital camera synchronously with the condensate film thickness measurement in order to confirm correspondence with the condensate behavior.

4.2.2 Condensate behavior and variation in condensate film thickness

Simultaneous measurement of condensate film thickness and observation of condensate behavior during the formation of the initial drops commenced after the departing drops swept by, under four ethanol vapor mass fractions c of 0.11, 0.21, 0.30, and 0.40, and a constant vapor velocity of 0.7 m/s. Figures 10 (a) and (b) show the images of condensate behavior and the variations of condensate thickness as measured from light transmittance for c = 0.30. Figure 10 (a) shows the aspect of condensate appearing after departing drops swept by, and figure 10 (b) shows variations in condensate thickness immediately after the departing drops swept by. Here, t = 0 was determined arbitrarily, and the variation in liquid thickness and the image coincide with each other. The relatively large dark region on the right-hand side of each image denotes the departing drop, and the light-colored area is thin condensate film at t = 0 ms. Condensate thickness is non-uniform; i.e., the droplets grew rapidly from a flat thin film that appeared after the departing drops swept by. The measuring position irradiated by the laser beam is highlighted as a circle measuring about 30 μm in diameter. After the measuring position was covered by the departing drop at t = 0 ms, the thin film observed as the minimum condensate thickness immediately after sweeping at t = 6 ms by the movement of the departing drop towards the lower right-hand corner. At t = 28 ms, the thickness was measured at the drop having diameter of less than 0.2 mm. Thus, good correspondence between the variation of measured value and the aspect of condensate is confirmed. In this case, the measured minimum condensate thickness was μm. These results confirm that a thin liquid film having a thickness of around 1 μm remains after sweeping and that the heat transfer surface was always covered by liquid in Marangoni dropwise condensation.

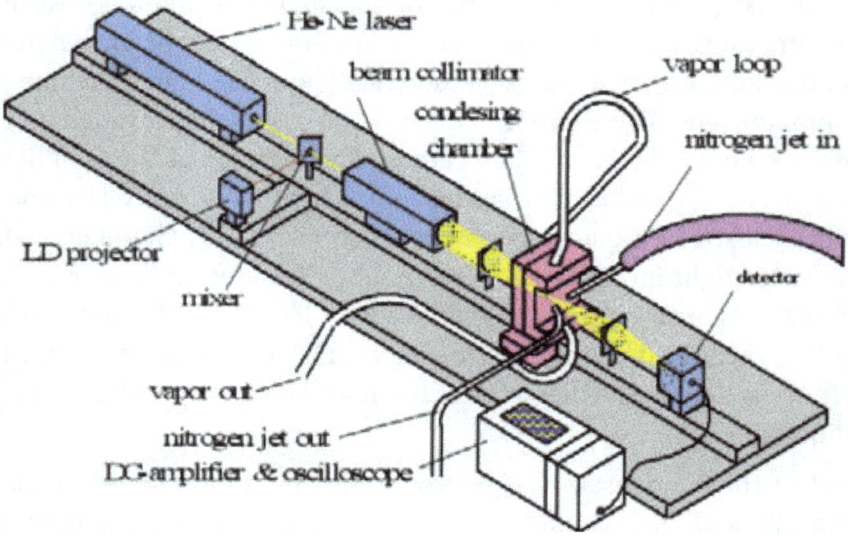

Fig. 9. Experimental apparatus for measuring condensate film thickness

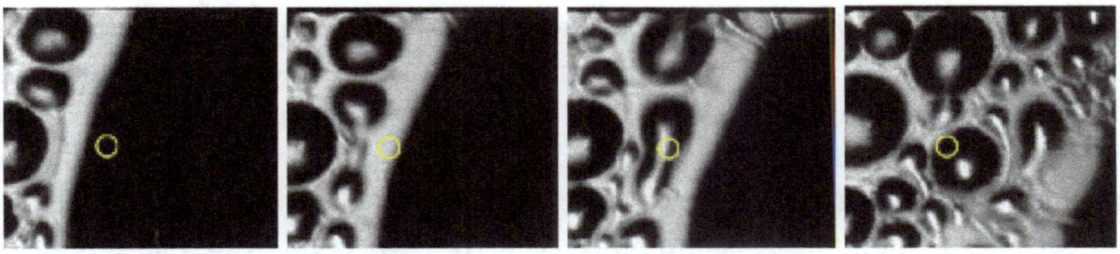

(a) Variation of condensate aspect 0.2 mm

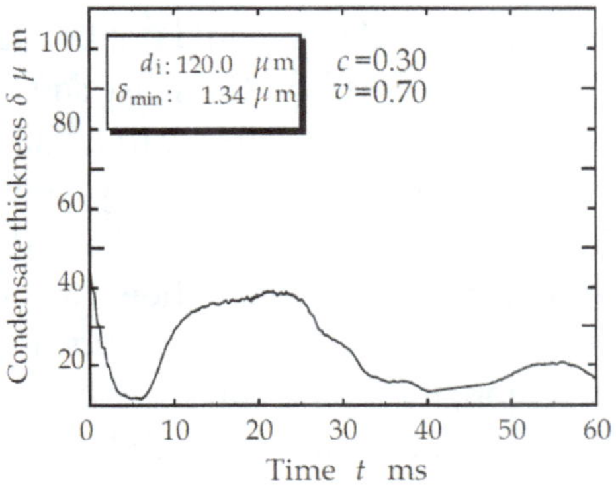

(b) Variation of condensate thickness

Fig. 10. Aspect of condensation and condensate thickness for c = 0.30 and δmin = 1.34 μm

Figures 11 (a),(b),(c), and (d) show the relations between the minimum condensate film thickness δmin appearing immediately after departing drops swept by in the Marangoni condensation cycle and the initial condensate thickness measured from the simultaneously obtained images for four ethanol vapor concentrations, i.e.; c = 0.11, 0.21, 0.30, and 0.42. Initial drop distance was adopted as the averaged value for at least several drop departures. For all vapor concentrations, minimum condensate thickness decreased as initial drop distances decreased in association with increasing surface subcooling. For the higher ethanol vapor concentrations of 0.30 and 0.42 shown in Figs. 11 (c) and (d), respectively, the minimum drop distances decreased linearly with decreasing initial drop distance in association with increased surface subcooling. Afterwards, the minimum thickness became minimum. Both the initial drop distance and the minimum condensate thickness increased with further increase in cooling intensity. Since, as shown in section 4.1, the maximum heat transfer coefficient appeared in a region near the minimum initial drop distance, and the return portion of the curve; i.e., the region near the minimum initial drop distance and the smallest minimum condensate thickness, should correspond to the maximum heat transfer point in the condensation characteristic curve. The minimum condensate thickness for the vapor ethanol mass fractions of 0.30 and 0.42 was very thin; that is, 1.2 μm and 1.5 μm, respectively. Minimum condensate thickness showed a tendency to decrease with decreasing vapor ethanol concentration. The minimum values of minimum condensate thickness for the mass fraction of ethanol vapor of 0.11 and 0.21 were about 1 μm, because of the limit of cooling intensity. In the present study, larger surface subcooling in the thinner ethanol concentration region associated with very large heat flux was difficult to realize, because quartz glass and the jet of evaporated liquid nitrogen gas were used as a heat transfer surface and a coolant, respectively. Therefore, minimum condensate thickness could not be measured near the region of minimum value. Linear curves fitted by the least squares method to the data showing the linear changes had slopes of 0.0087, 0.011, 0.016, and 0.016 for ethanol mass fractions of 0.11, 0.21, 0.30, and 0.42, respectively, and tended to increase with ethanol concentration. The dotted lines in Figs. 11 (a) and (b) were predicted on the basis of the characteristic features of the curves shown in Figs. 11 (c) and (d) and the previous results shown in section 4.1. Although results do not lend themselves to detailed quantitative discussion, minimum thickness can be predicted to decrease with increasing ethanol concentration and to reach less than 0.5 μm.

The results of this section and section 4.1 clarified that close coincidence exists among the features of heat transfer variation in the condensation characteristic curves and their characteristic quantities, such as initial drop distance, departing drop diameter, and minimum condensate thickness, whose minimum is attained at the surface subcooling near maximum heat transfer coefficient. Similar features of these characteristic quantities in relation to ethanol concentration were also obtained. Therefore, the variations in heat transfer coefficient in the condensation characteristic curves were shown to correspond to the changes in the distributions of surface tension differences, which are the driving force of condensate irregularity in Marangoni condensation. Most likely, the maximum driving force appears in the vicinity of peak value of heat transfer coefficient and the thinning of the condensate thin film. Therefore, the heat transfer is enhanced near the maximum heat transfer coefficient point when the thinning of the condensate film with the augmentation of surface tension difference and the decrease in diffusion resistance, which is discussed in next section, occur.

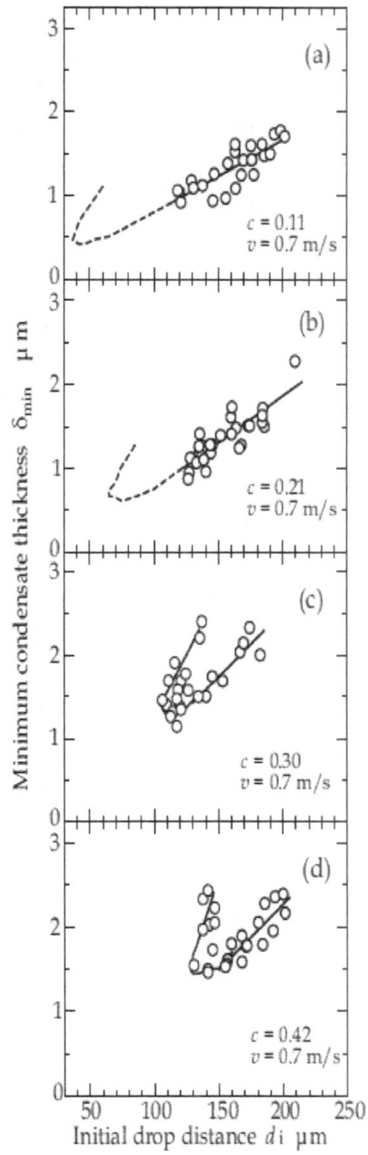

Fig. 11. Variation of minimum film thickness against initial drop distances after sweeping by departing drops

4.3 Characteristics of vapor-side diffusion

For Marangoni condensation, the condensate resistance based on the condensate aspect and the Mass diffusion resistance on the vapor side is essentially significant in regulating condensation properties. This phenomena involves the fluctuating process in which the condensate thickness and vapor concentration alter due to changes in the condensate mode. An examination of the vapor-side diffusion resistance is necessary to elucidate the intricate mechanism. The correlation between the condensate aspect and the condensation feature was already discussed in earlier sections. A model was examined that approximates the process immediately following the sweeping effect of a departing droplet in a standard condensation process, alongside the diffusion resistance (Wang & Utaka, 2005). Rapid condensation transpires upon the initiation of the heat transfer surface being traversed by a leaving droplet. The concentration distribution evolves as a result of rapid condensation from the initial state, where the heat transfer surface is obscured by a big droplet exhibiting relatively low heat transmission and an almost uniform concentration distribution. The thermal conductance change of the vapor-side is examined by numerically solving a one-dimensional unsteady diffusion equation derived from the aforementioned idealized process, focusing on rapid condensation from uniform concentration. The vapor-liquid interface temperature is specified as a boundary condition to account solely for the conductance in the vapor phase. Furthermore, given that the condensate film thickness was minimal, approximately 1 μm, as detailed in section 4.2, within the surface subcooling range adjacent to the maximum heat transfer coefficient, the thermal resistance of the condensate film is negligible, resulting in the surface temperature of the condensate liquid being close to that of the heat transfer surface. The fundamental equation, the boundary conditions, and the initial condition can be expressed as:

$$\frac{\partial C}{\partial t} + V\frac{\partial C}{\partial y} = D\frac{\partial^2 C}{\partial y^2} \tag{2}$$

$$y = 0 : C = C_I \tag{3}$$

$$(\rho D\frac{\partial C}{\partial y})_I = (C_{IL} - C_I)m \tag{4}$$

$$m = \rho V \tag{5}$$

$$y = \infty : C = C_\infty \tag{6}$$

$$t = 0 : C = C_\infty \tag{7}$$

where C is the ethanol mass fraction of the water-ethanol vapor mixture, D is the diffusivity between water and ethanol, V is the velocity of the vapor flowing into the heat transfer surface due to the condensation, ρ is the vapor density, m is the condensation rate, and CI , CIL and C∞ are the ethanol concentrations of vapor and liquid in the vapor-liquid interface and the bulk vapor concentration, respectively. The thermal conductance H and the non- dimensional temperature difference ΔT^* in the vapor phase are defined respectively as:

$$H = Lm/(T_S - T_I) \tag{8}$$

$$\Delta T^* = (T_S - T_I)/(T_S - T_L) \tag{9}$$

where L is the latent heat of condensation, T_S and T_L are the dew point and boiling point tempetatures, respectively, corresponding to the vapor concentration C_∞ in the phase equilibrium diagram of water and ethanol, and T_I is the temperature of the vapor-liquid interface.

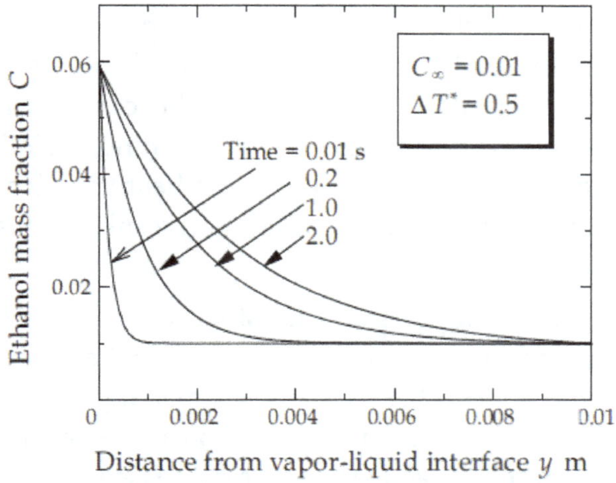

Fig. 12. Variation of distribution of mixture vapor concentration with time

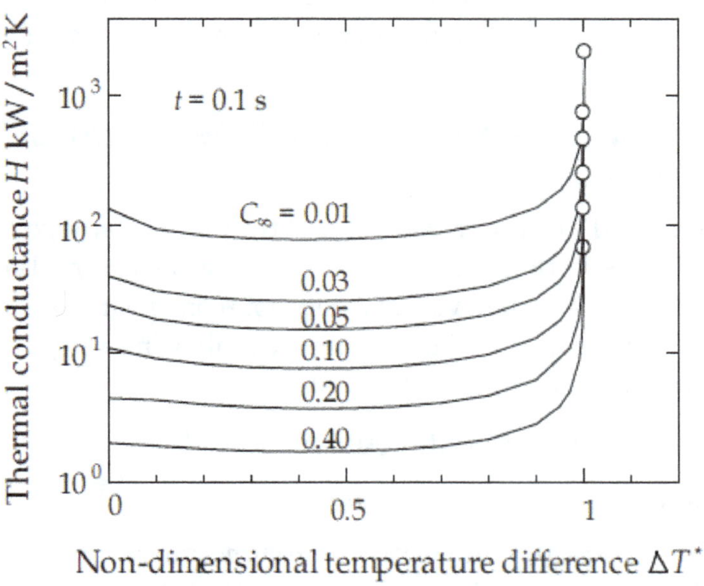

Fig. 13. Variation of thermal conductance with respect to non-dimensional temperature difference in the vapor layer (t = 0.1 s)

Figure 12 shows the time variation of ethanol mass fraction distribution in the vapor phase. It is seen that the thickness of different concentration region becomes thicker with time. Figure 13 shows the variation of thermal conductance with the non-dimensional temperature difference in the vapor phase for the different bulk vapor mass fractions, calculated when the time equals 0.1 seconds. Assuming a similar distribution of the condensate shape, the period of frequency between drop departures is inversely proportional to the heat flux. Therefore the characteristic time is dependent on that period. Also, the maximum heat flux of such a diffusion process is different for different vapor concentrations. However, for convenience, the thermal conductance at a fixed time is compared in order to examine the character of the variation in the condensation characteristic curve and its characteristic points. Since the sweeping period by the departing drop is approximately 0.1~0.2 s at the maximum heat transfer coefficient on the condensation characteristic curves, the calculation result for this time was adopted herein.

As a characteristic of the vapor-side diffusion resistance, the heat flux tends to increase rapidly for any vapor concentration when the non-dimensional temperature difference approaches 1.0 (i.e. the vapor-liquid interface temperature approaches the boiling point), because the compositions of the condensation and the bulk vapors also approach each other. The decrease in the diffusion resistance in the vapor-side due to the decrease in ethanol concentration improved the heat transfer performance. This is thought to be one of the reasons why a relatively high maximum heat transfer coefficient appears in the low-vapor- concentration region.

5. HEAT TRANSMISSION PROPERTIES OF MARANGONI CONDENSATION

The factors affecting the characteristics of heat transfer in Marangoni condensation is investigated in this section. That is, the effects of given external conditions such as vapor concentration, vapor velocity and non-condensing gas concentration. Also, the effect of surface subcooling varying the cooling intensity in detail. Impacts of Surface Subcooling and Vapor Concentration

5.1 Condensation characteristic curves

(a) Condensation characteristic curves

The condensation characteristic curves under atmospheric pressure and at vapor velocitie *U* of 1.5 m/s for a considerable range of ethanol concentrations are shown in Figs. 14 (a) and (Utaka & Wang, 2004). Since film condensation appeared in all subcooling regions for pure water vapor, monotonical change, which is a feature of ordinary film condensation heat transfer, was exhibited. In contrast, a common trend was observed for ethanol vapor mass fractions larger than 0.05%. That is, the qualitative features of each curve showed a common characteristic change having heat flux and heat transfer coefficient maxima as shown in section 2.

From these characteristic curves, excellent heat transfer were confirmed, whereby the heat transfer coefficient maxima shifted to the smaller subcooling region and the values at the maxima were very high at low ethanol concentrations, with the exception of the extremely low ethanol concentrations of 0.05% and 0.1%. In addition, in the low ethanol concentration range, the dependencies of the condensation heat transfer characteristics on the vapor concentration and the surface subcooling become greater, and the gradient of the characteristic curves at the smaller subcooling range increases.

As indicated in a previous report (Utaka & Terachi, 1995a), the realization of excellent heat transfer characteristics is likely due to the following. Since the temperature difference between the boiling point line and dew point line decreases as the ethanol concentration of the vapor mixture decreases, the subcooling region of condensation controlled by the diffusion resistance in the vapor-side becomes narrower due to the nature of vapor-liquid equilibrium relation. Moreover, since the difference between ethanol concentrations of the vapor in bulk and at the vapor-liquid interface is small and the variation of surface tension with respect to the concentration is large, higher heat transfer conductance is realized.

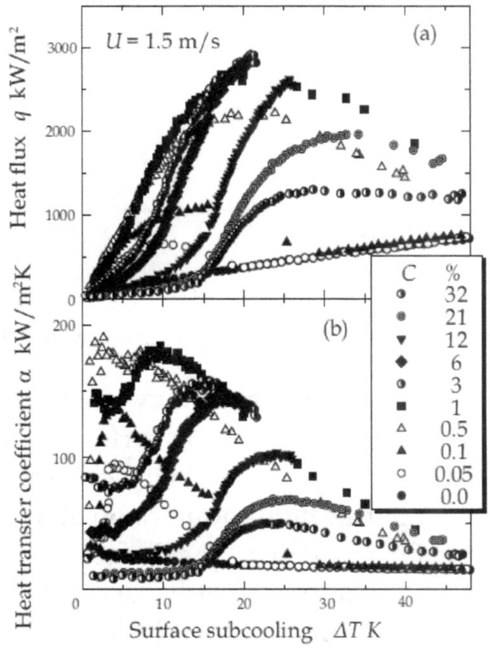

Fig. 14. Condensation heat transfer characteristic curves (U = 1.5 m/s)

In order to evaluate the condensation characteristic curves quantitatively, we examined the heat transfer quantities at the commencement points of the steep increase in heat transfer, the maximum heat transfer coefficient and maximum heat flux, as well as the condensing surface subcooling. Figure 15 shows the variation of surface subcooling with respect to vapor ethanol concentration at the commencement points of the steep increase in heat flux and the heat transfer coefficient for vapor velocity of 0.4 and 1.5 m/s. The steep increase in heat transfer is confirmed to have begun approximately when the temperature of the heat transfer surface reached the dew point line temperature. The surface subcooling at the commencement point of the steep increase almost coincided with the temperature difference between the boiling point line and the dew point line in the vapor-liquid equilibrium relation. However, the surface subcooling at the steep increase point is slightly higher than the temperature difference between the dew point and the bubble point in the vapor-liquid phase equilibrium, and is almost constant in the low-concentration range of less than 3% ethanol.

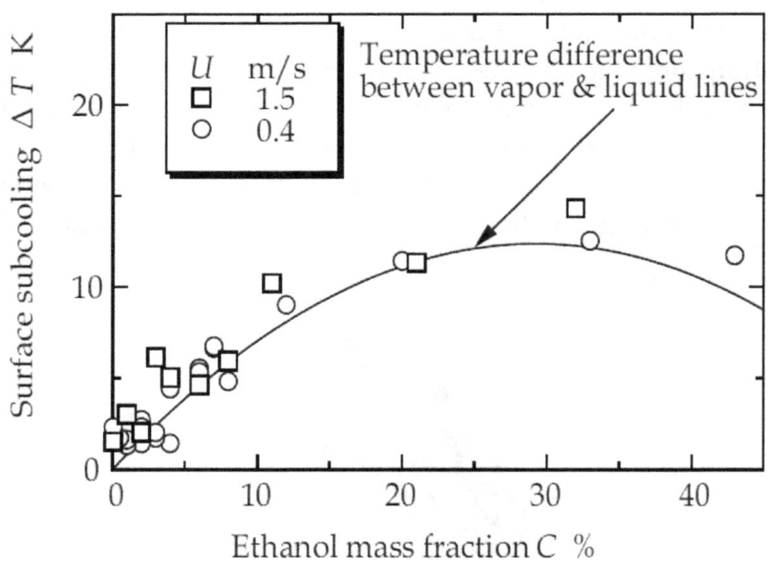

Fig. 15. Surface subcooling at the steep increase point of heat transfer

Figures 16 (a) and (b) show the changes in the maximum heat flux and heat transfer coefficient, respectively, with the condensing surface subcooling, the maximum values of which appear in the region of low ethanol concentration. When the vapor velocity was 0.4 m/s, a maximum heat flux of approximately 1.7 MW/m2 appeared at an ethanol concentration of nearly $C = 6\%$. The surface subcooling at the maximum value of heat flux increased monotonically with respect to the vapor concentration, and the rate of increase was relatively high in the low-concentration range. When the vapor velocity was 1.5 m/s, a maximum heat flux could not be directly obtained due to insufficient cooling intensity and excessive heat flux. Assuming a similar trend in behavior as for the vapor velocity of 0.4 m/s, the maximum heat flux at vapor velocity of 1.5 m/s was estimated to be approximately 3 MW/m2, as indicated by the dotted line in Fig. 15 (a). As shown in Fig. 15 (b), the maximum heat transfer coefficients appeared at an ethanol concentration of approximately C = 1%, and were 0.12 MW/m2K and 0.18 MW/m2K for vapor velocities of 0.4 m/s and 1.5 m/s, respectively. In addition, the subcooling at the maximum heat transfer coefficient showed some variation, with a peak that appeared at a vapor concentration of approximately 20%.

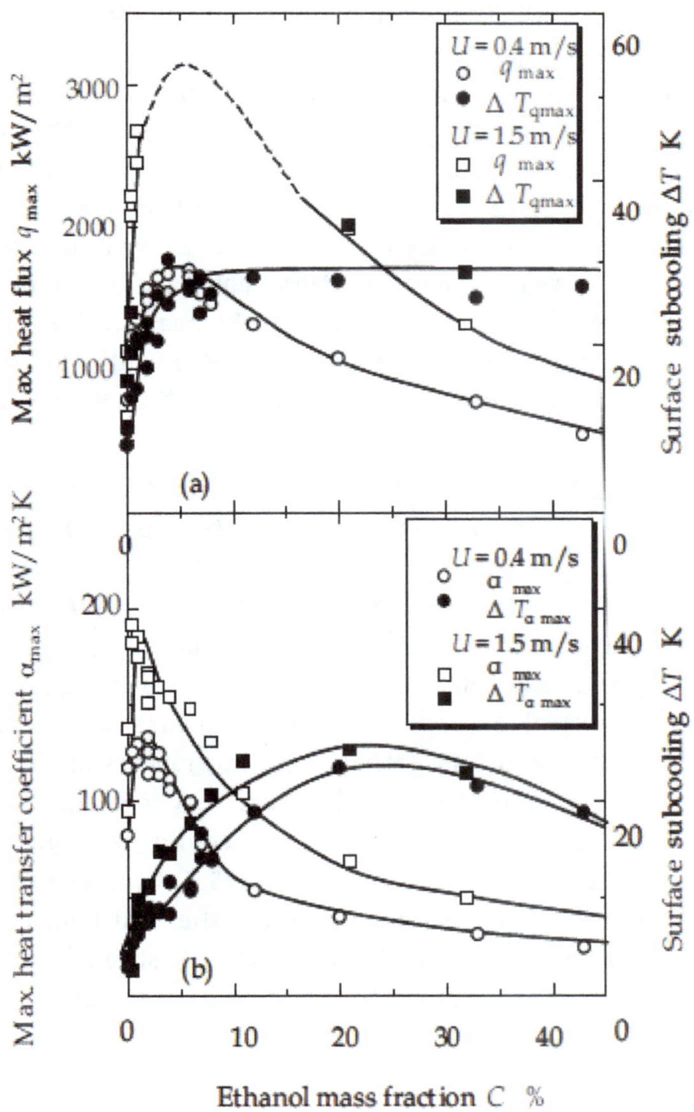

Fig. 16. Characteristics of maximum points of heat transfer (a) Maximum heat flux, (b) Maximum heat transfer coefficient

(b) Promotion of steam condensation by addition of ethanol

We examine the promotion of condensation heat transfer by adding ethanol to water. Figure 17 shows the variation of the condensation heat transfer coefficient of the vapor mixture normalized by that of pure steam with respect to the subcooling, for a vapor velocity of 0.4 m/s. For all vapor concentrations, the maximum shifted slightly from that of the characteristic curve toward larger subcooling. This is due to the lowering of the heat transfer coefficient for pure steam with the increase in subcooling. The condensation heat transfer was improved over almost the entire measured subcooling region for ethanol concentrations of less than approximately 6%. On the other hand, when the ethanol concentration of the vapor mixture was higher than 12%, the ratio of condensation heat transfer of the vapor mixture is smaller than that of steam in the low-subcooling region, but is higher in the subcooling region larger than that of the commencement point of steep increase. As discussed in section 3.4, the mutual relationship of the Marangoni driving force on the condensate and the diffusion resistance in the vapor phase in the condensation of binary vapor mixtures determines the condensation characteristics. The decrease in condensation heat transfer coefficient due to the diffusion resistance of the vapor-side is relatively small in the case of $C \leq 6\%$ for low ethanol concentration. In the case of $C \geq 12\%$, as the diffusion resistance is the controlling factor, the condensation heat transfer coefficient was reduced even in the dropwise mode.

Figure 18 shows the relationships between the ethanol concentration of the vapor mixture and the maximum value of the ratio of the heat transfer coefficient of the vapor mixture, α_{peak}, to steam, α0, with respect to the surface subcooling. For an ethanol concentration of approximately 1%, the addition of ethanol improves the condensation heat transfer coefficient of the mixture, compared to steam, by a factor of approximately six at a vapor velocity of 0.4m/s, and by a factor of approximately eight at a vapor velocity of 1.5 m/s. In addition, for vapor concentrations of 0.05% and 0.1% (ethanol concentrations of the liquid mixture were 0.005% and 0.01%, respectively), which are very low, the ratios of the heat transfer coefficients reached 3.5~5.5, and the subcooling is in the relatively small temperature difference range of approximately 3~5 K. Thus, the addition of a very small amount of ethanol is expected to promote the heat transfer process for a comparatively small temperature difference and at sustained high heat transfer coefficient regardless of the nature of the condensation surface.

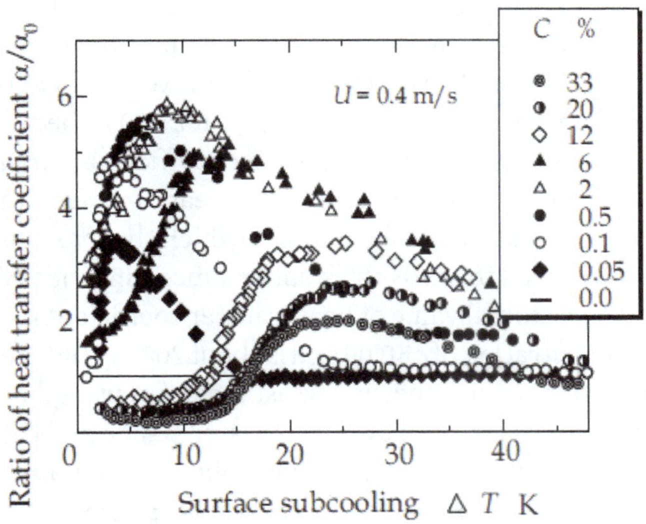

Fig. 17. Ratio of the condensation heat transfer coefficient of the mixture vapors to that of pure steam, with respect to surface subcooling

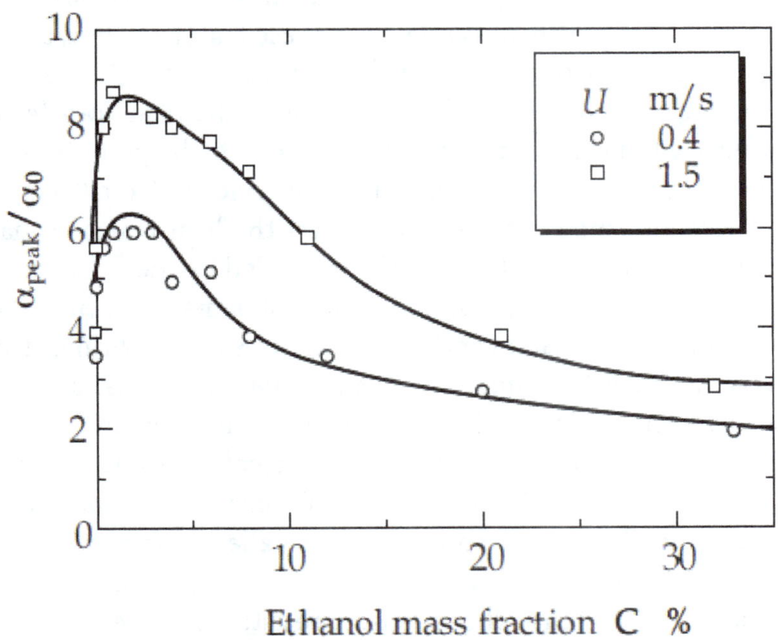

Fig. 18. Variation of the peak ratio of the condensation heat transfer coefficient of the mixture vapors to that of pure steam with respect to the ethanol concentration in the mixture vapors

5.2 Effect of vapor velocity

Condensation characteristic curves were measured for three ethanol mass fractions of water- ethanol vapor mixture c = 0.09, 0.32, 0.53. The results for each of the concentrations are shown in Figs. 19(a) and (b) for $c = 0.09, 0.32$ (Utaka & Kobayashi, 2003). The vapor velocity was varied from 0.3 to 45 m/s for $c = 0.32$ and 0.53 and from 0.3 to 68 m/s for $c = 0.09$. The region near the maximum heat tlux could not be measured when the vapor velocity was over 18 m/s for $c = 0.09$ because of the limited cooling intensity. The condensation heat transfer coefficient increased remarkably throughout whole subcooling region with the increase in vapor velocity under all conditions. Moreover, the surface subcooling in which the maximum heat transfer coefficient appeared moved towards the smaller subcooling-side with the increase in vapor velocity. The maximal value of heat transfer coefficient is very high for the thinnest ethanol concentration of $c = 0.09$ and is about 200 kW/m^{2}K for vapor velocity over 18 m/s. As describe before, thermal resistance of Marangoni condensation is composed of the resistance layers of diffusion in steam-side and the condensate. As shown in Table 1, with increasing surface subcooling, the dominant heat transfer resistances shifts from diffusion layer in the vapor phase to condensate layer. There are two effects of increase in vapor velocity, that is, weakening of the vapor-side resistance and condensate resistance due to the decrease in the largest droplet size in dropwise condensation region. Therefore, the characteristic measured from the above-mentioned experiment is related to these features. First, the increase in heat transfer coefficient with the vapor velocity is remarkable in the region where surface subcooling is relatively small. For example, the heat transfer coefficient increased by nearly tenfold at the steep increase point (the point B in Fig. 3) from 5 kW/m^{2}K to 45 kW/m2K with the change in vapor velocity from 0.8 m/s to 45 m/s for $c = 0.32$. Therefore, the vapor velocity affects the heat transfer characteristic at the steep increase point which is mainly controlled by the diffusion resistance, because the effect of the vapor flow influences steam-side diffusion resistance. Next, the maximum heat trasfer coefficient increases nearly 2 times from 45 kW/m^{2}K to 80 kW/m^{2}K under the same condition. Similar tendencies are also found in the other two concentration conditions. In general, since the diffusion resistance in vapor layer is smaller than the condensate thermal resistance near the surface subcooling at the point of maximum heat transfer coefficient, and the condensate resistance is also smaller compared to that at smaller surface subcooling region because of the appearance of clear dropwise mode. Especially, when the vapor velocity increases, the decrease of condensate resistance caused by a reduction in the size of the largest drop on the heat transfer surface becomes remarkable and is similar to that in so- called dropwise condensation process on a lyophobic surface. With increased vapor velociity, maximum heat transfer coefficient point shifts to smaller subcooling. This is due to lower condensate surface temperature caused by reduced average condensate thickness.

Figures 20 (a) and (b) show the relations among the maximum heat transfer coefficient as a representative point showing the feature of condensation characteristic curve, the vapor velocity and the departing drop diameter. In this region, since thermal resistance of the condensate is the main factor over the vapor-side thermal resistance, it is worth comparing the effect of vapor velocity with the result (Tanasawa et al. 1973) of dropwise condensation on a lyophobic surface where the effect of drop conduction is important. While the relation between maximum heat transfer coefficient and vapor velocity in Marangoni condensation is similar for all of the conditions, those are greatly different from that of dropwise condensation. Thus, the difference in the dependecies of departing drop size to the vapor velocity observed between two phenomena is due not only to the droplet shape but also to the condition of drop peripheral. That is, while there is a three phase interface of solid, vapor and liquid in dropwise condensation on a lyophobic surface, a continuous condensate film exists among the drops so that there is no boundary between drop and condensate film in Marangoni dropwise condensation. Figure 20 (b) shows the change of the maximum heat transfer coefficient against the departing drop diameter. The line fitted by using least square method in the figure was applied to the above measuring points for vapor velocity greater than 0.8 m/s. Since each gradient of the curves almost agrees with that for dropwise condensation on a lyophobic surface, it can be seen that the effect of departing drop diameter on the heat transfer coefficient is similar to each other. Similar trends of dependencies on departing drop diameters both for the maximum heat transfer coefficient in Marangoni condensation and the heat transfer coefficient of dropwise condensation is considered here.

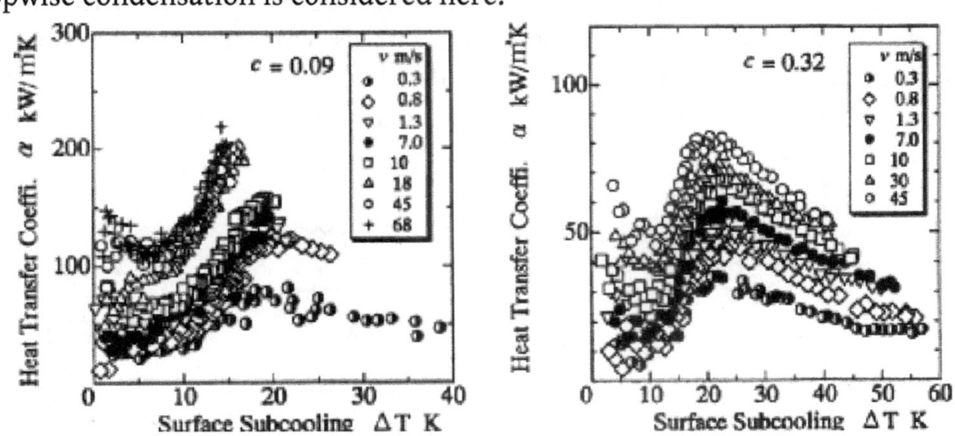

a) $c = 0.09$ (b) $c = 0.32$

Fig. 19. Condensation characteristic curves

It has been clarified that comparatively large drops on condensing surface in dropwise condensation on a lyophobic surface works as a hindrance to heat transfer and that the major part of heat transfer occurs through small drops. Therefore, considering that the condensate resistance is dominant and that the dropwise mode appears at the point of maximum heat transfer coefficient, it can be understood why the exponents agree with one another because relatively large drops work as heat resistance and the thin condensate between the dorps are taking on the major heat transfer.

5.3 Effect of non-condensing gas

To achieve the minimal non-condensing gas concentration, the leak-proof vapor loop, illustrated in Fig. 5, was initially isolated from the atmosphere (Wang & Utaka, 2005). A structure was implemented to include nitrogenous gas into the vapor mixture, hence altering the concentration of non-condensing gas. The quantity and stability of the introduced non-condensing gas were assessed by measuring the flow of nitrogen gas with a flow meter.Figures 20 (a) and (b) shows the condensation characteristic curves of heat transfer coefficient under ethanol mass fractions of the vapor mixture of 0.01 and 0.45, respectvely. The condensation heat transfer performance improves with the decrease in the concentration of the non-condensing gas for vapor mixtures of all ethanol concentrations. In particular, the condensation heat transfer performance of the vapor mixture can be improved by further reducing the concentration of the non-condensing gas, even in cases of very small non-condensing gas concentration.

As shown in Fig. 3, Marangoni condensation characteristic curves can be divided into four domains: the domain controlled by the diffusion resistance of the vapor-side, the rapid decrease domain of the diffusion resistance, and the domains controlled by dropwise and filmwise condensations. For each domain, since the main factor governing the heat transfer is different, the degree of the effect of the non-condensing gas on the condensation heat transfer should also be different.

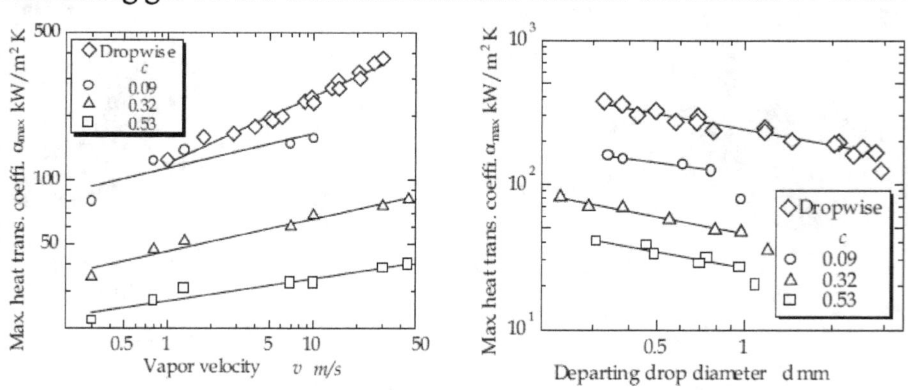

(a) (b)

Fig. 20. Heat transfer characteristics at point of maximum heat transfer coefficient; (a) Variation of maximum heat transfer coefficient against vapor velocity; (b) Variation of maximum heat transfer coefficient against departing drop diameter

First, in the domain controlled by the diffusion resistance of the vapor-side (domain A-B in Fig. 3), the diffusion resistance of the vapor-side is large and the amount of condensate is comparatively small due to the small heat flux. Therefore, the effect of the non-condensing gas is comparatively small. Particularly, for the case in which the vapor ethanol concentration is high (Fig. 21 (b)), this tendency is clear. Next, in the rapid decrease domain of the diffusion resistance (domain B-C), as the diffusion resistance of the vapor-side decreases rapidly while the condensate is maintained dropwise, the condensation heat flux increases suddenly. Corresponding to this change, the effect of the non-condensing gas increases suddenly. As seen in Fig. 21, with the increase of the non-condensing gas concentration, the condensation characteristic curve showed the tendency to decrease from the commencement point of the steep increase, and the drop became large with the increase in surface subcooling. Moreover, in the domain controlled by the dropwise condensation (domain C- E), since the condensate maintains the dropwise aspect and the diffusion resistance on the vapor-side is small, the heat transfer coefficient showed was high. Hence, the domain is a region in which the effect of non-condensing gas on Marangoni condensation appears easily. These figures show that the decrease in heat transfer coefficient or heat flux due to the existence of the non-condensable gas becomes remarkable around the peak value of the condensation characteristic curve, and is comparatively large. For example, in the case of the ethanol concentration in vapor mixture of $Ce = 0.01$, when the non-condensable gas concentration was increased from 15×10^{-6} to 494×10^{-6}, the peak value of the heat transfer coefficient was reduced by about 35% and the peak value of the heat flux was reduced by about 40%. Even at other vapor ethanol concentrations, the decrease of the peak value of the heat transfer coefficient or the heat flux due to the increase in the non-condensing gas concentration to the degree mentioned above is between approximately 30%-50%. The reasons for this are as follows. 1) In this domain, since the condensation heat flux is comparatively high, the non-condensing gas accumulates easily, therefore the diffusion resistance layer of the non-condensing gas form easily. 2) In this domain, both the heat transfer resistance in the condensate and the diffusion resistance due to the low-boiling- point component in the vapor mixture are small. In other words, the diffusion resistance caused by the non-condensing gas becomes relatively large in comparison with other heat transfer resistance in the condensation process. Furthermore, for this domain, as the driving force that makes the condensate irregular, the temperature difference formed on the surface of the condensate decreases again after an increase and gradually approaches zero (the condensate becomes filmwise). As a result, the variation of the heat transfer resistance in the condensate with respect to the surface subcooling having a tendency to decrease in the initial stage and increase again after reaching the minimum around the point of peak heat transfer coefficient. Hence, the effect of non-condensing gas on Marangoni condensation becomes greatest around the subcooling at which the peak heat transfer coefficient. Finally, in the domain controlled by the filmwise condensation (the area to the right of point E), as the condensate becomes filmwise, the diffusion resistance formed by the non-condensing gas and the low-boiling-point component in the vapor mixture is relatively small in the condensation process, the effect of the non-condensable gas is not readily apparent.

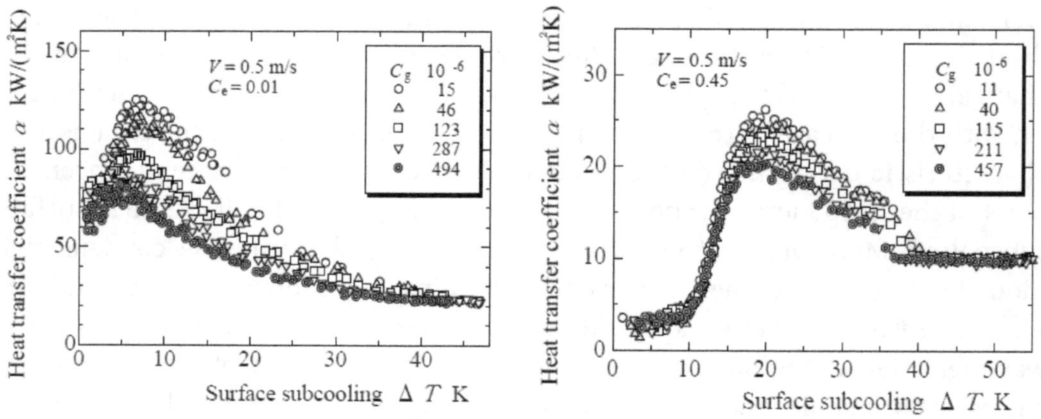

(a) $C_e = 0.01$ (b) $C_e = 0.45$

Fig. 21. Effect of non-condensing gas in Marangoni condensation curve

These figures also indicate that the effect of the non-condensing gas is different under different ethanol concentrations of vapor mixture. Specifically, in the case of the low ethanol concentration of vapor mixture, the heat flux characteristic curve of which has a strongly nonlinear shape, the nonlinearity became weak and the heat flux peak becomes indistinct with the increase in non-condensing gas concentration. The effect of the non-condensable gas becomes stronger because the diffusion resistance of the low-boiling-point component in the condensation process is small when the ethanol concentration of the vapor mixture is low. On the other hand, in the case of the larger ethanol concentration, the effect of non-condensing gas is weaker because of the existence of larger vapor-side heat transfer resistance due to much non-volatile (ethanol) concentration as seen in Fig. 21 (b).

6. CONCISE SUMMARY

This study investigates the mechanisms and heat transfer characteristics of Marangoni condensation phenomena, which exhibit an exceptionally high heat transfer coefficient in the dropwise condensation regime, arising from surface tension instability of the condensate in the condensation of a binary vapor mixture of a positive system, primarily based on the research conducted by the author's group. The discussion encompassed the general description of the mechanisms and characteristics of Marangoni condensation phenomena, the Marangoni dropwise condensation cycle, prevalent experimental apparatus and measurement methods, factors influencing the mechanisms of Marangoni condensation (including the relationship between initial drop distance and heat transfer, condensate film thickness, and vapor-side diffusion characteristics), as well as the heat transfer characteristics of Marangoni condensation (considering the effects of surface subcooling, vapor concentration, vapor velocity, and non-condensing gas).

REFERENCES

Ford, J.D. and Missen, R.W. (1968). On the Conditions for Stability of Falling Films Subject to Surface Tension Disturbances; the Condensation of Binary Vapor, *Can. J. Chem. Eng.*, Vol. 48, pp. 309-312.

Fujii, T., Osa, N., Koyama, S. (1993). Free Convective Condensation of Binary Vapor Mixtures on a Smooth Horizontal Tube: Condensing Mode and Heat Transfer Coefficient of Condensate, *Proc. US Engineering Foundation Conference on Condensation and Condenser Design*, St. Augustine, Florida, ASME, pp. 171-182.

Hijikata, K., Fukasaku, Y., Nakabeppu, O. (1996). Theoretical and Experimental Studies on the Pseudo-Dropwise Condensation of a Binary Vapor Mixture, *Journal of Heat Transfer*, Vol. 118, pp. 140-147.

Hovestreijdt, J. (1963). The Influence of the Surface Tension Difference on the Boiling of Mixture, *Chem. Eng. Sci.*, Vol. 18, pp. 631-639.

Mirkovich, V.V. and Missen, R.W. (1961). Non-Filmwise Condensation of Binary Vapor of Miscible Liquids, *Can. J. Chem. Eng.*, Vol. 39, pp. 86-87.

Morrison, J.N.A. and Deans, J. (1997). Augmentation of Steam Condensation Heat Transfer by Addition of Ammonia, *International Journal of Heat and Mass Transfer*, Vol. 40, pp. 765–772.

Murase, T., Wang, H.S., Rose, J.W. (2007). Marangoni condensation of steam-ethanol mixtures on a horizontal tube, *International Journal of Heat and Mass Transfer*, Vol. 50, pp. 3774-3779.

Tanasawa, I., Ochiai, J., Utaka, Y. and Enya, S. (1976). Experimental Study on Dropwise Condensation Process (Effect of Departing Drop Size), *Trans. JSME*, Vol. 42, No. 361, pp. 2846-2853.

Utaka, Y. and Terachi, N. (1995a). Measurement of Condensation Characteristic Curves for Binary Mixture of Steam and Ethanol Vapor, *Heat Transfer-Japanese Research*, Vol. 24, pp. 57-67.

Utaka, Y. and Terachi, N. (1995b). Study on Condensation Heat Transfer for Steam-Ethanol Vapor Mixture (Relation between Condensation Characteristic Curve and Modes of Condensate), *Transactions of Japan Society of Mechanical Engineers*, Series B, Vol. 61, No. 588, pp. 3059-3065.

Utaka, Y., Kenmotsu, T., Yokoyama, S. (1998). Study on Marangoni Condensation (Measurement and Observation for Water and Ethanol Vapor Mixture), *Proceedings of 11th International Heat Transfer Conference*, Vol. 6, pp. 397-402.

Utaka, Y. and Kobayashi, H. (2003). Effect of Vapor Velocity on Condensation Heat Transfer for Water-Ethanol Binary Vapor Mixture, *Proceedings of 6th ASME-JSME Thermal Engineering Conference.*

Utaka, Y. and Nishikawa, T. (2003a). An Investigation of Liquid Film Thickness during solutal Marangoni Condensation Using a Laser Absorption Method (Absorption Property and Examination of Measuring Method) , *Heat Transfer - Asian Research*, Vol.32, No.8, pp.700-711.

Utaka, Y. and Nishikawa, T. (2003b). Measurement of Condensate Film Thickness for Solutal Marangoni Condensation Applying Laser Extinction Method, *Journal of Enhanced Heat Transfer*, Vol. 10, No. 1, pp. 119-129.

Utaka, Y. and Wang, S. (2004). Characteristic Curves and the Promotion Effect of Ethanol Addition on Steam Condensation Heat Transfer, *International Journal of Heat and Mass Transfer*, Vol. 47, pp. 4507-4516.

Wang, S. and Utaka, Y. (2005). An Experimental Study on the Effect of Non-condensable Gas for Solutal Marangoni Condensation Heat Transfer, *Experimental Heat Transfer*, Vol. 18, No. 2, pp. 61-79.

INDEX